AF356380

# Multifield and Multiscale Coupling of Rocks in Deep Energy Exploitation

# Multifield and Multiscale Coupling of Rocks in Deep Energy Exploitation

Editors

**Chun Zhu**
**Manchao He**
**Junlong Shang**

Basel • Beijing • Wuhan • Barcelona • Belgrade • Novi Sad • Cluj • Manchester

*Editors*
Chun Zhu
Hohai University
Nanjing
China

Manchao He
China University of Mining
and Technology
Beijing
China

Junlong Shang
University of Glasgow
Glasgow
UK

*Editorial Office*
MDPI AG
Grosspeteranlage 5
4052 Basel, Switzerland

This is a reprint of articles from the Special Issue published online in the open access journal *Energies* (ISSN 1996-1073) (available at: https://www.mdpi.com/journal/energies/special_issues/ rocks_deep_energy_exploitation).

For citation purposes, cite each article independently as indicated on the article page online and as indicated below:

Lastname, A.A.; Lastname, B.B. Article Title. *Journal Name* **Year**, *Volume Number*, Page Range.

**ISBN 978-3-7258-2415-1 (Hbk)**
**ISBN 978-3-7258-2416-8 (PDF)**
**doi.org/10.3390/books978-3-7258-2416-8**

# Contents

*Editorial*

# Advances in Multifield and Multiscale Coupling of Rock Engineering

Chun Zhu [1,2,3,*], Jiabing Zhang [2,4], Junlong Shang [5], Dazhong Ren [1] and Manchao He [2,4]

[1] Shaanxi Key Laboratory of Advanced Stimulation Technology for Oil & Gas Reservoirs, College of Petroleum Engineering, Xi'an Shiyou University, Xi'an 710065, China; dzren@xsyu.edu.cn

[2] School of Earth Sciences and Engineering, Hohai University, Nanjing 210098, China; manchaohecumtb@163.com (M.H.)

[3] State Key Laboratory of Mining Response and Disaster Prevention and Control in Deep Coal Mines, Anhui University of Science and Technology, Huainan 232001, China

[4] State Key Laboratory for Geomechanics & Deep Underground Engineering, China University of Mining and Technology (Beijing), Beijing 100083, China

[5] James Watt School of Engineering, University of Glasgow, Glasgow G12 8QQ, UK; junlong.shang@glasgow.ac.uk

* Correspondence: zhu.chun@hhu.edu.cn

**Citation:** Zhu, C.; Zhang, J.; Shang, J.; Ren, D.; He, M. Advances in Multifield and Multiscale Coupling of Rock Engineering. *Energies* **2023**, *16*, 4004. https://doi.org/10.3390/en16104004

Received: 4 May 2023
Accepted: 6 May 2023
Published: 10 May 2023



## 1. Introduction

In deep rock engineering, the stability of the rock is a key factor. Rocks are composed of different internal structures at different scales. Both microstructures (such as minerals, porosity, and fractures) and the geological environment (such as geostress and being underground) lead to an explosive complex coupling system. Multifield and multiscale coupling modeling provides a very powerful framework to understand the behavior of rocks in natural and engineered states, incorporating theoretical, numerical, and experimental analysis.

The past few years have witnessed the advances and applications of multifield and multiscale coupling in rock mechanics and rock engineering. Multifield addresses the interaction of different physical behaviors, such as in pores, fractures, and fluids. Multiscale refers to the scale of rock mechanics and engineering and may range from the nano-scale to the meter scale for material characterization purposes, and to hundreds of kilometers for geological applications.

This Special Issue aims to provide a platform for publishing original articles and reviews on recent numerical and experimental advances and applications on multiscale and multi-physics couplings in rock mechanics and engineering. Since the launch of the Special Issue, a total of 30 well-known scholars have submitted their research work to this Special Issue. Following strict quality screening, thirteen high-level papers have been published in the *Energies* journal, and the acceptance rate is 43%.

## 2. Special Issue Content

Gao et al. [1] calculated and examined the stability of perilous rock masses on the surface of the coal mine before and after mining, and the movement law of the overlying strata in the goaf, the movement and deformation law, and the failure mode of perilous rock were analyzed. This study provides a theoretical basis for the treatment of unstable rock and coal seam mining, and has important guiding significance for safe and efficient production within the mine. The results demonstrate the following: (1) The perilous rock is basically in a stable state without the influence of mining. Through theoretical analysis and the construction of the collapse model of perilous rock, it was judged that perilous rocks W1, W3, W4, and W7 were basically stable, perilous rocks W2 and W5 were in an unstable state, and perilous rock W6 was stable without heavy rainfall. (2) As a result of the mining of the C3 coal seam, the cracks in the upper strata begin to develop and rise to the surface, and the longitudinal separation cracks gradually appear between the surface perilous rock and the rock matrix. Due to the existence of these cracks, the perilous rock had a downward shear force. In addition, due to the heavy rainfall in the Guizhou area, the transient saturated zone of perilous rock expanded and the strength of perilous rock was reduced. The seepage increases the sliding force of the perilous rock and aggravates the opening of cracks at any time. (3) The stability of the surface perilous rock mass is largely affected by the mining of underground coal mines. The simulation analysis was repeated using the method of setting coal pillars. When 45 m permanent protection coal pillars are set at both ends, and 15 m local protection coal pillars are set at 60 m, the safety of coal mining can be ensured without affecting the surface and perilous rock.

Gao et al. [2] performed triaxial tests under cyclic multi-level loading at different rates using an MTS-815 Rock Mechanics Testing System. The strain characteristics, elastic modulus, and energy evolution were obtained in order to explore the effects of the mechanism of loading rate on the evolution of deformation and energy parameters of tectonic coal. The results showed that the irreversible strain and plastic energy increased exponentially with the increase in the deviatoric stress, but the growth rate decreased with the increase in loading rate. Furthermore, the elastic strain increased linearly and the growth rate was essentially unaffected by the loading rate. During the compaction stage, the variation in each parameter was not sensitive to the loading rate; during the elastic and damage stage, the rate increase inhibited secondary defect propagation and improved rock strength. In addition, the stepwise and cumulative energy ratio was defined in order to describe the energy distribution during cyclic loading and unloading. It was found that the decrease in the loading rate was beneficial to the transformation of the total energy into plastic energy. The elastic modulus was the most sensitive to sample damage, but the energy density evolution could be used to describe the deformation damage process of tectonic coal in more detail. These findings provide important theoretical support for the tectonic coal deformation law and action mechanism in the damage process that occurs under complex stress conditions.

Santos et al. [3] studied the wave anelasticity (attenuation and velocity dispersion) of a periodic set of three flat porous layers saturated by two immiscible fluids. The fluids are very dissimilar in regard to properties, namely gas, oil, and water, and, at most, three layers are required to study the problem from a general point of view. The sequence behaves as viscoelastic and transversely isotropic (VTI) at wavelengths much longer than the spatial period. Wave propagation causes fluid flow and slow P modes, inducing anelasticity. The fluids are characterized by capillary forces and relative permeabilities, which allow for the existence of two slow modes and the presence of dissipation, respectively. The methodology to study the physics is based on a finite-element upscaling approach to compute the complex and frequency-dependent stiffnesses of the effective VTI medium. The results of the experiments indicate that there is higher dissipation and anisotropy compared to the widely used model based on an effective fluid that ignores the effects of surface tension (capillarity) and viscous flow interference between the two fluid phases.

Hu et al. [4] carried out uniaxial compression tests on granite specimens damaged by cyclic loading using the digital speckle correlation method, so as to reveal the strength characteristics and crack damage fracture laws after rock pre-damage. The experimental results indicate that the mechanical properties of pre-damaged specimens show large damage differences for different cycles. The damage variable of the pre-damaged specimens increases with the increase in cycle number and confining pressure. The specimen damage is primarily due to the strength weakening effect caused by cycle numbers, and the confining pressure restriction effect is not obvious. The evolution laws of uniaxial compression damage propagation in the pre-damaged specimens show differences and an obvious localization phenomenon. Pre-damaged specimens experienced three failure modes in the uniaxial compression test, namely tensile shear failure (Mode I), quasi-coplanar shear failure (Mode II), and stepped path failure (Mode III), and under different pre-damage stress environments with high confining pressures, the failure modes are dominated by Mode II and Mode III, respectively.

Wang et al. [5] carried out the in situ ground pressure monitoring and numerical simulation calculation using the FLAC2D software, so as to solve the problem of surrounding rock stability control for this roadway type. The influence laws of the surrounding rock lithology, the vertical and horizontal distance between the roadway and overlying working face, the positional relationship between the roadway and the overlying working face, and the support form and strength of the rock surrounding an oblique straddle roadway were obtained. Within the range of mining influence, the properties of the rock surrounding the roof and floor were very different, and the deformation of the rock surrounding the two sides exhibited regional differences. The influence range of the mining working face on the rock floor of the roadway was approximately 30–40 m, and that of horizontal mining was approximately 50–60 m. The mining influence on the rock surrounding the side roadway of the working face is large, but the mining influence on the roadway below is small. Using FLAC2D, the stress and displacement characteristics of the rock surrounding the obliquely straddled roadway were compared and analyzed when the bolt support, combined bolt and shed support, and bolt–shotcreting–grouting support were adopted. Then, the proposed support scheme of bolting and shotcreting was successfully applied. The deformation of the rock surrounding the roadway was satisfactorily controlled, and the results were useful as a reference for similar roadway maintenance projects.

Yu et al. [6] selected Yuanbao Bay Coal Mine, Shuozhou, Shanxi, to study the collapse of the overlying coal pillars on the longwall face and to reveal the mechanism of the pillar collapse and the disaster-causing mechanism caused by strong ground pressure. The results showed that the dynamic collapse process of coal pillars is relatively complicated. First, the coal pillars on both sides of the goaf are destroyed and destabilized, followed by the adjacent coal pillars, which eventually cause a large-scale collapse of the coal pillars. This results in a large-scale cut-off movement of the overlying strata, and the large impact load that acts on the longwall face causes an unmovable longwall face support. Moreover, the roof weighting is severe when strong ground pressure occurs on the longwall face, causing local support jammed accidents. Furthermore, the data of each measurement point of the strata movement inside the ground borehole significantly increase, and the position of the borescope peeping error holes in the ground drill hole rise steeply. The range of movement of the overlying strata increases instantaneously, and all strata begin to move. Research on the mechanism of strong ground pressure can effectively prevent mine safety accidents, thus avoiding huge economic losses.

Liu et al. [7] conducted dynamic flattened Brazilian disc (FBD) tensile tests on naturally saturated sandstone under static pre-tension using a modified split-Hopkinson pressure bar (SHPB) device. Combining high-speed photographs with digital image correlation (DIC) technology, we can observe the variation in strain applied to specimens' surfaces, including the central crack initiation. The experimental results indicate that the dynamic tensile strength of naturally saturated specimens increases with an increase in loading rate, but with the pre-tension increase, the dynamic strength at a certain loading rate decreases accordingly. Moreover, the dynamic strength of naturally saturated sandstone is found to be lower than that of natural sandstone. The fracture behavior of naturally saturated and natural specimens is similar, and both exhibit obvious tensile cracks. The comprehensive micromechanism of water effects, concerning the dynamic tensile behavior of rocks with static preload, can be explained by the weakening effects of water on mechanical properties, the water wedging effect, and the Stefan effect.

Han et al. [8] built the numerical model between different perforation spacing by the extended finite element method (XFEM), so as to reveal the variation in the stress shadow with perforation spacing. The variation in stress shadows was analyzed from the stress of two perforation centers, the fracture path, and the ratio of fracture length to spacing. The simulations showed that the reservoir rock at the two perforation centers is always in a state of compressive stress, and the smaller the perforation spacing, the higher the maximum compressive stress. Moreover, the compressive stress value can directly reflect the size of the stress shadow effect, which changes with the fracture propagation. When the fracture length extends to 2.5 times the perforation spacing, the stress shadow effect is the strongest. In addition, small perforation spacing leads to the backward-spreading of hydraulic fractures, and the smaller the perforation spacing, the greater the deflection degree of hydraulic fractures. Additionally, the deflection angle of the fracture decreases with the expansion of the fracture. Furthermore, the perforation spacing has an important influence on the initiation pressure, and the smaller the perforation spacing, the greater the initiation pressure. At the same time, there is also a perforation spacing which minimizes the initiation pressure. However, when the perforation spacing increases to a certain value (the result of this work is about 14 m), the initiation pressure will not change. This study will be useful in guiding the design of programs related to simultaneous fracturing.

Xiao et al. [9] analyzed the physical and mechanical behaviors of granite treated with different thermal shocks by non-destructive (P-wave velocity test) and destructive tests (uniaxial compression test and Brazil splitting test). The results show that the P-wave velocity (VP), uniaxial compressive strength (UCS), elastic modulus (E), and tensile strength (st) of specimens all decrease with the treatment temperature. Compared with air cooling, water cooling causes greater damage to the mechanical properties of granite. Thermal shock induces thermal stress inside the rock due to the inhomogeneous expansion of mineral particles, and further causes the initiation and propagation of microcracks which alter the mechanical behaviors of granite. Rapid cooling aggravates the damage degree of specimens. The failure pattern gradually transforms from longitudinal fracture to shear failure with temperature. In addition, there is a good fitting relationship between P-wave velocity and the mechanical parameters of granite after different temperature treatments, which indicates that P-wave velocity can be used to evaluate rock damage and predict rock mechanical parameters. The research results can provide guidance for high-temperature rock engineering.

Peng et al. [10] used physical simulation to study the formation and evolution of a normal fault from a strain energy perspective. Based on the similarity principle, we designed and conducted three repeated physical simulation experiments according to the normal fault in the Yanchang Formation of Jinhe Oilfield, Ordos Basin, China, and obtained dip angle, fault displacement, and strain energy via the velocity profile recorded by high-resolution particle image velocimetry (PIV). As a result, the strain energy is mainly released in the normal fault line zone, and can thus serve as a channel for oil/gas migration and escape routes connected to the earth's surface, destroying the already formed oil/gas reservoirs. Thus, one might need to avoid drilling near the fault line. Moreover, a significant amount of strain energy remaining in the hanging wall is the reason as to why the normal fault continues to evolve after the normal fault formation until the antithetic fault forms. Our findings provide important insights into the formation and evolution of normal faults from a strain energy perspective, which plays an important role in oil/gas exploration, prediction of the shallow-source earthquake, and post-disaster reconstruction site selection.

Bao et al. [11] proposed a new method based on the inverse logistic function, considering inverse distance weighting (IDW) to predict the displacement of landslides. Quantitative standards of dividing the deformation stages and determining the critical sliding time were put forward. The proposed method is applied in some landslide cases according to the displacement monitoring data, which shows that the new method is effective. Moreover, long-term displacement predictions are applied for two landslides. Finally, summarized with the application in other landslide cases, the value of displacement acceleration, 0.9 mm/day$^2$, is suggested as the first early warning standard of sliding, and the fitting function of the acceleration rate with the volume or length of the landslide can be considered as the secondary critical threshold function of landslide failure.

Xiong et al. [12] derived a novel procedure for the coupled simulation of thermal and fluid flow models (NPCTF) to model heat flow and thermal energy absorption characteristics in rough-walled rock fractures. The perturbation method is used to calculate the pressure and flow rate in connected wedge-shaped cells at a pore scale, and an approximate analytical solution of temperature distribution in wedge-shaped cells is obtained, which assumes an identical temperature between the fluid and fracture wall. The proposed method is verified in Barton and Choubey (1985) fracture profiles. The maximum deviation of temperature distribution between the proposed method and heat flow simulation is 13.2% and flow transmissivity is 1.2%, which indicates the results from the proposed method are in close agreement with those obtained from simulations. By applying the proposed NPCTF to real rock fractures obtained by a 3D stereotopometric scanning system, its performance was tested against heat flow simulations from a COMSOL code. The mean discrepancy between them is 1.51% for all cases of fracture profiles, meaning that the new model can be applicable for fractures with different fracture roughness. Performance analysis shows that small fracture aperture increases the deviation of NPCTF, but this decreases for a large aperture fracture. The accuracy of the NPCTF is not sensitive to the size of the mesh.

Peng et al. [13] used a combination of experiment and infrared detection, where the strength, deformation, and infrared temperature evolution behavior of marble with elliptical holes under uniaxial compression were studied. The test results showed that as the vertical axis b of the ellipse increased, the peak intensity first decreased and then increased, and the minimum value appeared when the horizontal axis was equal to the vertical axis. The detection results of the infrared thermal imager showed that the maximum temperature, minimum temperature, and average temperature of the observation area in the loading stage showed a downward trend, and the range of change was between 0.02 °C and 1 °C. It was mainly due to the accumulation of energy in the loading process of the rock sample that caused the surface temperature of the specimen to decrease. In the brittle failure stage, macroscopic cracks appeared on the surface of the rock sample, which caused the energy accumulated inside to dissipate, thereby increasing the maximum temperature and average temperature of the rock sample. The average temperature increase was from about 0.05 °C to about 0.19 °C. The evolution of infrared temperature was consistent with the mechanical characteristics of rock sample failure, indicating that infrared thermal imaging technology can provide effective monitoring for the study of rock mechanics. The research in this paper provides new ideas for further research on the basic characteristics of rock failure under uniaxial compression.

## 3. Summarize Remarks

The papers presented in this Special Issue cover the important aspects of the latest progress in multifield and multiscale coupling of rock engineering. Even though the multifield and multiscale coupling of rock engineering is an extensive topic, this small contribution could stimulate the community to conduct further research, and thus stimulate progress in this area. Therefore, we believe that the presented papers will have practical importance for future development within rock engineering sector. Finally, together with the other co-Guest Editors, Prof. Junlong Shang and Prof. Manchao He, we wish to thank the authors who contributed their works to this Special Issue.

**Funding:** This research received no external funding.

**Acknowledgments:** This research was supported by the Natural Science Basic Research Plan in Shaanxi Province of China (Grant No. 2022JQ-304), the Opening Fund of State Key Laboratory of Mining Response and Disaster Prevention and Control in Deep Coal Mines (Grant No. SKLMRDPC21KF04). We would like to acknowledge the authors who contributed to this Special Issue.



## References

1. Gao, L.; Kang, X.T.; Gao, L.; Ma, Z.Q. Study of the Stability of the Surface Perilous Rock in a Mining Area. *Energies* **2022**, *15*, 1542. [CrossRef]
2. Gao, D.Y.; Sang, S.X.; Liu, S.Q.; Geng, J.S.; Wang, T.; Sun, T.M. Investigation of the Energy Evolution of Tectonic Coal under Triaxial Cyclic Loading with Different Loading Rates and the Underlying Mechanism. *Energies* **2021**, *14*, 8124. [CrossRef]

3. Santos, J.E.; Carcione, J.M.; Gao, L.; Ba, J. Two-Phase Flow Effects on Seismic Wave Anelasticity in Anisotropic Poroelastic Media. *Energies* **2021**, *14*, 6528. [CrossRef]
4. Hu, J.H.; Zeng, P.P.; Yang, D.J.; Wen, G.P.; Xu, X.; Ma, S.W.; Zhao, F.W.; Xiang, R. Experimental Investigation on Uniaxial Compression Mechanical Behavior and Damage Evolution of Pre-Damaged Granite after Cyclic Loading. *Energies* **2021**, *14*, 6179. [CrossRef]
5. Wang, P.; Zhang, N.; Kan, J.G.; Wang, B.; Xu, X.L. Stabilization of Rock Roadway under Obliquely Straddle Working Face. *Energies* **2021**, *14*, 5759. [CrossRef]
6. Yu, D.; Yi, X.Y.; Liang, Z.M.; Lou, J.F.; Zhu, W.B. Research on Strong Ground Pressure of Multiple-Seam Caused by Remnant Room Pillars Undermining in Shallow Seams. *Energies* **2021**, *14*, 5221. [CrossRef]
7. Liu, X.Y.; Dai, F.; Liu, Y.; Pei, P.D.; Yan, Z.L. Experimental Investigation of the Dynamic Tensile Properties of Naturally Saturated Rocks Using the Coupled Static–Dynamic Flattened Brazilian Disc Method. *Energies* **2021**, *14*, 4784. [CrossRef]
8. Han, W.G.; Cui, Z.D.; Zhu, Z.G. The Effect of Perforation Spacing on the Variation of Stress Shadow. *Energies* **2021**, *14*, 4040. [CrossRef]
9. Xiao, P.; Zheng, J.; Dou, B.; Tian, H.; Cui, G.D.; Kashif, M. Mechanical Behaviors of Granite after Thermal Shock with Different Cooling Rates. *Energies* **2021**, *14*, 3721. [CrossRef]
10. Peng, X.F.; Deng, H.C.; He, J.H.; Chen, H.D.; Zhang, Y.Y.; Kashif, M. Research on the Evolution and Damage Mechanism of Normal Fault Based on Physical Simulation Experiments and Particle Image Velocimetry Technique. *Energies* **2021**, *14*, 2825. [CrossRef]
11. Bao, L.L.; Zhang, G.C.; Hu, X.L.; Whang, S.S.; Liu, X.D. Stage Division of Landslide Deformation and Prediction of Critical Sliding Based on Inverse Logistic Function. *Energies* **2021**, *14*, 1091. [CrossRef]
12. Xiong, F.; Zhu, C.; Jiang, Q.H. A Novel Procedure for Coupled Simulation of Thermal and Fluid Flow Models for Rough-Walled Rock Fractures. *Energies* **2021**, *14*, 951. [CrossRef]
13. Peng, Y.Y.; Lin, Q.C.; He, M.C.; Zhu, C.; Zhang, H.J.; Guo, P.F. Experimental Study on Infrared Temperature Characteristics and Failure Modes of Marble with Prefabricated Holes under Uniaxial Compression. *Energies* **2021**, *14*, 713. [CrossRef]

*Article*

# Experimental Study on Infrared Temperature Characteristics and Failure Modes of Marble with Prefabricated Holes under Uniaxial Compression

Yanyan Peng [1,2], Qunchao Lin [1,2], Manchao He [2,3], Chun Zhu [1,2,4,*], Haijiang Zhang [1,2] and Pengfei Guo [1,2]

1    School of Civil Engineering, Shaoxing University, Shaoxing 312000, China; pengyy@usx.edu.cn (Y.P.); starryskylz@163.com (Q.L.); haijiang1086@126.com (H.Z.); gpf2018@usx.edu.cn (P.G.)

2    Key Laboratory of Rock Mechanics and Geohazards of Zhejiang Province, Shaoxing 312000, China; mche@cashq.ac.cn

3    State Key Laboratory for Geomechanics and Deep Underground Engineering, China University of Mining and Technology (Beijing), Beijing 100083, China

4    School of Earth Sciences and Engineering, Hohai University, Nanjing 210098, China

*    Correspondence: zhuchuncumtb@163.com

**Abstract:** In rock engineering, it is of great significance to study the failure mechanical behavior of rocks with holes. Using a combination of experiment and infrared detection, the strength, deformation, and infrared temperature evolution behavior of marble with elliptical holes under uniaxial compression were studied. The test results showed that as the vertical axis b of the ellipse increased, the peak intensity first decreased and then increased, and the minimum value appeared when the horizontal axis was equal to the vertical axis. The detection results of the infrared thermal imager showed that the maximum temperature, minimum temperature, and average temperature of the observation area in the loading stage showed a downward trend, and the range of change was between 0.02 °C and 1 °C. It was mainly due to the accumulation of energy in the loading process of the rock sample that caused the surface temperature of the specimen to decrease. In the brittle failure stage, macroscopic cracks appeared on the surface of the rock sample, which caused the energy accumulated inside to dissipate, thereby increasing the maximum temperature and average temperature of the rock sample. The average temperature increase was about 0.05 °C to about 0.19 °C. The evolution of infrared temperature was consistent with the mechanical characteristics of rock sample failure, indicating that infrared thermal imaging technology can provide effective monitoring for the study of rock mechanics. The research in this paper provides new ideas for further research on the basic characteristics of rock failure under uniaxial compression.

**Keywords:** infrared thermal imaging; uniaxial compression experiment; prefabricated hole marble; stress–strain; temperature field

**Citation:** Peng, Y.; Lin, Q.; He, M.; Zhu, C.; Zhang, H.; Guo, P. Experimental Study on Infrared Temperature Characteristics and Failure Modes of Marble with Prefabricated Holes under Uniaxial Compression. *Energies* **2021**, *14*, 713. https://doi.org/10.3390/en14030713

Academic Editor: Manoj Khandelwal

Received: 30 December 2020

Accepted: 27 January 2021

Published: 30 January 2021



## 1. Introduction

Deformation and failure of rock is a complex mechanical behavior, and signals such as light, sound, electricity, and magnetism are generated during the failure process. Many scholars have used digital images [1–3], acoustic emission [4–6], CT [7–9], and other technical means to conduct a lot of in-depth research on the process of rock deformation and failure and have obtained a lot of research results.

The process of rock deformation and failure is also a process of energy release. Infrared thermal imaging technology can perform high-frequency, real-time monitoring of the infrared radiation temperature of the rock surface, providing an effective method for rock mechanics' research. In 1986, Brand and Roswell [10] began to use optical spectroscopy to study the optical radiation characteristics of basalt and granite during uniaxial compression failure, hoping to predict the failure of rocks through related studies on rock electromagnetic wave radiation. At the end of the 20th century, Luong [11–13] carried out a series of studies using infrared thermal imaging technology. According to the principle of thermal-mechanical coupling, the infrared heat radiation characteristics of concrete in the process of fatigue damage and stress damage were analyzed. Infrared thermal imaging technology was used to study the process of crack initiation, development, and propagation and to evaluate the damage of concrete. At the beginning of the 21st century, Wu Lixin and his team [14–19] carried out a series of systematic rock mechanics' tests based on infrared thermal imaging technology and tested the discontinuous combination fault rupture, the double-shear stick-slip fault, the stick-slip of the intersection fault, and the rock pressure. The thermal infrared radiation law of shear fracture has been studied and analyzed in detail, and the infrared radiation characteristics and influencing factors of the process of rock stick-slip, rock frictional slip, and rock low-speed impact were also studied. The infrared radiation laws and thermal infrared precursor characteristics of granite, gneiss, quartz sandstone, and other rocks during the stress failure process were explored.

Subsequently, Gong Weili et al. [20–22] studied the deformation and failure characteristics of deep soft rock roadways based on infrared thermal imaging technology and revealed the friction and slip phenomenon between rock layers during the deformation and failure process of roadways through infrared images. Zhang Yanbo et al. [23,24] first observed the whole process of uniaxial compression failure of a round hole rock sample with an infrared thermal imaging camera, studied the infrared radiation characteristics of the failure process of the porous rock, and analyzed that regional zonal heating is an important precursor to rock failure. Infrared thermal imaging and acoustic emission technology were used to jointly monitor the failure process of granite samples, and the damage evolution process of the rock was studied from the two perspectives of surface temperature changes and internal acoustic emission characteristics. Wu Xianzhen et al. [25,26] proposed the concept of "Infrared Temperature Variation Field (ITVF)" and conducted an in-depth study on the characteristics of the instantaneous change of the infrared temperature field during the fracture process of saturated siltstone, providing a new method for exploring the precursors of rock fracture. Ma Liqiang et al. [27,28] proposed a new quantitative analysis index of infrared radiation, "differential infrared variance", and studied the abnormal characteristics of infrared radiation in the process of coal destruction. Zhou Zilong et al. [29] carried out uniaxial compression tests on granite samples under different loading rates, used infrared thermal imaging cameras to monitor the whole process, and studied the infrared radiation characteristics during the failure of the samples. Junwei Ma et al. [30] used infrared thermal imaging technology combined with ground laser scanning and particle tracking velocimetry to study the stability of landslides and showed that the decrease of the average temperature change can be used as a precursor to landslide damage.

In summary, infrared thermal imaging technology has been widely used in rock mechanics' experiments such as roadways, slopes, and jointed rock masses [31–35]. However, the current infrared experiments are mainly focused on the research of complete models and specimens. The rock mass is usually rich in joints, cracks, and holes, which play a vital role in the stability of the rock mass [36]. Through a series of indoor experiments, researchers have conducted a lot of in-depth research on the failure process of prefabricated joints, cracks, and holes in the rock [37–44], which is of great significance for exploring the failure mechanism of rocks. In this paper, the size of the horizontal axis a of the rock sample elliptical hole remained unchanged and the vertical axis b gradually increased. The ratio of the vertical axis to the horizontal axis k had a significant impact on the failure strength of the rock. An FLIR X8501sc infrared thermal imaging camera was used as the experimental observation equipment to conduct synchronous, high-frequency, real-time observation of the uniaxial compression test of marble to study the evolution characteristics of the temperature field during the failure process of the sample.

## 2. Experimental

### 2.1. Testing Machine

This experiment used cylindrical marble as the experimental sample, with a diameter of 50 mm and a height of 50 mm. The non-parallelism and non-perpendicularity of the end faces were both less than 0.02 mm, which met the basic requirements of ISRM. There were four different prefabricated elliptical holes and one round hole in the geometric center of the sample side. The size of the elliptical hole was 12 mm × 4 mm (horizontal axis a × vertical axis b), 12 mm × 8 mm, 12 mm × 16 mm, and 12 mm × 20 mm. The size of the circular hole was 12 mm × 12 mm. There were three samples per group. The uniaxial compression test was carried out with test system for deep rock gas escape under temperature-pressure coupling condition. The loading method was selected as displacement loading, and the loading rate was 0.1 mm/min. At the same time, an infrared thermal imaging camera observation system was set up for synchronous, high-frequency, real-time observation, and the observation rate was 15 frames/s. The sample picture is shown in Figure 1. Uniaxial compression test system-infrared thermal imaging camera monitoring system equipment is shown in Figure 2 below.

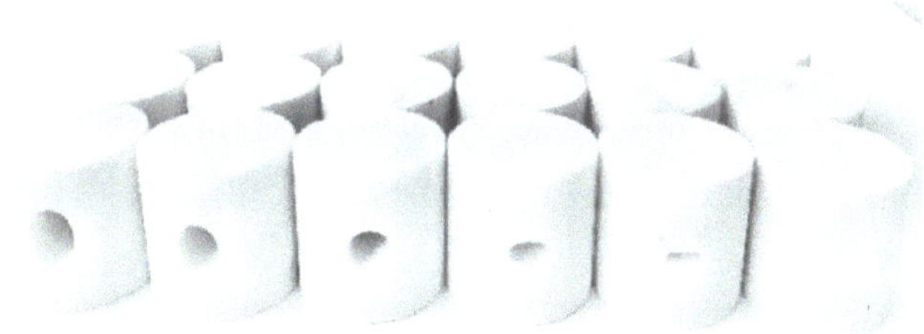

**Figure 1.** Marble rock samples.

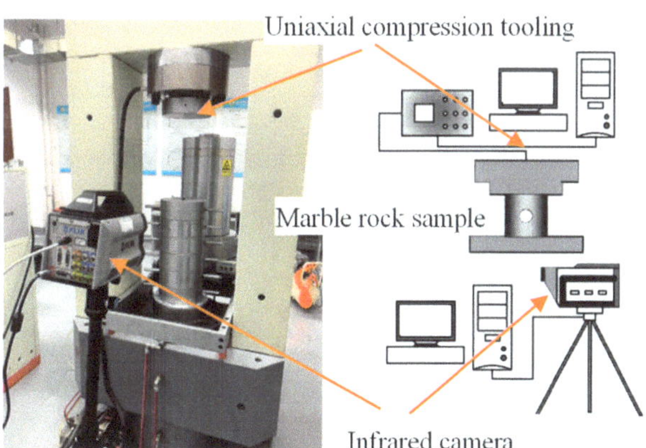



(**a**) Laboratory equipment     (**b**) schematic diagram of experimental arrangement

**Figure 2.** Test system for deep rock gas escape under Temperature-Pressure coupling condition.

From Figure 2 above, it can be seen that the infrared camera could observe the whole process of the experiment during the experiment. The prefabricated hole marble specimen in Figure 1 was placed under the loading head of the equipment to complete the uniaxial compressive strength test.

The FLIR X8501sc infrared thermal imaging camera used in this study had a temperature measurement range of $-20\,°C \sim 1500\,°C$, a thermal sensitivity of $0.02\,°C$, an integration time adjustment range of 0.3 us~30 ms, and a pixel clock of 355 MHz. Infrared heat radiation is easily affected by the surrounding environment. However, the FLIR X8501sc infrared thermal imaging camera uses a refrigerated photon detector and uses indium antimonide (InSb) as the refrigeration material, which is not easily affected by the environment.

### 2.2. Testing Program

(1) Fix the FLIR X8501 sc thermal imaging camera on a tripod. Then adjust the tripod to a suitable height and a suitable distance from the observation surface of the specimen. The camera has high temperature measurement accuracy, strong resistance to environmental interference, and can record temperature changes during the entire experiment.

(2) Connect the relevant lines. Then open the software operation interface. Connect the infrared thermal imager. Select the appropriate temperature measurement range. Set the emissivity, atmospheric temperature, storage path and other related parameters. Adjust the camera lens focal length to make the image clear.

(3) Since the infrared thermal imaging camera is very sensitive to heat sources, the external environment and personnel activities will have a great impact on the test results. In order to make the test data more realistic, the experiment was carried out in a closed environment without the influence of people walking and air flow, so as to minimize the environmental impact.

(4) Simultaneously, when the uniaxial compression test is loaded, the data will be collected synchronously at a frame rate of 15 frames/s.

(5) After the test is over, the data collection is completed, and the relevant software is used to analyze the infrared radiation data during the destruction of the sample.

### 2.3. Principle of Infrared Thermal Imaging

Any object in nature above absolute zero is constantly radiating electromagnetic waves through its own molecular motion. The infrared spectrum is distributed between visible light and microwave, and the wavelength is between 0.75 μm and 1000 μm. Infrared thermal imaging technology actually converts the difference in infrared radiation on the surface of an object into infrared images with different temperature distributions. The infrared thermal imager cannot directly measure the temperature of the surface of the object but measures the infrared radiation energy projected on the detector. The radiant temperature is calculated by the function of radiant energy and temperature. That is, the temperature detected by the infrared camera is actually the radiation temperature.

The working principle of the infrared camera is that the optical system collects the infrared radiation on the sensitive surface of the detector. The scanner matches the small field of view of the detector with the large field of view of the optical system. The signal processing circuit converts infrared radiation into electrical signals and amplifies them. Finally, it is displayed by the monitor in the form of an infrared thermal image.

## 3. Analysis of Experimental Results

### 3.1. Stress–Strain Curve

Because the stress–strain curve of DL2 is similar to that of DL1, the stress–strain curve of DL3 is similar to that of DL4 and the stress–strain curve of DL6 is similar to that of DL5. Therefore, only the stress–strain curves of DL1, DL3, and DL5 during uniaxial compression failure are shown in Figure 3. The uniaxial compressive strength data of all specimens are shown in Table 1, and the variation trend of the uniaxial compressive strength of the sample with k (ratio of horizontal and vertical axis a:b) is shown in Figure 4.

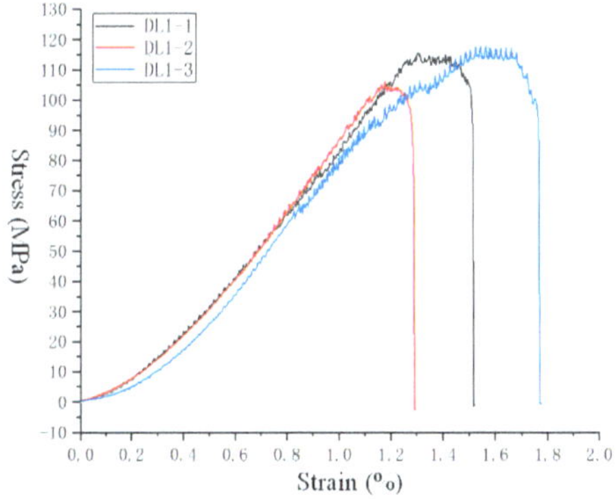

(**a**) Stress–strain curve of DL1 group specimens DL1

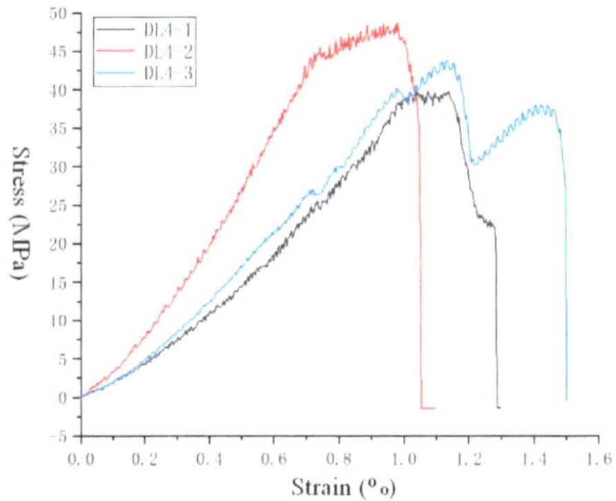

(**b**) Stress–strain curve of DL4 group specimens

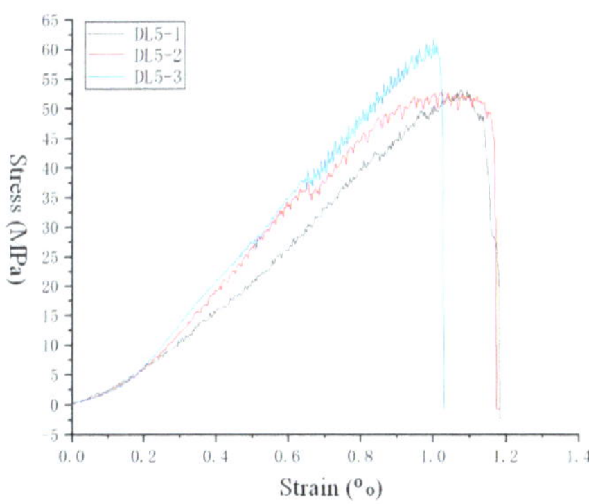

(**c**) Stress–strain curve of DL5 group specimens.

**Figure 3.** Stress–strain curves of marble specimens.

**Table 1.** Uniaxial compressive strength of all samples.

| Number | Picture | Schematic Diagram | k (a/b) | Uniaxial Compressive Strength (MPa) | Difference Value (MPa) | Average UCS (MPa) |
|---|---|---|---|---|---|---|
| DL1-1 | | | | 116.5 | 3.2 | |
| DL1-2 | | — | | 105.8 | −7.5 | 113.3 |
| DL1-3 | | | | 117.5 | 4.2 | |
| DL2-1 | | | | 78.9 | 8.9 | |
| DL2-2 | | | 3 | 63.7 | −6.3 | 70.0 |
| DL2-3 | | | | 67.5 | −2.5 | |
| DL3-1 | | | | 61.8 | 4.9 | |
| DL3-2 | | | 1.5 | 51.7 | −5.2 | 56.9 |
| DL3-3 | | | | 57.4 | 0.5 | |
| DL4-1 | | | | 40 | −4.2 | |
| DL4-2 | | | 1 | 48.9 | 4.7 | 44.2 |
| DL4-3 | | | | 43.7 | −0.5 | |
| DL5-1 | | | | 53.2 | −2.7 | |
| DL5-2 | | | 0.75 | 52.9 | −3 | 55.9 |
| DL5-3 | | | | 61.7 | 5.8 | |
| DL6-1 | | | | 57.8 | 6 | |
| DL6-2 | | | 0.6 | 50.6 | −1.2 | 51.8 |
| DL6-3 | | | | 47.1 | −4.7 | |

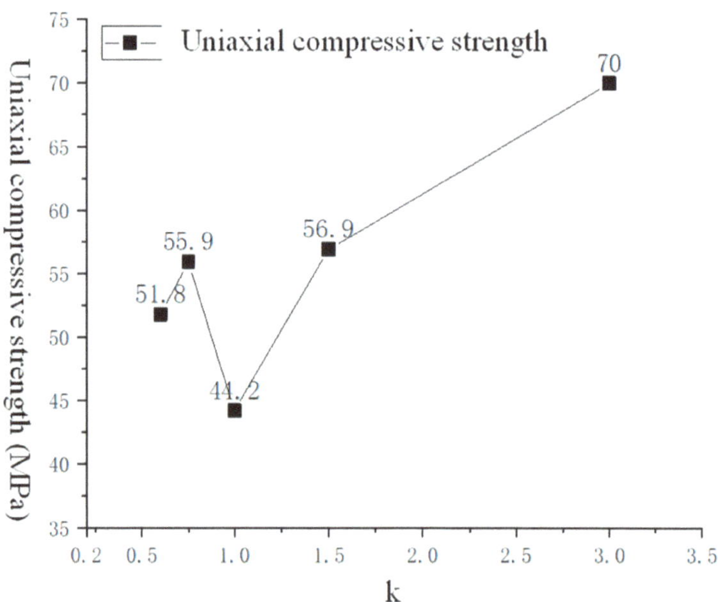


**Figure 4.** Trend chart of uniaxial compressive strength changing with k.

From Figure 3, it can be seen that the stress–strain curve of the prefabricated hole marble sample can be roughly divided into two types:

(1) With the gradual increase of the axial strain, the axial stress also gradually increased. When the yield strength is reached, a cliff-like decline and termination will occur, which is determined by the physical properties of the brittle rock itself.

(2) As the axial strain increases, the axial stress also increases and there will be a sudden drop after reaching the yield strength. However, it will rise slowly as the strain increases and then drop sharply to a certain extent. There will be two terminations after the second yield or even multiple yields.

From the data in Figure 3a and Table 1, we can see that the average uniaxial compressive strength of the complete rock sample was 113.3 MPa. It can be seen from the curve in Figure 3a that the stress–strain curve of the complete rock sample conformed to the standard uniaxial compressive strength curve. Combined with the data in Figure 3b and Table 1, it can be seen that when the rock sample had a prefabricated circular hole (the hole diameter was 12 mm), the average uniaxial compressive strength of the rock sample was reduced to 44.2 MPa. That is, the average uniaxial compressive strength of DL4 was less than half of DL1, indicating that the pores had a great influence on the strength of the rock. Combining the data in Figure 3c and Table 1, it can be seen that when the k value of the prefabricated cavities of the rock sample was less than 1, the average uniaxial compressive strength of the rock sample had an increasing trend, indicating that the shape of the cavities had an effect on the strength of the rock.

*3.2. Strength Analysis*

Due to the difference of the shape and size of the holes and the heterogeneity of the rock itself, the peak stress of the sample was different, but it also had certain regularity. From Table 1 and Figure 4, the specific data and trend of the uniaxial compressive strength of the prefabricated marble samples can be obtained. Due to the difference of processing factors and the physical properties of the rock itself, the uniaxial compressive strength of the sample with the same shape and size of the prefabricated hole was different. The k value in Table 1 and Figure 4 represents the ratio of the horizontal axis a to the vertical axis b of the prefabricated holes in the rock sample, that is, k = a/b. Since the horizontal axis a of our prefabricated hole remained unchanged at 12 mm and the vertical axis b increased from 0 mm to 20 mm, the value of k gradually changed from large to small.

It can be seen from Figure 4 that the uniaxial compressive strength changed with the change of the k. However, the overall analysis showed that k = 1 was the dividing line. When k > 1, as the value of k decreased, the uniaxial compressive strength of the marble sample with prefabricated elliptical holes gradually decreased. It shows that when the horizontal axis a is larger than the vertical axis b, as the vertical axis b increases, the uniaxial compressive strength gradually decreases. When k < 1, as the value of k decreases, the uniaxial compressive strength of marble samples with prefabricated elliptical holes will first increase and then decrease. From an overall point of view, when k = 1, that is, when the horizontal and vertical axes are equal, the uniaxial compressive strength of the marble specimens with prefabricated holes is the lowest.

## 4. Evolution Characteristics of Temperature Field

The FLIR X8501sc thermal imaging camera can record the evolution of the temperature field in the observation area during the uniaxial compression failure process of the sample in high frequency and real time. The calculation and analysis software of the infrared system can also obtain the average temperature, maximum temperature, and minimum temperature change data in the observation area, as well as the temperature distribution data of all pixels in the observation area at any time during the loading process. Figure 5 shows infrared images of the specimen DL1-3 during the destruction process. Because of this study, a total of six groups of 18 uniaxial compression experiments were done. It was not possible to show the infrared image of the destruction process of each specimen. Therefore, the infrared images of two typical specimens of DL1-3 and DL5-3 were selected for comparative analysis to demonstrate the influence of prefabricated holes on the mechanical properties of rocks.

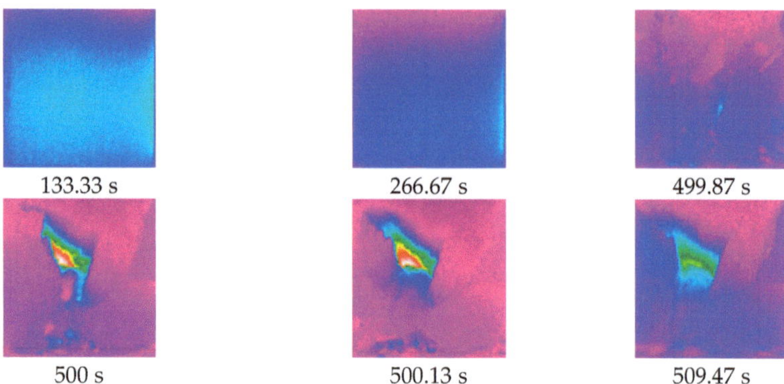

133.33 s  266.67 s  499.87 s

500 s  500.13 s  509.47 s

**Figure 5.** Cloud diagram of temperature field change during DL1-3 specimen failure.

### 4.1. Analysis of Infrared Results of Typical Specimens

Figure 5 is a cloud diagram of the temperature field evolution of the DL1-3 specimen during uniaxial compression failure. Figure 6 shows the temperature–time change curve of the sample failure process. Figure 7 shows the temperature oscillogram of the observation area at the critical moment during the destruction of the sample.

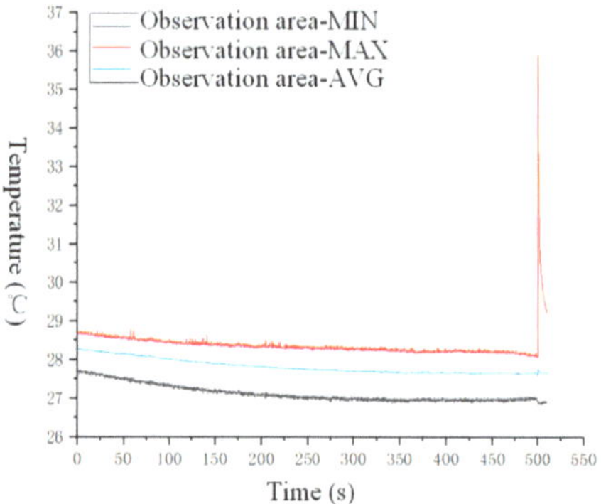


**Figure 6.** Temperature—time change curve of DL1-3 sample destruction process.

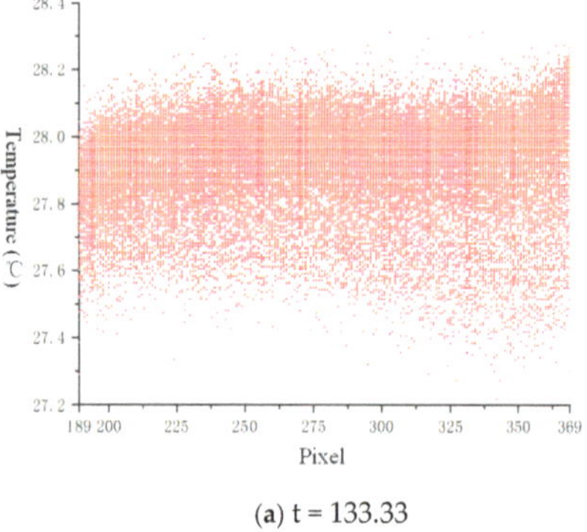


(a) t = 133.33

**Figure 7.** *Cont.*

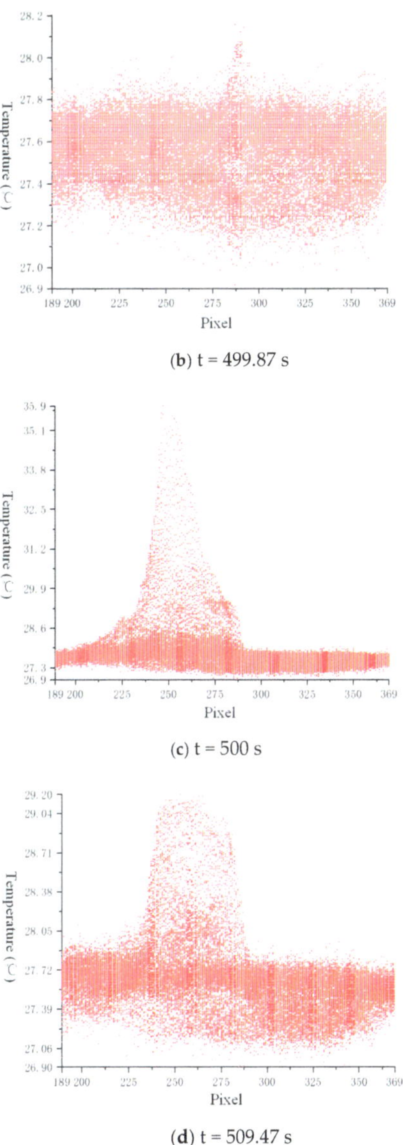

(**b**) t = 499.87 s

(**c**) t = 500 s

(**d**) t = 509.47 s

**Figure 7.** The oscillogram of the temperature in the observation area at the critical moment of the DL1-3 sample destruction.

From Figure 5, we can see the evolution of the temperature field in the observation area during the uniaxial compression test of the DL1-3 sample. From the temperature field cloud diagram from 133.3 to 266.67 s, the color temperature distribution of the temperature field cloud diagram changed little. From the temperature field cloud diagram from 499.87 s to 509.47 s, the sample was brittle failure under continuous loading pressure. The dense particles inside the sample generated a lot of heat energy under the action of intense friction, forming an obvious local heating zone.

Combining with the temperature–time variation curve of the observation area in the uniaxial compression failure process of the sample in Figure 6, it can be seen that the maximum temperature, the minimum temperature, and the average temperature in the observation area changed little during the time period from 0 s to 499.87 s when the sample was loaded. The maximum temperature slowly dropped from 28.66 °C to 28.08 °C. The minimum temperature gradually dropped from 27.68 °C to about 27.01°C. The average temperature gradually dropped from 28.25 °C to about 27.62 °C. After 499.87 s, due to the propagation and penetration of internal cracks in the sample, macroscopic failure occurred. The maximum temperature of the observation area increased sharply, the average temperature also increased, and the minimum temperature decreased. It shows that the brittle failure process of DL1-3 sample will form a local heating zone and a local cooling zone. The range and extent of the temperature increase in the heating zone were greater than those in the cooling zone, so the overall performance was temperature rises. In the brittle failure stage of the sample, the minimum temperature decreased from 27.04 °C to 26.83 °C, a decrease of about 0.2 °C. The maximum temperature rose sharply from 28.17 °C in 499.87 s to 35.87 °C in 500 s, an increase of to 7.7 °C. The average temperature rose from 27.56 °C to 27.75 °C, an increase of about 0.19 °C. The minimum and average temperature changes were relatively small, and the maximum temperature changes were relatively large.

Figure 7 shows the temperature distribution oscillogram of the observation area at any time during the sample loading process, where each red dot represents a pixel. The oscillogram at the early stage of loading is shown in 133.3 s. The temperature distribution was relatively uniform. A large number of distributions were between about 27.4 °C~28.2 °C and a few were between 28.2 °C~28.4 °C and 27.2 °C~27.4 °C. It can be seen from the oscillogram of 449.87 s that the trend of local aggregation in the high-temperature region began to appear. At 500 s, it can be seen that the temperature distribution of some pixels was 28.2 °C~35.9 °C, forming a local heating zone. It can be seen from the oscillogram of 509.47°C that the maximum temperature in the local heating zone dropped from 35.9 °C to 29.2 °C. The temperature in the local heating zone first rose rapidly and then dropped relatively slowly.

Figure 8 below is a cloud diagram of temperature field changes during uniaxial compression failure of DL5-3 sample. Figure 9 shows the temperature–time change curve of the sample failure process. Figure 10 shows the temperature oscillogram of the observation area at the critical moment during the destruction of the sample.

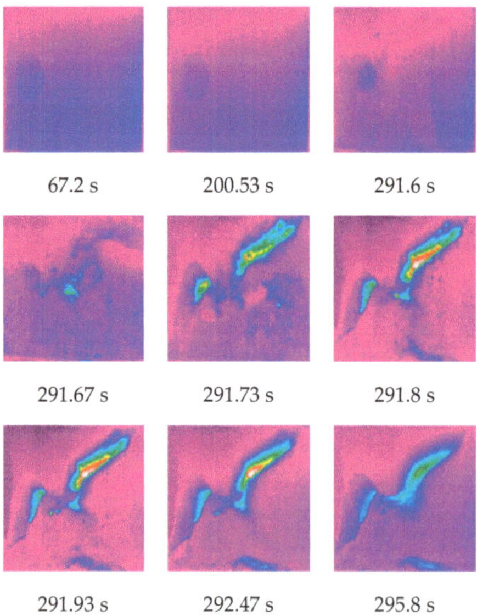


**Figure 8.** Cloud diagram of temperature field change during DL5-3 specimen failure.

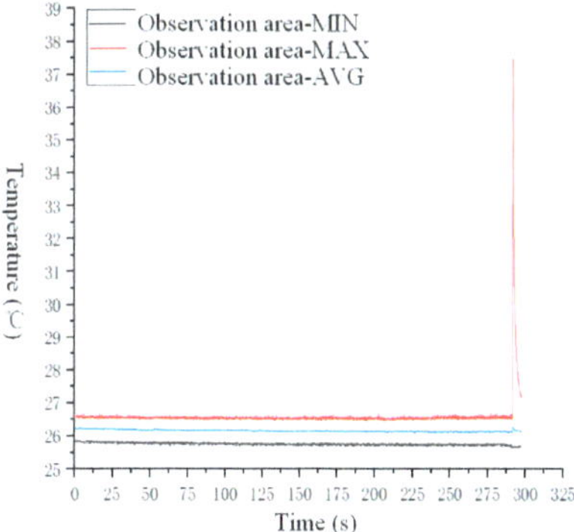

**Figure 9.** Temperature—time change curve of DL5-3 sample destruction process.

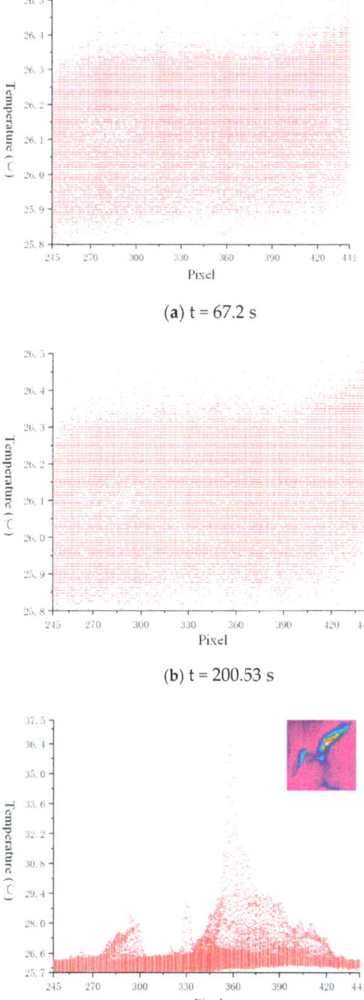

(**a**) t = 67.2 s

(**b**) t = 200.53 s

(**c**) t = 291.8 s

**Figure 10.** *Cont.*

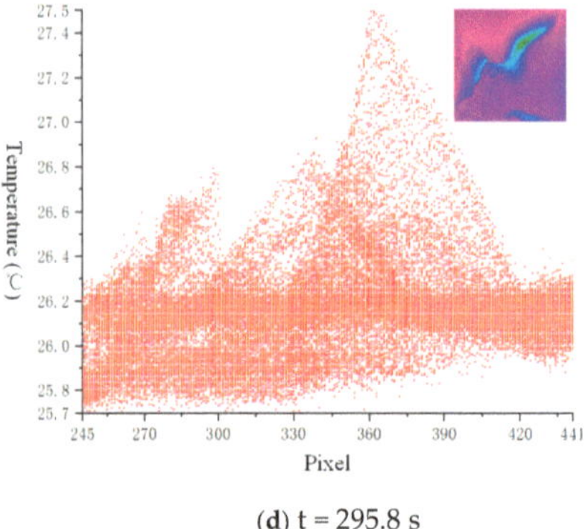

(**d**) t = 295.8 s

**Figure 10.** The oscillogram of the temperature in the observation area at the critical moment of the DL5-3 sample destruction.

From Figure 8, we can see the evolution of the temperature field in the observation area during the uniaxial compression test of the DL5-3 sample. In the time period of 0~291.67 s, the color temperature distribution of the temperature cloud image in the observation area of the sample was uniform and the amount of change was small. At 291.67 s, the temperature-changing area gradually appeared. The temperature-changing area gradually increased with the destruction of the sample and finally formed a semi-"X" deformation temperature area.

It can be seen from Figure 9 that the maximum temperature, minimum temperature, and average temperature in the observation area of the sample had very small changes during the period of 0 to 291.67 s, and the overall trend was horizontal. The maximum temperature gradually dropped from 26.61 °C to 26.51 °C. The minimum temperature dropped from 25.83 °C to 25.71 °C. The average temperature slowly dropped from 26.23 to 26.13 °C. At the moment of 291.67 s, the maximum temperature, the minimum temperature, and the average temperature all changed abruptly, which proved that brittle failure began to appear at 291.67 s. The minimum temperature decreased from 25.72 °C to 25.62 °C, a decrease of 0.1 °C, with a relatively small change. The maximum temperature rose sharply from 26.53 °C to 37.48 °C, with a particularly large change, rising 10.95 °C. The average temperature rose from 26.11 °C to 26.30 °C, an increase of 0.19 °C, and the range of change was relatively small. From the temperature change curve of the observation area during the sample loading process, it can be seen that during the brittle failure stage of the sample, the particles inside the sample released a large amount of heat energy due to intense friction, forming a local heating zone in the observation area. The location of the local heating zone coincided with the crack penetration area, and the crack penetration caused energy release to form a local cooling zone. Therefore, the overall performance of the sample observation area was increased and then gradually decreased to ambient temperature.

It can be seen from Figure 10 that the temperature distribution of all pixels in the sample observation area at 67.2 s and 200.53 s was extremely similar and the temperature was evenly distributed between 25.8 °C and 26.5 °C. The sample suffered brittle failure at 291.67 s, and some pixels began to heat up. As can be seen from the 291.8-s temperature pixel curve, the maximum temperature rose sharply by 37.5 °C and then quickly began to fall. The temperature distribution of all pixels in the observation area of the sample changed dramatically in a short time. During the brittle failure stage, the local heating zone showed a trend of sudden temperature rise and drop, and the whole observation zone also showed this trend. The variation range and distribution area of the local cooling zone were relatively small and the overall impact was small.

Figures 7 and 10 are graphs of temperature field changes during the loading process of two different specimens. From the two pictures, it can be seen that the temperature field pictures of the two specimens before failure were similar. However, there was a clear difference in the temperature field between the two in the process of specimen failure. The main reason is that the specimen in Figure 10 had prefabricated holes, while Figure 7 was a complete specimen, which led to different temperature field changes when the two were broken.

### 4.2. Temperature Change Trend Analysis

Through the analysis of temperature change data in the observation area during the uniaxial compression failure process of the prefabricated marble sample obtained by the infrared thermal imaging camera, we obtained the maximum temperature, minimum temperature, and average temperature change ($\triangle t_{max}$, $\triangle t_{min}$, $\triangle t_{avg}$) in the loading stage and brittle failure stage of the sample, as shown in Table 2.

**Table 2.** Temperature variation of all rock samples during the failure process.

| Number | Picture | Loading Stage | | | | Brittle Failure Stage | | | |
|---|---|---|---|---|---|---|---|---|---|
| | | $\triangle t_{min}$ (°C) | $\triangle t_{max}$ (°C) | $\triangle t_{avg}$ (°C) | $\overline{\triangle t_{avg}}$ (°C) | $\triangle t_{min}$ (°C) | $\triangle t_{max}$ (°C) | $\triangle t_{avg}$ (°C) | $\overline{\triangle t_{avg}}$ (°C) |
| DL1-1 | | −0.36 | −0.39 | −0.37 | | −0.15 | 6.51 | 0.16 | |
| DL1-2 | | −0.23 | −0.43 | −0.29 | | −0.13 | 5.82 | 0.17 | |
| DL1-3 | | −0.67 | −0.58 | −0.63 | | −0.2 | 7.70 | 0.19 | |
| DL2-1 | | −0.24 | −0.22 | −0.23 | | −0.11 | 2.44 | 0.14 | |
| DL2-2 | | −0.27 | −0.16 | −0.25 | | −0.09 | 3.35 | 0.12 | |
| DL2-3 | | −0.31 | −0.27 | −0.29 | | −0.15 | 4.76 | 0.11 | |
| DL3-1 | | −0.49 | −0.42 | −0.47 | −0.29 | −0.12 | 5.23 | 0.13 | 0.12 |
| DL3-2 | | −0.34 | −0.32 | −0.33 | | −0.07 | 1.64 | 0.07 | |
| DL3-3 | | −0.11 | −0.11 | −0.11 | | −0.08 | 1.68 | 0.05 | |
| DL3-1 | | −0.47 | −0.43 | −0.46 | | −0.11 | 2.31 | 0.08 | |
| DL3-2 | | −0.39 | −0.33 | −0.37 | | −0.14 | 5.76 | 0.14 | |
| DL3-3 | | −0.33 | −0.29 | −0.32 | | −0.09 | 6.33 | 0.15 | |
| DL5-1 | | −0.18 | −0.17 | −0.17 | | −0.13 | 3.21 | 0.09 | |
| DL5-2 | | −0.23 | −0.21 | −0.22 | | −0.08 | 4.57 | 0.10 | |
| DL5-3 | | −0.12 | −0.10 | −0.10 | | −0.1 | 10.96 | 0.19 | |
| DL6-1 | | −0.23 | −0.17 | −0.21 | | −0.14 | 2.34 | 0.09 | |
| DL6-2 | | −0.09 | −0.07 | −0.08 | | −0.09 | 1.57 | 0.08 | |
| DL6-3 | | −0.35 | −0.30 | −0.33 | | −0.07 | 2.11 | 0.06 | |

It can be seen from Table 2 that the prefabricated hole marble sample was loaded from the beginning to the brittle failure stage. The maximum temperature, minimum temperature, and average temperature of the observation area of the sample decreased, and the range of decrease was 0.02~1 °C. In the brittle failure stage, the maximum temperature in the observation area of the sample increased sharply, with a large amount of change, and the maximum increase observed reached 10.96 °C. The minimum temperature dropped sharply, and the drop range was about 0.07 °C~0.2 °C. The average temperature increased sharply, rising by about 0.05 °C to 0.19 °C.

From Table 2 we can know the temperature change rate of all rock samples during the loading process, including the minimum temperature, maximum temperature, and average temperature. During the loading stage, all the temperatures of the rock sample showed a downward trend, mainly due to the accumulation of energy in the loading process, which caused the surface temperature of the sample to decrease. In the brittle failure stage, macroscopic cracks appeared on the surface of the specimen, which led to the dissipation of the accumulated energy inside, which increased the maximum temperature and average temperature of the specimen. Among them, the maximum temperature increase of the specimen was relatively large, mainly concentrated at the moment when the specimen ruptured, because the temperature increase at the moment of the specimen rupture was mainly due to the change of the ambient temperature. Therefore, the average temperature of the specimen should be the main research object.

After statistical calculation, the average value of the average temperature change of all specimens during the loading stage was $-0.29$ °C and the standard deviation was 0.14 °C. The standard deviation reflects the degree of dispersion of the value relative to the average value, indicating that the dispersion of the loading stage is greater. During the brittle failure stage, the average temperature change of all rock samples was 0.12 degrees and the standard deviation was 0.04 °C. It shows that the dispersion of the average temperature change in the destruction stage was small. The comparison between the standard deviation and the average value shows that the temperature change rate in the loading stage was unstable, while the temperature change rate in the failure stage was relatively small, which is statistically significant.

## 5. Conclusions

Aiming at the scientific problem of detecting the mechanical behavior of rock masses in the engineering environment, the marble with different prefabricated holes was used as the research object and uniaxial compression experiments and infrared detection research were carried out. Through the processing and analysis of the infrared image, the average infrared temperature change revealed the changing law of the rock mass under the action of external load, and the infrared image represented the structural response of the rock mass and the degree of damage of the rock mass. This research is of great significance to the understanding of the mechanical behavior and failure mechanism of complex rock mass structures. Infrared thermal imaging technology has broad application prospects.

(1) Prefabricated holes will affect the strength of the marble. When $k > 1$, with the decrease of $k$, the uniaxial compressive strength of the sample gradually decreases. It shows that as the vertical axis $b$ increases, the uniaxial compressive strength gradually decreases. When $k < 1$, as the $k$ decreases, the uniaxial compressive strength of the sample will first increase and then decrease. When $k = 1$, the uniaxial compressive strength of the marble specimens with prefabricated holes is the lowest. It shows that ratio of horizontal and vertical axis $k$ has a very obvious effect on the uniaxial compressive strength and the strength of marble is the lowest in the case of circular holes.

(2) During the loading stage, the color temperature distribution of the temperature field cloud map changed little. In the brittle failure stage, obvious local temperature change areas were formed. The randomly distributed infrared temperature represented the elastic deformation of the rock mass, and the large-scale concentrated infrared temperature represented the plastic deformation of the rock mass. The temperature distribution of infrared images revealed that, under the action of external load, there was energy accumulation inside the rock, which eventually led to brittle failure of the rock.

(3) During the loading stage, the maximum temperature, minimum temperature, and average temperature of the observation area showed a downward trend. It was mainly due to the accumulation of energy in the loading process of the rock sample that caused the surface temperature of the specimen to decrease. In the brittle failure stage, macroscopic cracks appeared on the surface of the specimen, which led to the dissipation of the accumulated energy inside, which increased the maximum temperature and average temperature of the specimen. Since the temperature increase at the moment of specimen rupture was mainly the change of ambient temperature, the average temperature of the specimen should be the main research object.

In this paper, the whole process of the uniaxial compression test of marble with prefabricated holes was observed in real time through the synchronous, high-frequency infrared thermal imaging camera. The experimental results obtained the evolution characteristics of the infrared heat radiation temperature field in the observation area during the sample loading process and the infrared heat radiation temperature information at any point in the observation area at any time. However, due to the experimental conditions, only one surface could be observed by one infrared thermal imaging camera. If multiple infrared thermal imaging cameras can be used for simultaneous and omni-directional observation, it will be of more research significance.

**Author Contributions:** Conceptualization, C.Z.; methodology, M.H.; software, Q.L.; validation, H.Z., P.G. and Y.P.; data curation, Q.L.; writing—original draft preparation, Q.L.; writing—review and editing, Y.P.; funding acquisition, Y.P. and C.Z. All authors have read and agreed to the published version of the manuscript.

**Funding:** This research was funded by the funding of National Natural Science Foundation of China, grant number 41702381; Zhejiang Public Welfare Technology Research Program/Social Development, grant number LGF20D020002; Key Laboratory of Rock Mechanics and Geohazards of Zhejiang Province, grant number ZJRMG-2020-02 and Fundamental Research Funds for the Central Universities, grant number B210201001.

**Data Availability Statement:** The data are available and explained in this article; readers can access the data supporting the conclusions of this study.

**Conflicts of Interest:** Authors declare no conflict of interest. The manuscript is approved by all authors for publication. I would like to declare on behalf of my coauthors that the work described is original and has not been previously published.

# References

1. Huon, V.; Cousin, B.; Wattrisse, B.; Maisonneuve, O. Investigating the thermo-mechanical behaviour of cementitious materials using image processing techniques. *Cem. Concr. Res.* **2009**, *39*, 529–536. [CrossRef]
2. Ma, S.-P.; Wang, L.-G.; Zhao, Y.-H. Experimental study on deformation field evolution during failure procedure of a rock borehole structure. *Yantu Lixue (Rock Soil Mech.)* **2006**, *27*, 1082–1086.
3. Yong, R.; Ye, J.; Li, B.; Du, S. Determining the maximum sampling interval in rock joint roughness measurements using Fourier series. *Int. J. Rock Mech. Min. Sci.* **2018**, *101*, 78–88. [CrossRef]
4. He, M.; Miao, J.; Feng, J. Rock burst process of limestone and its acoustic emission characteristics under true-triaxial unloading conditions. *Int. J. Rock Mech. Min. Sci.* **2010**, *47*, 286–298. [CrossRef]
5. Li, A.; Zhang, R.; Ai, T.; Gao, M.; Zhang, Z.; Jing, X. Acoustic emission space-time evolution rules and failure precursors of granite under uniaxial compression. *Chin. J. Geotech. Eng.* **2016**, *38*, 306–311.
6. Yang, S.-Q.; Tian, W.-L.; Huang, Y.-H.; Ma, Z.-G.; Fan, L.-F.; Wu, Z.-J. Experimental and discrete element modeling on cracking behavior of sandstone containing a single oval flaw under uniaxial compression. *Eng. Fract. Mech.* **2018**, *194*, 154–174. [CrossRef]
7. Li, S.; Li, T.; Wang, G.; Bai, S. CT real-time scanning tests on rock specimens with artificial initial crack under uniaxial conditions. *Yanshilixue Yu Gongcheng Xuebao/Chin. J. Rock Mech. Eng.* **2007**, *26*, 484–492.
8. Sun, X.; Li, X.; Zheng, B.; He, J.; Mao, T. Study on the progressive fracturing in soil and rock mixture under uniaxial compression conditions by CT scanning. *Eng. Geol.* **2020**, *279*, 105884. [CrossRef]
9. Wang, Y.; Xiao, Y.; Hou, Z.; Li, C.; Wei, X. In situ X-ray computed tomography (CT) investigation of crack damage evolution for cemented paste backfill with marble waste block admixture under uniaxial deformation. *Arab. J. Geosci.* **2020**, *13*, 1–16. [CrossRef]
10. Brand, B.T.; Roswell, G.A. Laboratory investigation of the electrodynamics of rock fracture. *Nature* **1986**, *321*, 488–492.
11. Luong, M. Infrared thermographic scanning of fatigue in metals. *Nucl. Eng. Des.* **1995**, *158*, 363–376. [CrossRef]
12. Luong, M.P. *Infrared Thermography of Macrostructural Aspects of Thermoplasticity. Micro-and Macrostructural Aspects of Thermoplasticity*; Springer: Dordrecht, The Netherlands, 2006; pp. 437–446.
13. Luong, M.P.; Parganin, D.; Loizeau, J. Infrared Thermography of Thermomechanical Couplings in Solids. In *Iutam Symposium on Advanced Optical Methods and Applications in Solid Mechanics*; Springer: Berlin/Heidelberg, Germany, 2000; Volume 82, pp. 297–304.
14. Shanjun, L.; Lixin, W.; Huanping, W.; Yuhua, W.; Tao, C.; Guohua, L. Quantitative study on the thermal infrared radiation of dark mineral rock in condition of uniaxial loading. *Chin. J. Rock Mech. Eng.* **2002**, *21*, 1585–1589.
15. Shan-jun, L.; Li-xin, W.; Chuan-ying, W.; Daqing, G.; Yuhua, W. Remote sensing-rock mechanics (VIII): TIR omens of rock fracturing. *Chin. J. Rock Mech. Eng.* **2004**, *23*, 1621–1627.
16. Shanjun, L.; Lixin, W.; Yuhua, W.; Yongqiang, L. Remote sensing-rock mechanics (V)—Analysis on the factors affecting thermal infrared radiation in process of rock viscosity sliding. *Chin. J. Rock Mech. Eng.* **2004**, *23*, 730–735.
17. Liu, S.-J.; Wu, L.-X.; Zhang, Y.-B. Temporal-spatial evolution features of infrared thermal images before rock failure. *J. Northeast. Univ. Nat. Sci* **2009**, *30*, 1034–1038.
18. Wu, L.; Liu, S.; Wu, Y.; Li, Y. Remote sensing-rock mechanics (II)—Laws of thermalinfrared radiation from viscosity-sliding of bi-shearedfaults and its meanings for tectonic earthquake omens. *Chin. J. Rock Mech. Eng.* **2004**, *23*, 192–198.

19. Lixin, W.; Shanjun, L.; Yunhua, W. Remote-sensing-rock mechanics (IV)—Laws of thermal infrared radiation from compressively-sheared fracturing of rock and its meanings for earthquake omens. *Chin. J. Rock Mech. Eng.* **2004**, *23*, 539–544.

20. Gong, W.; Peng, Y.; Sun, X.; He, M.; Zhao, S.; Chen, H.; Xie, T. Enhancement of low-contrast thermograms for detecting the stressed tunnel in horizontally stratified rocks. *Int. J. Rock Mech. Min. Sci.* **2015**, *74*, 69–80. [CrossRef]

21. Gong, W.; Peng, Y.; He, M.; Wang, J. Thermal image and spectral characterization of roadway failure process in geologically 45° inclined rocks. *Tunn. Undergr. Space Technol.* **2015**, *49*, 156–173. [CrossRef]

22. Gong, W.; Peng, Y.; He, M.; Xie, T.; Zhao, S. An overview of the thermography-based experimental studies on roadway excavation in stratified rock masses at CUMTB. *Int. J. Min. Sci. Technol.* **2015**, *25*, 333–345. [CrossRef]

23. Zhang, Y.-b.; Liu, S. Thermal radiation temperature field variation of hole rock in loading process. *Rock Soil Mech.* **2011**, *32*, 1013–1017.

24. Zhang, Y.; Wu, W.; Yao, X.; Liang, P.; Tian, B.; Huang, Y.; Liang, J. Acoustic emission-infrared characteristics and damage evolution of granite under uniaxial compression. *Rock Soil Mech* **2020**, *41*, 139–146.

25. Wu, X.; Gao, X.; Liu, X.; Zhao, K. Abnormality of infrared temperature mutation in the process of saturated siltstone failure. *J. China Coal Soc.* **2015**, *40*, 328–336.

26. Wu, X.; Gao, X.; Zhao, K.; Liu, J.; Liu, X. Abnormality of transient infrared temperature field (ITF) in the process of rock failure. *Chin. J. Rock Mech. Eng.* **2016**, *35*, 1578–1594.

27. Ma, L.; Zhang, Y.; Sun, H.; Wang, S.; Najeem, A. Experimental study on dependence of infrared radiation on stress for coal fracturing process. *J. China Coal Soc.* **2017**, *42*, 140–147.

28. Ma, L.; Zhang, D.; Guo, X.; Sun, H.; Najeem, A.; Zhang, Y. Characteristics on the variance of differential infrared image sequence during coal failures under uniaxial loading. *Chin. J. Rock Mech. Eng.* **2017**, *36*, 3927–3934.

29. Zhou, Z.; Chang, Y.; Cai, X. Experimental study of infrared radiation effects of rock with different loading rates. *J. Cent. South Univ. (Sci. Technol.)* **2019**, *50*, 1127–1134.

30. Ma, J.; Niu, X.; Liu, X.; Wang, Y.; Wen, T.; Zhang, J. Thermal Infrared Imagery Integrated with Terrestrial Laser Scanning and Particle Tracking Velocimetry for Characterization of Landslide Model Failure. *Sensors* **2019**, *20*, 219. [CrossRef]

31. Teena, M.; Manickavasagan, A. Thermal Infrared Imaging. In *Imaging with Electromagnetic Spectrum*; Manickavasagan, A., Jayasuriya, H., Eds.; Springer: Berlin/Heidelberg, Germany, 2014; pp. 147–173.

32. Mujtaba, B.; De Lima, J.L.M.P. Laboratory testing of a new thermal tracer for infrared-based PTV technique for shallow overland flows. *Catena* **2018**, *169*, 69–79. [CrossRef]

33. Morello, R. Potentialities and limitations of thermography to assess landslide risk. *Measurement* **2018**, *116*, 658–668. [CrossRef]

34. Sobrino, J.A.; Del Frate, F.; Drusch, M.; Jimenez, J.C.; Manunta, P.; Regan, A. Review of Thermal Infrared Applications and Requirements for Future High-Resolution Sensors. *IEEE Trans. Geosci. Remote. Sens.* **2016**, *54*, 2963–2972. [CrossRef]

35. Frodella, W.; Gigli, G.; Morelli, S.; Lombardi, L.; Casagli, N. Landslide Mapping and Characterization through Infrared Thermography (IRT): Suggestions for a Methodological Approach from Some Case Studies. *Remote Sens.* **2017**, *9*, 1281. [CrossRef]

36. Xu, B.; Dong, S.; Yin, S.; Li, S.; Xu, Y.; Dai, Z. Analysis of Crack Initiation and Propagation Thresholds of Inclined Cracks under High-Pressure Grouting in Ordovician Limestone. *Energies* **2021**, *14*, 360. [CrossRef]

37. Wang, Y.; Tang, J.; Dai, Z.; Yi, T. Experimental study on mechanical properties and failure modes of low-strength rock samples containing different fissures under uniaxial compression. *Eng. Fract. Mech.* **2018**, *197*, 1–20. [CrossRef]

38. Yang, S.; Huang, Y.; Jing, H.; Liu, X. Discrete element modeling on fracture coalescence behavior of red sandstone containing two unparallel fissures under uniaxial compression. *Eng. Geol.* **2014**, *178*, 28–48. [CrossRef]

39. Shi, G.; Yang, X.; Yu, H.; Zhu, C. Acoustic emission characteristics of creep fracture evolution in double-fracture fine sandstone under uniaxial compression. *Eng. Fract. Mech.* **2019**, *210*, 13–28. [CrossRef]

40. Li, Z.; Liu, S.; Ren, W.; Fang, J.; Zhu, Q.; Dun, Z. Multiscale Laboratory Study and Numerical Analysis of Water-Weakening Effect on Shale. *Adv. Mater. Sci. Eng.* **2020**, *2020*, 1–14. [CrossRef]

41. Meng, Q.; Wang, H.; Cai, M.; Xu, W.; Zhuang, X.; Rabczuk, T. Three-dimensional mesoscale computational modeling of soil-rock mixtures with concave particles. *Eng. Geol.* **2020**, *277*, 105802. [CrossRef]

42. Wang, Y.; Zhang, B.; Gao, S.; Li, C. Investigation on the effect of freeze-thaw on fracture mode classification in marble subjected to multi-level cyclic loads. *Theor. Appl. Fract. Mech.* **2021**, *111*, 102847. [CrossRef]

43. Li, Z.; Liu, H.; Dun, Z.; Ren, L.; Fang, J. Grouting effect on rock fracture using shear and seepage assessment. *Constr. Build. Mater.* **2020**, *242*, 118131. [CrossRef]

44. Meng, Q.-X.; Wang, H.; Xu, W.-Y.; Chen, Y.-L. Numerical homogenization study on the effects of columnar jointed structure on the mechanical properties of rock mass. *Int. J. Rock Mech. Min. Sci.* **2019**, *124*, 104127. [CrossRef]

*Article*

# A Novel Procedure for Coupled Simulation of Thermal and Fluid Flow Models for Rough-Walled Rock Fractures

Feng Xiong [1,2,*], Chu Zhu [3] and Qinghui Jiang [2]

1   Faculty of Engineering, China University of Geosciences, Wuhan 430074, China
2   School of Civil Engineering, Wuhan University, Wuhan 430072, China; jqh1972@whu.edu.cn
3   School of Earth Sciences and Engineering, Hohai University, Nanjing 210098, China; zhu.chun@hhu.edu.cn
*   Correspondence: fengxiong@cug.edu.cn; Tel.: +86-27-6877-2221

**Abstract:** An enhanced geothermal system (EGS) proposed on the basis of hot dry rock mining technology has become a focus of geothermal research. A novel procedure for coupled simulation of thermal and fluid flow models (NPCTF) is derived to model heat flow and thermal energy absorption characteristics in rough-walled rock fractures. The perturbation method is used to calculate the pressure and flow rate in connected wedge-shaped cells at pore-scale, and an approximate analytical solution of temperature distribution in wedge-shaped cells is obtained, which assumes an identical temperature between the fluid and fracture wall. The proposed method is verified in Barton and Choubey (1985) fracture profiles. The maximum deviation of temperature distribution between the proposed method and heat flow simulation is 13.2% and flow transmissivity is 1.2%, which indicates the results from the proposed method are in close agreement with those obtained from simulations. By applying the proposed NPCTF to real rock fractures obtained by a 3D stereotopometric scanning system, its performance was tested against heat flow simulations from a COMSOL code. The mean discrepancy between them is 1.51% for all cases of fracture profiles, meaning that the new model can be applicable for fractures with different fracture roughness. Performance analysis shows small fracture aperture increases the deviation of NPCTF, but this decreases for a large aperture fracture. The accuracy of the NPCTF is not sensitive to the size of the mesh.

**Keywords:** rough fracture; coupled hydrothermal model; joint roughness coefficient; aperture; mesh size

**Citation:** Xiong, F.; Zhu, C.; Jiang, Q. A Novel Procedure for Coupled Simulation of Thermal and Fluid Flow Models for Rough-Walled Rock Fractures. *Energies* **2021**, *14*, 951. https://doi.org/10.3390/en14040951

Academic Editor: Jacek Majorowicz

Received: 23 December 2020
Accepted: 8 February 2021
Published: 11 February 2021

**Publisher's Note:** MDPI stays neutral with regard to jurisdictional claims in published maps and institutional affiliations.



## 1. Introduction

Geothermal energy, which is clean, renewable, and widespread, has been developed and utilized in many countries, and considerable attention has been given to exploiting thermal energy from hot dry rock (HDR) 3–10 km underground, using the enhanced geothermal system (EGS) to exchange and loop heat in/out of effectively connected systems of underground fractures [1–14]. For example, the Landau plant in Germany [15], the Soultz plant in France [16], and the recently commissioned Geodynamics Habanero pilot plant [17] demonstrate the feasibility of EGS running. Figure 1 shows the traditional heat extraction process from the geothermal system, where the vast rock fracture system is the main path for heat fluid flow. However, prediction of EGS service life and thermal energy extraction encounters considerable challenges due to limited knowledge of the heat exchange behavior between flowing fluid and high temperature rock [18–21].

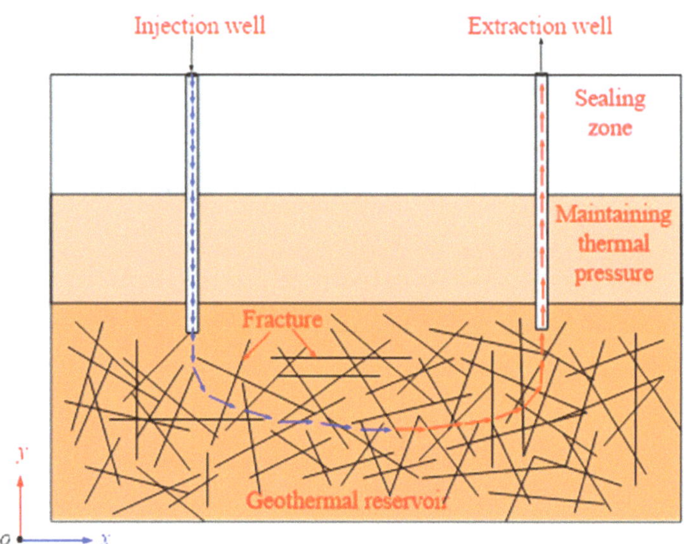


**Figure 1.** Schematic illustration of heat extraction from a geothermal reservoir.

Natural rock fractures are characterized by irregular shape, rough surface, and even contact asperities, which play important roles in the hydraulic and heat transfer properties of fractures. Because of the rough fracture surface, there may be a 1–2 order of magnitude error of permeability calculated by the widely used cubic law, which assumes the fracture is a smooth parallel plate [22–24]. Zou et al. [25] and Wang et al. [26] used a 3D Lattice Boltzmann method to predict the fluid flow behavior in fractures and found that the secondary roughness enhanced the local complexity of velocity distribution and led to nonlinear flow behavior. Similar studies were also reported by Yin et al. [27] and Xiong et al. [24]. In addition, the rough fracture surface also causes a decrease in heat transport intensity. Recently, many efforts [28–30] paid attention to the study of the effects of fracture roughness on heat transport behaviors in fractures, based on laboratory experiments and numerical simulations. For example, He et al. [28] found that heat transport characteristics mainly depended on fracture surface roughness, followed by aperture and flow rate, combining the experimental and numerical modeling approaches. Huang et al. [29] conducted seepage and convective heat transport experiments, and demonstrated that large roughness in the direction perpendicular to flow would increase the capacity of heat transport. Ma et al. [30] adopted numerical and experimental approaches to improve understanding of the heat transfer characteristics of water flowing through rough fractures, and results showed the total heat transfer coefficient increased with the increase in the value of the joint roughness coefficient (JRC). However, there are no analytical models describing the heat transport properties between fluid and high temperature rock, considering fracture surface geometry. In previous research studies, the parallel plate is traditionally used to simplify the real rough fracture surface, and many models can be observed in the literature, such as the Gehlin and Hellstrom model [2], Gringarten et al. model [31], Cheng et al. model [32], Martinez et al. model [33], Zhao model [34], Yost and Einstein model [35], and Yan and Jiao model [7] which have inevitable relative error comparing with the real heat transport process in rough-walled fractures.

In view of analytical heat transport models for rough fractures being rare, this paper aims at developing a proposed novel procedure for coupled simulation of thermal and fluid flow (NPCTF) models to understand the heat transport process in rough fractures, based on representing the measured fracture geometry with a series of connecting wedges formed using adjacent apertures in longitude direction along the fracture plane, which is a common approximation approach for fracture geometry [36]. The performance of the proposed NPCTF is assessed by comparing predictions with results from numerically solving the Navier–Stokes and energy equations, using COMSOL code in rock fracture profiles obtained by a 3D stereotopometric scanning system, which have different relative surface roughness and aperture.

## 2. Fluid Flow Model

Flow behavior in natural rock fractures has been studied through considering fracture walls as parallel plates, saw-tooth shaped walls, and sinusoidally varying walls [30]. Among these approximates, fractures are found to be well described by a series of connected wedges. Wang et al. [37,38] used a wedge-shaped fracture to obtain the perturbation solution of pressure $\Delta p$ and flowrate $Q$ under the pressure boundary condition. The perturbation parameter $\varepsilon$ is selected as the relative aperture variation along the wedge length l in his study. This section provides a brief description of the approach rather than a detailed derivation. A more detailed perturbation derivation can be viewed in Wang et al. [37,38]. By perturbation analysis, the pressure difference can be made dimensionless and expressed as expanded series with a small parameter $\varepsilon$:

$$\Delta P = \Delta P_0 + \epsilon \Delta P_1 + O\left(\epsilon^2\right) \tag{1}$$

$$\Delta P_0 = \frac{1}{\omega} \int_0^\omega \frac{1}{B^3} dX \tag{2}$$

$$\Delta P_1 = -\frac{1}{\omega} \int_0^\omega \frac{9RB'}{70B^3} dX \tag{3}$$

where $X$ is defined as $X = \varepsilon x / b_m$; $\omega$ is the dimensionless absolute aperture variation defined as $\omega = |a| / b_m$, where $a$ is the difference between the upper and lower wedge edge; $B$ is the dimensionless aperture defined as $B = c / c_m$, $c$ is the fracture local aperture, and $c_m$ is the mean aperture; $B'$ is the first derivatives of $B$ with respect to $X$; $R$ is the Reynolds number defined as $R = \rho Q / \mu$; $Q$ is flow rate per unit width of fracture and $\mu$ is fluid viscosity; $\Delta P$ is the dimensionless pressure difference given as $\Delta P = \Delta p / \Delta p_m$ and $\Delta p_m$ is the pressure difference of flow through a fracture with a uniform aperture defined from the cubic law as:

$$\Delta p_m = \frac{12\mu l Q}{b_m{}^3} \tag{4}$$

Noted that only the first two order terms in Equation (1) are used which is enough accuracy for nonlinear flow simulation. Because the quadratic polynomial form, called the Forchheimer law, can well describe nonlinear flow behavior in a single fracture [39,40], the highest order term is the second order term.

## 3. Heat Transport Model

The steady heat fluid transport differential equation in the fracture, including advection, conduction, and convection terms for the fracture walls, can be expressed as [34]:

$$v\frac{\partial T_f(x,y)}{\partial x} - \frac{K_w}{\rho_w c_w}\frac{\partial^2 T_f(x,y)}{\partial^2 x} - \frac{2K_r}{\rho_w c_w b}\frac{\partial T_r(x,y)}{\partial y}\Big|_{boundary} = 0 \tag{5}$$

where $\rho_w$ is the fluid density; $c_w$ is the specific heat of fluid; $K_w$ is the fluid thermal conductivity; $K_r$ is the rock thermal conductivity; $v$ is the steady flow velocity. $b$ is half aperture of the fracture; $T_f$ is the bulk temperature of the fluid; $T_r$ is the temperature of the reservoir rock matrix. In this equation, the temperature at the rock fracture surface is identified as that of the bulk fluid. The steady heat conduction in the rock is governed by [32,34]:

$$\frac{\partial T^2{}_r(x,y)}{\partial y^2} = 0 \tag{6}$$

which assumes that the heat conduction is two dimensional, perpendicular to the fracture plane.

The boundary conditions associated with the heat flow in the fracture are:

$$T_f(0) = T_{in} \tag{7}$$

$$T_f(\infty) = T_0 \tag{8}$$

$$T_r(x, R) = T_0 \tag{9}$$

where $R$ is the height of the rock. The identical temperature between the fluid and fracture surfaces is expressed as:

$$T_r(x, f_t(x)) = T_f(x) \tag{10}$$

$$T_r(x, f_l(x)) = T_f(x) \tag{11}$$

where $f_t(x)$ and $f_l(x)$ are the upper fracture wall and lower fracture wall, respectively. Zhao et al. [34] used this assumption to obtain a solution of temperature in the plane fracture. This solution has good agreement with the experiment results. However, there are not any temperature solutions for the fracture wedge, which is the basic element in rough fractures. Therefore, a solution for the fracture wedge would be developed based on the above heat fluid transport equations.

(1)  Symmetric fracture wedge

For the symmetric fracture wedge (Figure 2a), the wedge wall (upper fracture $f_t(x)$, lower fracture $f_l(x)$) can be described as:

$$f_t(x) = Ax + b_0 \tag{12}$$

$$f_l(x) = -Ax - b_0 \tag{13}$$

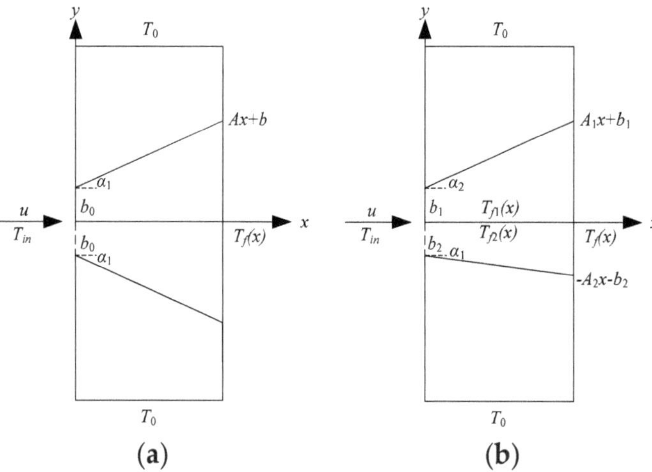


**Figure 2.** A conceptual diagram of heat flow through a rock fracture. (**a**) Symmetric fracture wedge; (**b**) Asymmetric fracture wedge.

The aperture $b$ can be determined as $f_t(x) - f_l(x)$ according to Equations (12) and (13). If thermal diffusion in the fluid is not considered, the boundary conditions in Equations (10) and (11) are substituted into Equations (5) and (6); a temperature distribution equation of fluid can be expressed as:

$$v\frac{\partial T_f(x,y)}{\partial x} - \frac{2K_r}{\rho_w c_w (Ax + b_0)}\frac{T_f(x) - T_0}{Ax + b_0 - R} = 0 \tag{14}$$

The solution to Equation (14) considering the temperature boundary in Equations (7)–(9) is:

$$T_f(x) = T_0 + (T_{in} - T_o)\exp\left(\frac{k_r ln(|\frac{b_0+R}{b_0+R+Ax}|)}{Ab_0 * c_w * \rho_w v}\right) \tag{15}$$

(2) Asymmetric fracture wedge

For the asymmetric fracture wedge (Figure 2b), the wedge wall can be described as:

$$f_t(x) = A_1 x + b_1 \tag{16}$$

$$f_l(x) = -A_2 x - b_2 \tag{17}$$

Similarly, the solutions to Equations (5) and (6) for the asymmetric wedge with boundaries in Equations (16) and (17) are:

$$T_{f1}(x) = T_0 + (T_{in} - T_0) * \exp\left(\frac{2k_r \ln(|\frac{b_1+R}{b_1+R+A_1 x}|)}{A_1 b_1 c_w p_w v - A_1 b_2 c_w p_w v}\right) \tag{18}$$

$$T_{f2}(x) = T_0 + (T_{in} - T_0) * \exp\left(\frac{2k_r \ln(|\frac{b_2-R}{b_2-R+A_2 x}|)}{A_2 b_1 c_w p_w v - A_2 b_2 c_w p_w v}\right) \tag{19}$$

$$T_f(x) = \frac{T_{f1}(x) + T_{f2}(x)}{2} \tag{20}$$

The temperature mathematic solutions for the fracture wedge were first proposed in the literature based on a few assumptions which have well-defined parameters.

## 4. Procedure for Coupled Simulation of Thermal and Fluid Flow Models (NPCTF)

### 4.1. Model Description

As presented above, the fluid flow model (Equation (1)) and heat transport model (Equations (18)–(20)) were obtained in the fracture wedge. Each fracture wedge was connected to another and constituted a rough-walled fracture (Figure 3a). According to the flowrate equilibrium principle and fluid flow model (Equation (1)), the flowrate and pressure distribution in the entire fracture can be determined (seen in Figure 3b). If the flowrate distribution is known, the temperature solution in each connected fracture wedge would be obtained using the heat transport model (as Figure 3c). When the pressure and temperature condition changes, the density of water ($\rho_w$) is no longer a constant value; this can be expressed as a function of pressure and temperature [6]:

$$\frac{1}{\rho_w} = 3.086 - 0.899017\left(647.25 - T_f\right)^{0.147166} - 0.39\left(658.15 - T_f\right)^{-1.6}(P - 225.5) + \delta_w \tag{21}$$

where $\delta_w$ is a function of $P$ and $T_f$, and the value of $\delta_w$ does not exceed 6% of $1/\rho_w$. The temperature of water also affects the kinematic viscosity of water ($\mu$). An empirical formula for kinematic viscosity is [6]:

$$\mu = \frac{0.01775}{1 + 0.033 T_f + 0.000221 T_f^2} \tag{22}$$

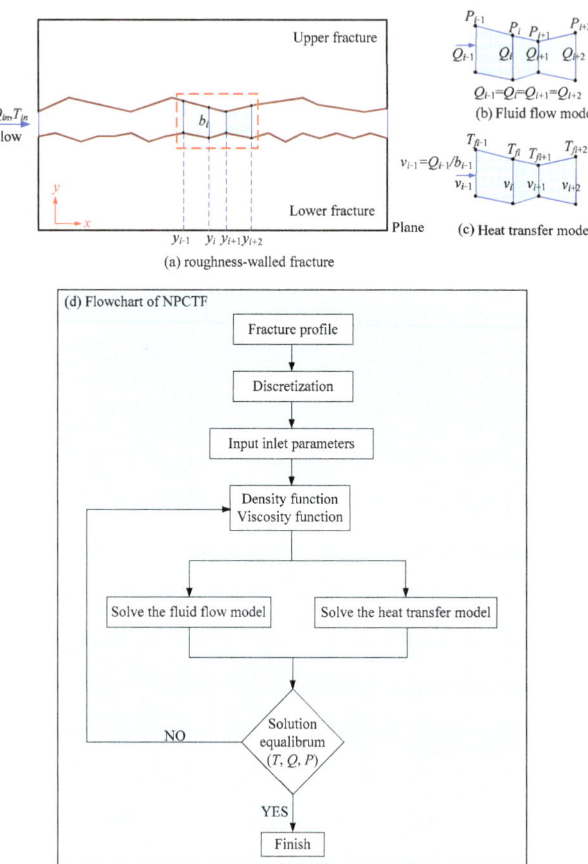



**Figure 3.** Computational flow chart of the proposed novel procedure for coupled simulation of thermal and fluid flow models (NPCTF). (**a**) Roughness-walled fracture; (**b**) Fluid flow model; (**c**) Heat transport model; (**d**) Flowchart of NPCT.

Therefore, the function of density and viscosity of fluid can relate the fluid flow model and heat transport model, and a procedure for coupled simulation of thermal and fluid flow (NPCTF) models in a two-dimension fracture profile was proposed. A detailed flowchart for the proposed simulation method is seen in Figure 3d. When fracture profile information $(x,y)$ is obtained, the fluid flow model is used to evaluate flow rate $Q$ and pressure $P$, and the heat transport model then predicts the temperature distribution $T_f(x)$. An iterative process is used to solve the fluid flow model and heat transport model by modifying the fluid's physical properties (fluid density $\rho_w$ and kinematic viscosity $\mu$) using new flowrate, pressure, and temperature before moving to the next step. The iterative process stops once the temperature and flowrate stabilize to a specified tolerance. The tolerances of flowrate and temperature are defined as $||Q_{j-1} - Q_j|| \leq 0.00001$ and $||Tf_{j-1}-Tf_j|| \leq 0.00001$, respectively. $j$ is each iteration step in the calculation process. This calculation method would increase calculated efficiency, compared to the previous finite element method or finite difference method.

## 4.2. Validation of NPCTF

Barton and Bandis (1985) [41] analyzed the roughness of 136 natural fracture surfaces. According to the roughness of the fracture surface, it is divided into 10 levels, of which values are in the range of 0–20. The roughness index is called the joint roughness coefficient

(JRC). The JRC parameter is widely used in rock mechanics and engineering. In order to valid the NPCTF, 5 JRC profiles were selected to calculate temperature and pressure distribution, and those results were compared to numerical results with *COMSOL*. The JRC values of the selected fracture profiles are 1, 5, 10, 15, and 20, respectively. Table 1 lists the thermal and flow parameters for predicting the temperature in JRC profiles.

**Table 1.** The used parameters for heat transport modeling.

| Parameters | $K_r$/W m$^{-1}$ K$^{-1}$ | $\rho_w$/kg m$^{-3}$ | $\mu$/pa s | $c_w$/J kg$^{-1}$ K$^{-1}$ | $T_{in}$/K | $T_0$/K |
|---|---|---|---|---|---|---|
| Value | 3.5 | 1000 | 0.001 | 4200 | 293.15 | 363.15 |

Figure 4 shows the comparison between the predicted results of the NPCTF and the calculated results of *COMSOL*. As illustrated in Figure 4, the calculated results of the NPCTF are consistent with the *COMSOL* results. With the increase in JRC value, the error becomes larger and a more obvious difference is observed at the nonlinear part of the temperature profile. However, the maximum error value is 12.3%. The reason is that the effect of thermal diffusion is not considered. With the increase in JRC, the thermal diffusion effect is more obvious, and enhances the differences in the nonlinear part of the temperature profile between the proposed NPCTF and *COMSOL* results. Figure 5 shows the temperature counter in the fracture profile with different JRC. In addition, the results of flow field calculated by proposed NPCTF and *COMSOL* are compared. The flow field is expressed in terms of transmissivity *Ta*, which is defined as follows:

$$Ta = \frac{QL}{P} \tag{23}$$

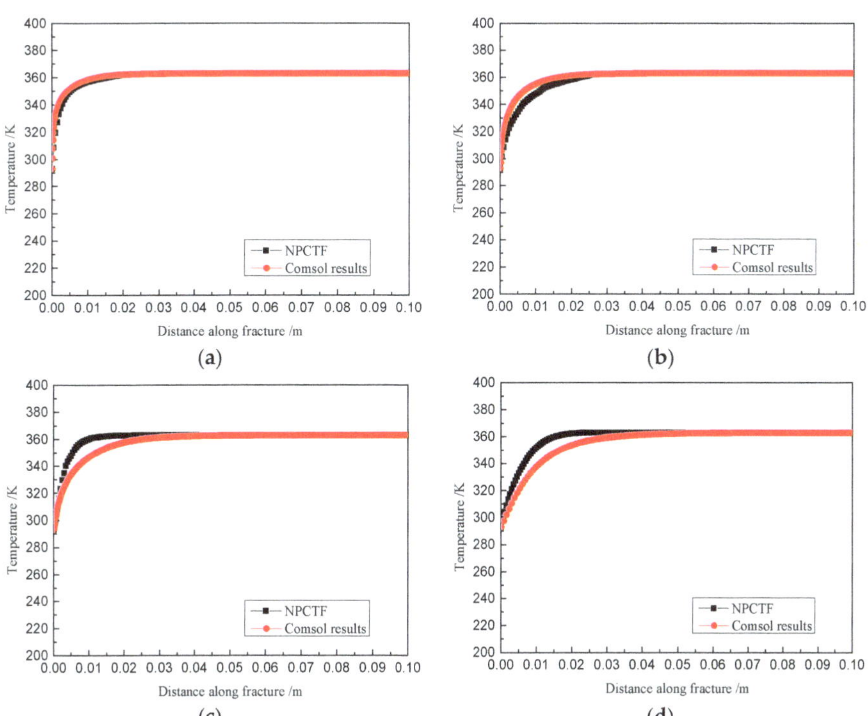


**Figure 4.** *Cont.*

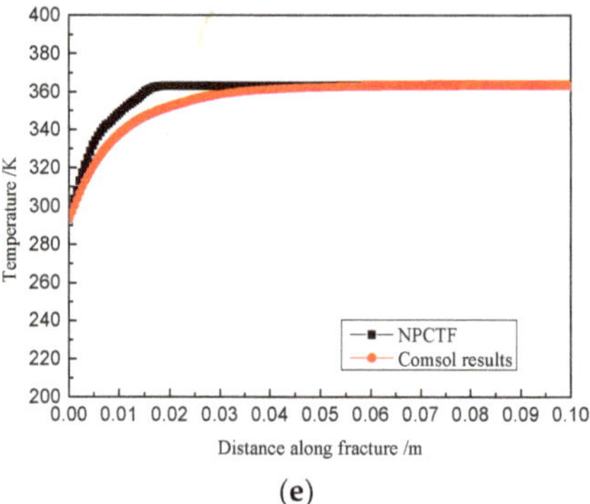



**Figure 4.** Temperature distribution along 5 fracture profiles calculated by the proposed NPCTF and COMSOL with different joint roughness coefficient (JRC) values. (**a**) JRC = 1; (**b**) JRC = 5; (**c**) JRC = 10; (**d**) JRC = 15; (**e**) JRC = 20.

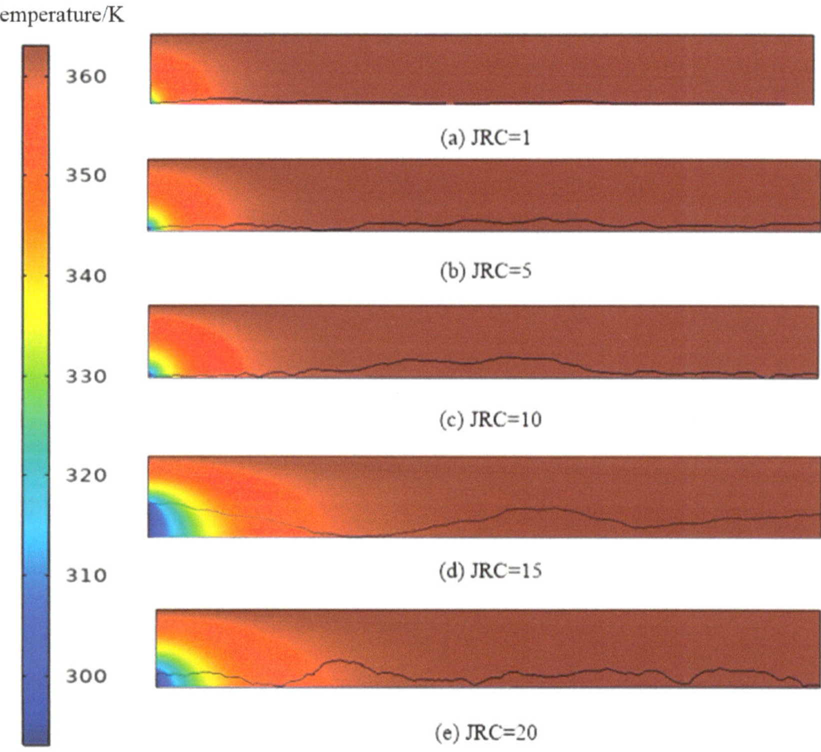



**Figure 5.** Temperature contour of fracture profile with different JRC (1–20).

Figure 6 shows the comparison of transmissivity $Ta$ of fractures under different JRC values. The transmissivities predicted by the proposed NPCTF are consistent with the *COMSOL* results. As the JRC value increases, the error increases, and the maximum error is 1.2%.

Therefore, the proposed NPCTF can predict temperature and flow velocity distribution in rough-walled fractures.

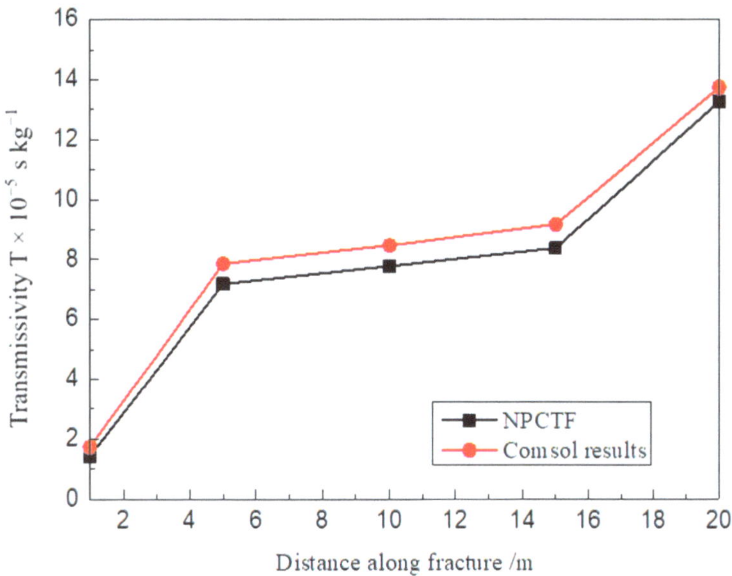



**Figure 6.** The transmissivity comparison calculated by the proposed NPCTF and COMSOL simulation in 5 fracture profiles.

## 5. Coupled Hydrothermal Simulation in Three-Dimensional Rock Fracture

### 5.1. Modelling Setup

To obtain the rock fractures, an intact granite block was split using the Brazilian indirect tensile test into two halves with dimensions of $150 \times 150 \times 75$ mm. This method was also used by Xiong et al. [24] to study fluid flow behavior in rock fractures.

To determine the surface information of the fracture, an advanced 3D CaMega stereotopometric scanning system (BoWeiHengXin Inc., Beijing, China) was used to perform a non-contact 3D scan of the rough fracture surfaces. Then, the coordinates of 22,801 points of the upper half of the fracture were obtained, as well as the lower half fracture. Tatone et al. [42] introduced a novel method to measure fracture geometry based on fracture surface data. Based on this method, a 3D fracture model was obtained, which is shown in Figure 7a.

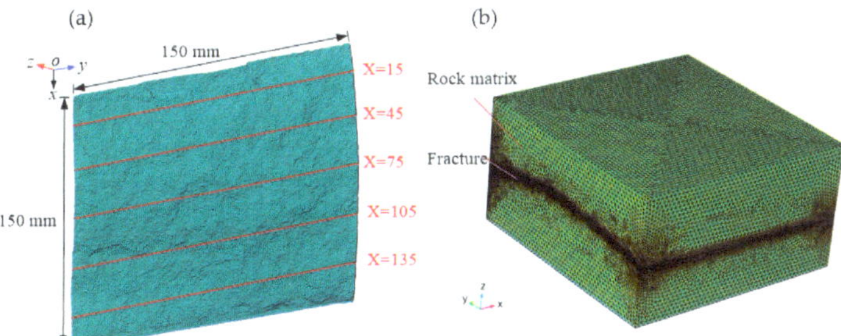



**Figure 7.** The 3D fracture model setup scanned by stereotopometric scanning system. (**a**) 3D fracture geometry model. (**b**) 3D fracture mesh model.

Five evenly spaced surface profiles parallel to the flow direction with lengths of 150 mm of the upper and lower half of the fracture were taken into consideration, namely slices $X = 15, 45, 75, 105,$ and $135$ mm, as shown in Figure 7a. The profile curves of the five slices were obtained by importing the coordinates (x, z) of every point of each slice to Auto CAD.

Correspondingly, different values of JRC are calculated according to Tse and Cruden [43], who proposed a relationship of JRC and the root mean square of the first derivative of profile $Z_2$:

$$JRC = 32.2 + 32.47 \log Z_2 \tag{24}$$

Table 2 lists the JRC and mean aperture of the fractures.

**Table 2.** Fracture geometry parameters.

| Fracture Position | JRC | Mean Aperture/mm |
|---|---|---|
| $X = 15$ | 14.86 | 1.2 |
| $X = 45$ | 16.34 | 1.5 |
| $X = 75$ | 13.26 | 1.5 |
| $X = 105$ | 12.26 | 1.8 |
| $X = 135$ | 11.96 | 1.6 |

The heat flow simulations in this work were conducted by solving a 3D Navier–Stokes equation and an energy equation with an advanced *COMSOL* Multiphysics code. The mesh scheme in the *COMSOL* mesh module was optimized to reduce the mesh-related effect on flow simulations. The fracture computational mesh was further refined, with denser mesh in the fracture part to capture the details of the flow. The final fracture mesh is shown in Figure 7b. The mean mesh sizes of the rock matrix and fracture are 35.0 and 5.0 μm, respectively. In general, over 3.2 million tetrahedron mesh elements were constructed for the fracture model. The temperature boundary conductions were the upper and lower surface of the rock matrix. No-flux and non-slip boundary conditions were set for the fracture walls. The inflow boundary was assigned a constant mass flow rate and a constant temperature, and the outflow boundary was set as a constant pressure condition.

*5.2. Temperature and Pressure Distribution Along Fracture*

To demonstrate the general thermal transport within the fractures, the temperature counter and distribution are plotted in Figure 8 along five fracture profiles tested in this study using *COMSOL* simulation results. The calculated parameters are listed in Table 1. The corresponding injection pressure was 0.01 Pa. As illustrated in Figure 8, the water has a strong trend to attain equilibrium temperature (363.15 K) close to the fracture inlet since the coupling between the water and the rock is strong and, therefore, the heat transport is more intense. Far away from the fracture inlet, the water temperature slowly attains equilibrium temperature. If the fracture is long enough, the temperature at the fracture outlet position would be same as the equilibrium temperature. Additionally, each temperature along the rough-walled fracture is not a smooth curve; the rougher fracture surface leads to more fluctuation for the temperature curve, indicating the fracture roughness affects the temperature distribution, as shown in Figure 8. It can also be seen that as fracture aperture increases, the temperature at which water is absorbed decreases. The difference between the temperature related to the aperture parameter is readily apparent when the change in temperature that occurred between fractures with $b_m = 1.2$ mm and $b_m = 1.8$ mm (the smallest and largest fracture) aperture is examined. A decrease in temperature of nearly 25% was calculated between the flow in these two fractures. This relatively large change in aperture relates well to the large simulated change in temperature. Hence, both fracture roughness and aperture have significant influences on temperature distribution.

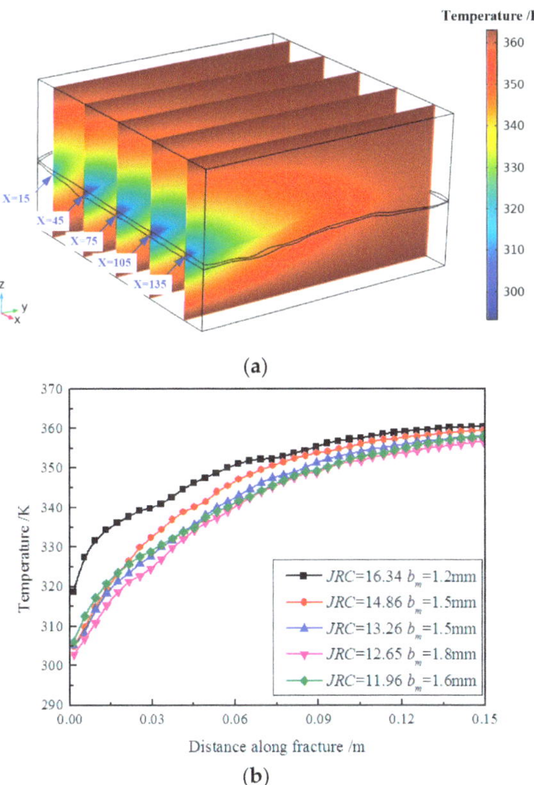

(a)

(b)

**Figure 8.** Temperature distribution of 3D fracture for (**a**) spatial temperature variation and (**b**) temperature change along fracture profiles.

The pressure distribution of the fracture is also plotted in Figure 9. As illustrated in Figure 9, from the inlet to the outlet, the fluid pressure decreases from 1 to 0 Pa. Both fracture roughness and aperture have small influences on pressure distribution.

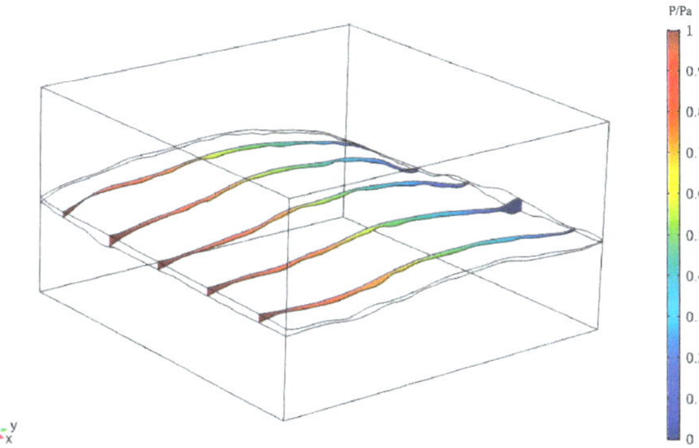

**Figure 9.** Pressure distribution of 3D fracture.

## 6. Model Performance

### 6.1. Comparison with Previous Models from the Literature

To demonstrate the robustness of the proposed NPCTF (the calculation process shown in Figure 3) in describing this thermal transport in realistic 3D rock fractures, the deviation $D$ is defined as the relative error in estimating the temperature distribution:

$$D = \frac{T_{i-model} - T_{i-real}}{T_{i-model}} \times 100\% \tag{25}$$

where $T_{i-model}$ is the fracture's local temperature at $i$ node obtained from the NPCTF and $T_{i-real}$ is the fracture local temperature at $i$ node obtained from numerical simulations. The temperature deviations $D$ calculated by Equation (25) for the five fracture profiles are plotted in Figure 10 at different fracture positions. In general, the two temperatures predicted by NPCTF and COMSOL simulations are in close agreement for all fracture profiles, with the deviation $D$ ranging from 0.44% to 8.10% with a mean of 1.51%. A highest mean deviation $<D>$ of 2.57% is observed for the fracture profile with JRC = 16.34. The fracture profile with JRC = 11.96 has the lowest $<D>$ of 0.75% in this case and the range of $D$ is from 0.48% to 3.98%. The root mean squares (RMS) of deviation $D$ were calculated. It is interesting that the fracture profile with the higher JRC has the higher RMS. The highest RMS of 0.0173 is observed for the fracture profile with JRC = 16.34. The fracture profile with JRC = 11.96 has the lowest RMS of 0.00858. The deviation $D$ increases with increases in JRC, also suggesting that the prediction accuracy of the proposed NPCTF is related to fracture roughness.

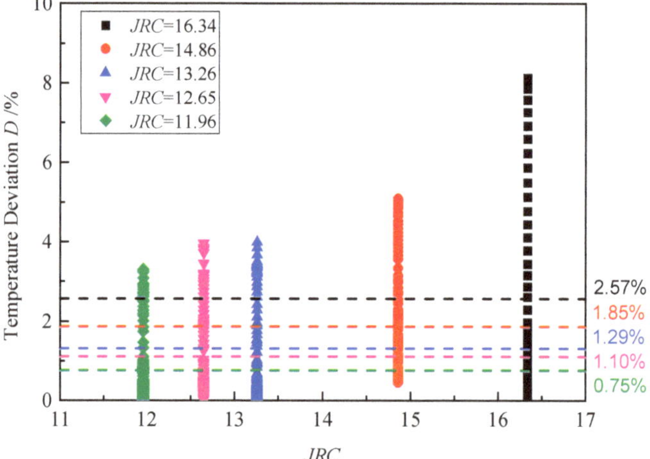


**Figure 10.** Deviation from heat flow simulations Ds for the proposed NPCTF, where the mean deviations of the model with different JRC are 2.57%, 1.85%, 1.29%, 1.10%, and 0.75%, respectively, as represented by dashed lines.

Further assessments of NPCTF were made by comparing the values of $D$ with those obtained from previously proposed heat transport models including the Gehlin and Hellstrom [2] and Zhao models [34]. A steady-state analytical solution describing the temperature distribution in the vertical fracture was developed by Gehlin and Hellstrom [2], as follows:

$$T_f(x) = T_o + (T_{in} - T_o)\exp(-\frac{\alpha}{c_w v}x) \tag{26}$$

where $\alpha$ is the heat transfer coefficient; the value of $900 \text{ W m}^{-2} \text{ K}^{-1}$ was used in the research of Shaik et al. [44]. $T_o$ is rock temperature and $T_{in}$ is water temperature in the inlet of the fracture. When the diverging angle or converging angle is 0, the proposed heat transport model can be degenerated as in the Zhao model [34]:

$$T_f(x) = T_o + (T_{in} - T_o)\exp(-\frac{k_r}{v\rho_w c_w bR}x) \tag{27}$$

For comparison, the heat transport model took the place of the Gehlin and Hellstrom and Zhao heat transport models in the NPCTF, when applying the Gehlin and Hellstrom [2] and Zhao [34] models to calculate the temperature distribution in all five fracture cases. The mean temperature deviations $<D>$ (see in Equation (25)) for all five fracture cases by comparing the NPCTF, the Gehlin and Hellstrom model [2], and the Zhao model [34] with *COMSOL* simulation are plotted in Figure 11. Overall, the highest deviation is observed for the Gehlin and Hellstrom model [2], with a mean of 23.27% and a maximum of 27.81%. The Zhao model [34] deviates by less than 10% of $<D>$ when JRC is less than 13; however, the deviation rises to 12.45%, 12.75%, and 14.16%, respectively, as JRC increases to 16.34. The poorer performance of the two models is due to not taking fracture roughness and aperture variation into consideration, compared to the proposed NPCTF. Therefore, it can be observed that the $D$ from the NPCTF is less than other models. Hence, the proposed NPCTF outperformed other models in different JRC with the lowest $<D>$ of 0.44%.

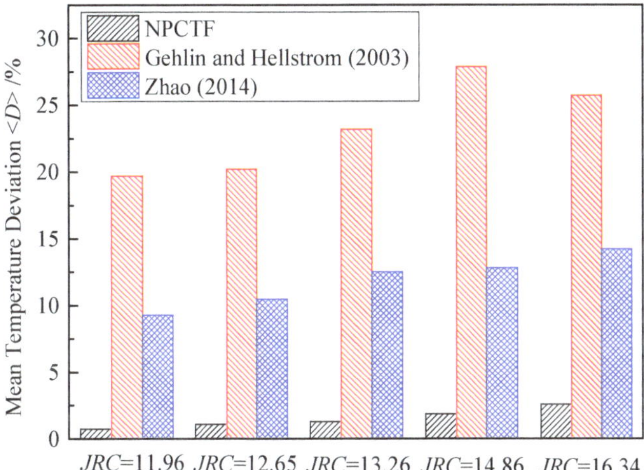


**Figure 11.** Mean deviation from heat flow simulations $<D>$ for the proposed NPCTF, Gehlin and Hellstrom (2003) [2], and Zhao (2014) [35] models.

*6.2. Sensitivity Analysis*

The accuracy of the proposed NPCTF is mainly affected by the fracture geometry (i.e., roughness and aperture variation) and mesh size (one dimensional element). The fracture roughness has been discussed above. Hence, the two factors (aperture and mesh size) need to be tested in the NPCTF.

The fracture profile with maximum roughness ($JRC = 16.34$ and $b_m = 1.5$ mm) was selected to test, and aperture $b_m$ was fixed at 0.6, 0.9, 1.2, 1.5, 1.8, and 2.1 mm by moving the distance of the lower and upper fracture surface. Figure 12 plots the effect of aperture on mean deviation $<D>$ (see in Equation (25)) from NPCTF). As shown in Figure 13, the mean deviation $<D>$ decreases as the value of $b_m$ increases. At JRC = 16.34, as the value of $b_m$ changes from 0.6 to 2.1 mm, the mean deviation $<D>$ decreases from 1.80 to 0.62. The deviation of cases for aperture less than 1.5 mm is larger than that of cases for an aperture value of 1.8 to 2.1 mm. It is evident that the effect of $b_m$ on $<D>$ is larger for a small aperture fracture and smaller for a larger aperture fracture.

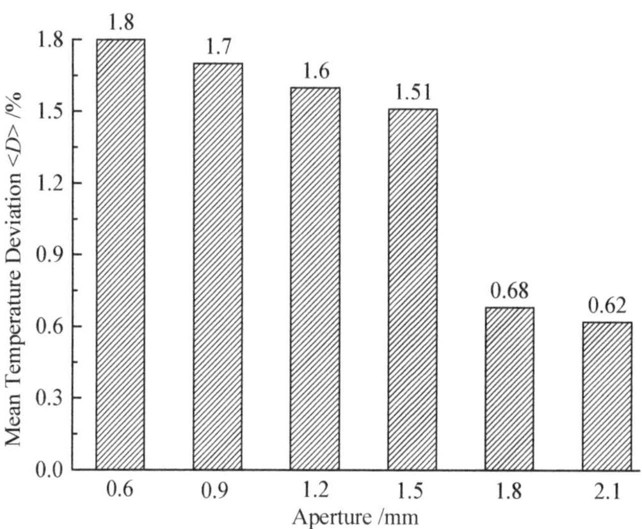

**Figure 12.** The effect of fracture aperture on mean deviation $<D>$ from NPCTF for JRC = 16.34 fracture.

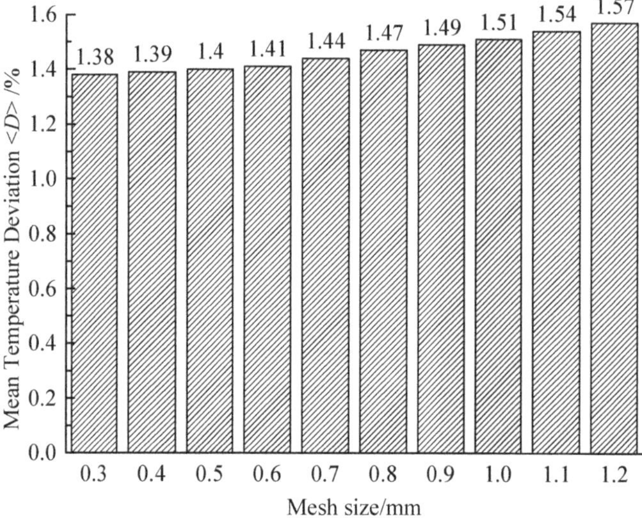

**Figure 13.** The effect of mesh size on mean deviation $<D>$ from NPCTF or JRC = 16.34 fracture.

In order to analyze the effect of mesh size, the mesh size is set as 0.3, 0.4, 0.5, 0.6, 0.7, 0.8, and 0.9 mm and 1.0, 1.1, and 1.2 mm for the fracture profile with JRC = 16.34 and $b_m$ = 1.5 mm, respectively. Figure 13 shows the effect of mesh size on mean deviation $<D>$ (see in Equation (25)) from the NPCTF. As shown in Figure 13, the mean deviation $<D>$ increases as the mesh size increases from 0.6 to 1.2 mm. The maximum deviation for a mesh size value of 1.2 mm is 1.57%. The results tend to converge when the simulation mesh size is less than 0.6 mm. The differences in deviation from using a mesh size of 0.3 and 0.6 mm are approximately 1.38% and 1.41%, respectively. Therefore, this indicates that the NPCTF is not sensitive to the size of mesh defined in the range of 0.2 to 0.8 mm.

### 6.3. Limitations of NPCTF

The main limitation of this proposed NPCTF from the heat transport model is neglecting the heat diffusion term. This assumption can be violated as the diverging or converging angle increases. This heat diffusion effect can only cause bigger deviation in the distance from the fracture inlet to equilibrium temperature position. Furthermore, the deviation from the proposed heat transport model is reasonable when presented a diverging wedge with an angle that is 63.47°. The feature of the diverging and converging angle is mostly considered and reported for rock fractures [36,37,45] and therefore, the validity of the assumption is warranted in most cases. Additionally, the validity of the used flow perturbation solution in the fracture lies in the assumptions that the variation of aperture along the fracture length direction is small. However, Wang et al. [37] discussed that the assumption can meet the geometry of most natural fractures.

### 7. Conclusions

We presented an NPCTF for rough-walled fractures, including a fluid flow model and a heat transport model. The fluid flow model describes the relationship of flowrate and pressure in the wedge-shaped cell, based on the perturbation method. The heat transport properties between the flowing fluid and rock fracture wall are captured by the proposed heat transport model, which assumes an identical temperature between the fluid and fracture wall. Both of them connected to each other in a series of connected wedge-shaped fractures. The proposed NPCTF was validated by a commercial *COMSOL* Multiphysics code in JRC profiles. The temperature relative deviation between them is less than 12.3% and the maximum transmissivity deviation is 1.2%. To examine the performance of the proposed NPCTF, heat flow simulations in real rock fractures with different relative surface roughness, obtained by a 3D stereotopometric scanning system, were performed. It has been shown that the results from the proposed NPCTF agree well with the *COMSOL* simulation results. In general, the absolute deviation of the proposed model from the simulation results, in terms of the estimated temperature, is in the range of 0.44% to 8.10% with a mean deviation of 1.51% for all cases of the fracture profiles examined. In addition, the NPCTF is sensitive to fracture aperture variation, but not to the size of the mesh. The effect of aperture on deviation of the NPCTF is larger for a small aperture fracture compared with a large aperture fracture.

The proposed NPCTF has good estimation of the heat transport effect of rough-walled fractures. Therefore, in the next work, we would take the proposed NPCTF to use in a two-dimension rock fracture network. On this basis, the heat energy exploitation process in EGS can be simulated.

**Author Contributions:** Conceptualization, F.X. and Q.J.; methodology, F.X.; software, C.Z.; validation, C.Z.; formal analysis, F.X.; investigation, F.X.; resources, Q.J.; data curation, C.Z.; writing—original draft preparation, F.X.; writing—review and editing, Q.J.; visualization, C.Z.; supervision, Q.J.; project administration, Q.J.; funding acquisition, Q.J. All authors have read and agreed to the published version of the manuscript.

**Funding:** This research was funded by the National Science Foundation of China, grant number 1679173.

**Institutional Review Board Statement:** No applicable.

**Informed Consent Statement:** No applicable.

**Data Availability Statement:** All data included in this study are available upon request by contact with the corresponding author.

**Acknowledgments:** This research was financially supported by the National Science Foundation of China (No. 51679173). This support is gratefully acknowledged.

**Conflicts of Interest:** The authors declared that they have no conflicts of interest to this work.

**Abbreviations**

| | |
|---|---|
| $\varepsilon$ | Perturbation parameter |
| $X$ | Dimensionless roughness |
| $\omega$ | Dimensionless absolute aperture variation |
| $a$ | Difference between the upper and lower wedge edge |
| $B$ | Dimensionless aperture |
| $c$ | Fracture local aperture |
| $c_m$ | Mean aperture |
| $B'$ | First derivatives of $B$ with respect to $X$ |
| $R$ | Reynolds number |
| $Q$ | Flow rate per unit width of fracture |
| $\mu$ | Fluid viscosity |
| $\Delta P$ | Dimensionless pressure difference |
| $\Delta p_m$ | Pressure difference of flow through a fracture with a uniform aperture defined from the cubic law |
| $\rho_w$ | fluid density |
| $c_w$ | Specific heat of fluid; |
| $K_w$ | Fluid thermal conductivity |
| $K_r$ | Rock thermal conductivity |
| $v$ | Steady flow velocity |
| $b$ | Half aperture of the fracture |
| $T_f$ | Bulk temperature of fluid |
| $T_r$ | Temperature of reservoir rock matrix |
| $T_o$ | rock temperature |
| $T_{in}$ | water temperature in inlet of fracture |
| $R$ | Height of rock |
| $\alpha$ | Inclined angle |
| $\delta_w$ | Function of $P$ and $T_f$ |
| $Ta$ | transmissivity |
| JRC | Joint roughness coefficient |
| $Z_2$ | root mean square of first derivative of profile |
| $D$ | Relative deviation between two variables |
| $<D>$ | Mean deviation |


## References

1. Kolditz, O.; Clauser, C. Numerical simulation of flow and heat transfer in fractured crystalline rocks: Application to the hot dry rock site in Rosemanowes (UK). *Geothermics* **1997**, *27*, 1–23. [CrossRef]
2. Gehlin, S.; Hellstrom, G. Influence on thermal response test by groundwater flow in vertical fractures in hard rock. *Renew. Energy* **2003**, *28*, 2221–2238. [CrossRef]
3. Zhao, Y.; Feng, Z.; Feng, Z.; Yang, D.; Liang, W. THM (Thermo-hydro-mechanical) coupled mathematical model of fractured media and numerical simulation of a 3D enhanced geothermal system at 573 K and buried depth 6000-7000 M. *Energy* **2015**, *82*, 193–205. [CrossRef]
4. Jiao, Y.Y.; Zhang, X.L.; Zhang, H.Q.; Li, H.B.; Yang, S.Q.; Li, J.C. A coupled thermo-mechanical discontinuum model for simulating rock cracking induced by temperature stresses. *Comput. Geotech.* **2015**, *67*, 142–149. [CrossRef]
5. Sun, Z.; Zhang, X.; Yao, J.; Wang, H.; Lv, S.; Sun, Z.; Huang, Y.; Cai, M.; Huang, X. Numerical simulation of the heat extraction in EGS with thermal-hydraulic-mechanical coupling method based on discrete fractures model. *Energy* **2017**, *120*, 20–33. [CrossRef]
6. Yao, J.; Zhang, X.; Sun, Z.; Huang, Z.; Liu, J.; Li, Y.; Xin, Y.; Yan, X.; Liu, W. Numerical simulation of the heat extraction in 3D-EGS with thermal-hydraulic-mechanical coupling method based on discrete fractures model. *Geothermics* **2018**, *74*, 19–34. [CrossRef]
7. Yan, C.; Jiao, Y.Y. FDEM-TH3D: A three-dimensional coupled hydrothermal model for fractured rock. *Int. J. Numer. Anal. Methods Geomech.* **2019**, *43*, 415–440. [CrossRef]
8. Ehyaei, M.; Ahmadi, A.; Rosen, M.A.; Davarpanah, A. Thermodynamic Optimization of a Geothermal Power Plant with a Genetic Algorithm in Two Stages. *Processes* **2020**, *8*, 1277. [CrossRef]
9. Davarpanah, A.; Shirmohammadi, R.; Mirshekari, B.; Aslani, A. Analysis of hydraulic fracturing techniques: Hybrid fuzzy approaches. *Arab. J. Geosci.* **2019**, *12*, 402. [CrossRef]
10. Sun, S.; Zhou, M.; Lu, W.; Davarpanah, A. Application of Symmetry Law in Numerical Modeling of Hydraulic Fracturing by Finite Element Method. *Symmetry* **2020**, *12*, 1122. [CrossRef]
11. Zhu, M.; Yu, L.; Zhang, X.; Davarpanah, A. Application of Implicit Pressure-Explicit Saturation Method to Predict Filtrated Mud Saturation Impact on the Hydrocarbon Reservoirs Formation Damage. *Mathematics* **2020**, *8*, 1057. [CrossRef]

12. Hu, X.; Xie, J.; Cai, W.; Wang, R.; Davarpanah, A. Thermodynamic effects of cycling carbon dioxide injectivity in shale reservoirs. *J. Pet. Sci. Eng.* **2020**, *195*, 107717. [CrossRef]
13. Xiong, F.; Jiang, Q.; Xu, C. Fast Equivalent Micro-Scale Pipe Network Representation of Rock Fractures Obtained by Computed Tomography for Fluid Flow Simulations. *Rock Mech. Rock Eng.* **2020**, 1–17. [CrossRef]
14. Xiong, F.; Wei, W.; Xu, C.; Jiang, Q. Experimental and numerical investigation on nonlinear flow behaviour through three dimensional fracture intersections and fracture networks. *Comput. Geotech.* **2020**, *121*, 103446. [CrossRef]
15. Heimlich, C.; Gourmelen, N.; Masson, F.; Schmittbuhl, J.; Kim, S.W.; Azzola, J. Uplift around the geothermal power plant of Landau (Germany) as observed by InSAR monitoring. *Geotherm. Energy* **2015**, *3*, 2. [CrossRef]
16. Tenzer, H.; Park, C.-H.; Kolditz, O.; McDermott, C.I. Application of the geomechanical facies approach and comparison of exploration and evaluation methods used at Soultz-sous-Forêts (France) and Spa Urach (Germany) geothermal sites. *Environ. Earth Sci.* **2010**, *61*, 853–880. [CrossRef]
17. Xu, C.; Dowd, P.; Tian, Z. A simplified coupled hydro-thermal model for enhanced geothermal systems. *Appl. Energy* **2015**, *140*, 135–145. [CrossRef]
18. Kolditz, O. Modelling flow and heat transfer in fractured rocks: Conceptual model of a 3-D deterministic fracture network. *Geothermics* **1995**, *24*, 451–470. [CrossRef]
19. Natarajan, N.; Xu, C.; Dowd, P.A.; Hand, M. Numerical modelling of thermal transport and quartz precipitation/dissolution in a coupled fracture–skin–matrix system. *Int. J. Heat Mass Transf.* **2014**, *78*, 302–310. [CrossRef]
20. Bai, B.; He, Y.; Li, X.; Hu, S.; Huang, X.; Li, J.; Zhu, J. Local heat transfer characteristics of water flowing through a single fracture within a cylindrical granite specimen. *Environ. Earth Sci.* **2016**, *75*, 1460. [CrossRef]
21. Li, Z.; Feng, X.; Zhang, Y.; Zhang, C.; Xu, T.; Wang, Y. Experimental research on the convection heat transfer characteristics of distilled water in manmade smooth and rough rock fractures. *Energy* **2017**, *133*, 206–218. [CrossRef]
22. Tsang, Y. The effect of tortuosity on fluid flow through a single fracture. *Water Resour. Res.* **1984**, *20*, 1209–1215. [CrossRef]
23. Brown, S. Fluid flow through rock Joints-the effect of surface roughness. *J. Geophys. Res.* **1987**, *92*, 1337–1347. [CrossRef]
24. Xiong, F.; Jiang, Q.; Ye, Z.; Zhang, X. Nonlinear flow behavior through rough-walled rock fractures: The effect of contact area. *Comput. Geotech.* **2018**, *102*, 179–195. [CrossRef]
25. Zou, L.; Jing, L.; Cvetkovic, V. Roughness decomposition and nonlinear fluid flow in a single rock fracture. *Int. J. Rock Mech. Min. Sci.* **2015**, *75*, 102–118. [CrossRef]
26. Wang, M.; Chen, Y.; Ma, G.; Zhou, J.; Zhou, C. Influence of surface roughness on nonlinear flow behaviors in 3D self-affine rough fractures: Lattice Boltzmann simulations. *Adv. Water Resour.* **2016**, *96*, 373–388. [CrossRef]
27. Yin, Q.; Ma, G.; Jing, H.; Wang, H.; Su, H.; Wang, Y.; Liu, R.C. Hydraulic properties of 3D rough-walled fractures during shearing: An experimental study. *J. Hydrol.* **2017**, *55*, 169–184. [CrossRef]
28. He, Y.; Bai, B.; Hu, S.; Li, X. Effects of surface roughness on the heat transfer characteristics of water flow through a single granite fracture. *Comput. Geotech.* **2016**, *80*, 312–321. [CrossRef]
29. Huang, Y.; Zhang, Y.; Yu, Z.; Ma, Y.; Zhang, C. Experimental investigation of seepage and heat transfer in rough fractures for enhanced geothermal systems. *Renew. Energy* **2019**, *135*, 846–855. [CrossRef]
30. Ma, Y.; Zhang, Y.; Yu, Z.; Huang, Y.; Zhang, C. Heat transfer by water flowing through rough fractures and distribution of local heat transfer coefficient along the flow direction. *Int. J. Heat Mass* **2018**, *119*, 139–147. [CrossRef]
31. Gringarten, A.C.; Witherspoon, P.A.; Ohnishi, Y. Theory of heat extraction from fractured hot dry rock. *J. Geophys. Res.* **1975**, *80*, 1120–1124. [CrossRef]
32. Cheng, A.; Ghassemi, A.; Detournay, E. Integral equation solution of heat extraction from a fracture in hot dry rock. *Int. J. Numer. Anal. Methods Geomech.* **2001**, *25*, 1327–1338. [CrossRef]
33. Martínez, A.R.; Roubinet, D.; Tartakovsky, D.M. Analytical models of heat conduction in fractured rocks. *J. Geophys. Res.* **2014**, *119*, 83–98. [CrossRef]
34. Zhao, Z. On the heat transfer coefficient between rock fracture walls and flowing fluid. *Comput. Geotech.* **2014**, *59*, 105–111. [CrossRef]
35. Yost, K.; Valentin, A.; Einstein, H. Estimating cost and time of wellbore drilling for Engineered Geothermal Systems (EGS)–Considering uncertainties. *Geothermics* **2015**, *53*, 85–99. [CrossRef]
36. Brush, D.J.; Thomson, N.R. Fluid flow in synthetic rough-walled fractures: Navier-Stokes, Stokes, and local cubic law simulations. *Water Resour. Res.* **2003**, *39*, 1085. [CrossRef]
37. Wang, Z.; Xu, C.; Dowd, P. Perturbation Solutions for Flow in a Slowly Varying Fracture and the Estimation of Its Transmissivity. *Transp. Porous Media* **2019**, *128*, 97–121. [CrossRef]
38. Wang, Z.; Xu, C.; Dowd, P.; Xiong, F.; Wang, H. A non-linear version of the Reynolds equation for flow in rock fractures with complex void geometries. *Water Resour. Res.* **2020**, *56*, e2019WR026149. [CrossRef]
39. Zimmerman, R.; Al-Yaarubi, A.; Pain, C.; Grattoni, C. Non-linear regimes of fluid flow in rock fractures. *Int. J. Rock Mech. Min. Sci.* **2004**, *41*, 163–169. [CrossRef]
40. Chen, Y.F.; Zhou, J.Q.; Hu, S.H.; Hu, R.; Zhou, C.B. Evaluation of Forchheimer equation coefficients for non-Darcy flow in deformable rough-walled fractures. *J. Hydrol.* **2015**, *529*, 993–1006. [CrossRef]
41. Barton, N.; Bandis, S.; Bakhtar, K. Strength, deformation and conductivity coupling of rock joints. *Int. J. Rock Mech. Min. Sci. Geomech. Abstr.* **1985**, *22*, 121–140. [CrossRef]

42. Tatone, B.; Grasselli, G. Quantitative measurements of fracture aperture and directional roughness from rock cores. *Rock Mech. Rock Eng.* **2012**, *45*, 619–629. [CrossRef]
43. Tse, R.; Cruden, D.M. Estimating joint roughness coefficients. *Int. J. Min. Sci. Geomech. Abstr.* **1979**, *16*, 303–307. [CrossRef]
44. Shaik, A.R.; Rahman, S.S.; Tran, N.H.; Tran, T. Numerical simulation of fluid-rock coupling heat transfer in naturally fractured geothermal system. *Appl. Therm. Eng.* **2011**, *31*, 1600–1606. [CrossRef]
45. Sisavath, S.; Al-Yaarubi, A.; Pain, C.C.; Zimmerman, R.W. A Simple Model for Deviations from the Cubic Law for a Fracture Undergoing Dilation or Closure. *Pure Appl. Geophys.* **2003**, *160*, 1009–1022. [CrossRef]

*Article*

# Stage Division of Landslide Deformation and Prediction of Critical Sliding Based on Inverse Logistic Function

**Liulei Bao, Guangcheng Zhang *, Xinli Hu, Shuangshuang Wu and Xiangdong Liu**

Faculty of Engineering, China University of Geosciences, Wuhan 430074, China; baoliulei@cug.edu.cn (L.B.); huxinli@cug.edu.cn (X.H.); wshuang@cug.edu.cn (S.W.); liuxiangdong@cug.edu.cn (X.L.)
* Correspondence: zhangguangc@cug.edu.cn

**Abstract:** The cumulative displacement-time curve is the most common and direct method used to predict the deformation trends of landslides and divide the deformation stages. A new method based on the inverse logistic function considering inverse distance weighting (IDW) is proposed to predict the displacement of landslides, and the quantitative standards of dividing the deformation stages and determining the critical sliding time are put forward. The proposed method is applied in some landslide cases according to the displacement monitoring data and shows that the new method is effective. Moreover, long-term displacement predictions are applied in two landslides. Finally, summarized with the application in other landslide cases, the value of displacement acceleration, $0.9 \ mm/day^2$, is suggested as the first early warning standard of sliding, and the fitting function of the acceleration rate with the volume or length of landslide can be considered the secondary critical threshold function of landslide failure.

**Keywords:** displacement-time curve; the deformation stage division; critical sliding prediction; inverse logistic curve; inverse distance weighted

**Citation:** Bao, L.; Zhang, G.; Hu, X.; Wu, S.; Liu, X. Stage Division of Landslide Deformation and Prediction of Critical Sliding Based on Inverse Logistic Function. *Energies* **2021**, *14*, 1091. https://doi.org/10.3390/en14041091

Academic Editor: Chun Zhu
Received: 10 January 2021
Accepted: 10 February 2021
Published: 19 February 2021

**Publisher's Note:** MDPI stays neutral with regard to jurisdictional claims in published maps and institutional affiliations.



## 1. Introduction

Judging the deformation stages and predicting the critical sliding time of a landslide is very significant to determine how and when the landslide will fail. The most common method of dividing the deformation stages and predicting the trend of landslides is generally based on the cumulative displacement-time curve [1,2]. From the perspective of the prediction time length, it can be divided into a long-term prediction (1 to 3 years) [3–7], medium-term prediction (3 to 12 months), short-term prediction (1 to 3 months) and critical sliding prediction (1 to 10 days) [8].

Saito [1,9] first proposed a landslide prediction model based on the three-stage creep theory of rock and soil materials and successfully predicted the Gaochangshan landslide in Japan. Since then, many different landslide prediction models have been developed [10–18]. Over the last decades, Wang and Nie [19] combined nonlinear regression analysis with the quadratic curve exponential smoothing method to predict the future displacement value of landslides. Crosta and Agliardi [20] discussed how to set velocity thresholds to achieve different levels for forecasting landslides. Li et al. [21] presented an application of a linear combination model with optimal weight in landslide displacement prediction. Bozzano et al. [22] described four years of continuous slope monitoring using an integrated platform, calibrating the empirical parameters of the Voight function to predict the landslide failure time. Lian et al. [23] established a set of predictors considering different environmental factors and proposed a switched method to select the appropriate individual predictor. Liao et al. [24] proposed and applied a step-like displacement prediction model based on a kernel extreme learning machine with grey wolf optimization (GWO-KELM). Krkač et al. [25] presented a methodology for the prediction of landslide movements using random forests. Meng et al. [26] used a vector autoregressive model to predict periodic displacement based on time series analysis. Landslide displacement can be divided into trend displacement, reflecting the long-term trend of landslides, which is the response of geologic structures, and periodic displacement, reflecting the volatility of landslides, which is mainly affected by external factors such as rainfall. Tang et al. [27] studied the evolution mechanism of the landslide considering the main influencing factors and providing an opinion for landslide prediction with consideration of the interior geology characteristics and external dynamic factors.

From the beginning of deformation to the final failure, landslide deformation has some inherent patterns. Terzaghi [28] first approached the correlation between the accelerating phase and landslide movements. Ter-Stepanian [29] and Tavenas and Leroueil [30] identified the presence of relevant creep deformations before failure. Furthermore, several authors discussed the possibility of defining alert threshold levels. Cruden and Masoumzadeh [31] proposed three velocity levels to correspond with three accelerating creep stages. Intrieri et al. [32] approached the task by studying the most critical periods of the entire dataset of the Torgiovannetto landslide. Moreover, according to the characteristics of the cumulative displacement and time curves of landslides, most scholars generally recognize the three-stage evolution mode of landslide deformation—the initial deformation stage [10,33–36], the constant velocity deformation stage, and the accelerated deformation stage. Some other scholars further subdivide the third stage according to the characteristics of acceleration variation. Xu et al. [33] proposed dividing the third stage into the initial acceleration, medium acceleration, and accelerated acceleration stages as shown in Figure 1.

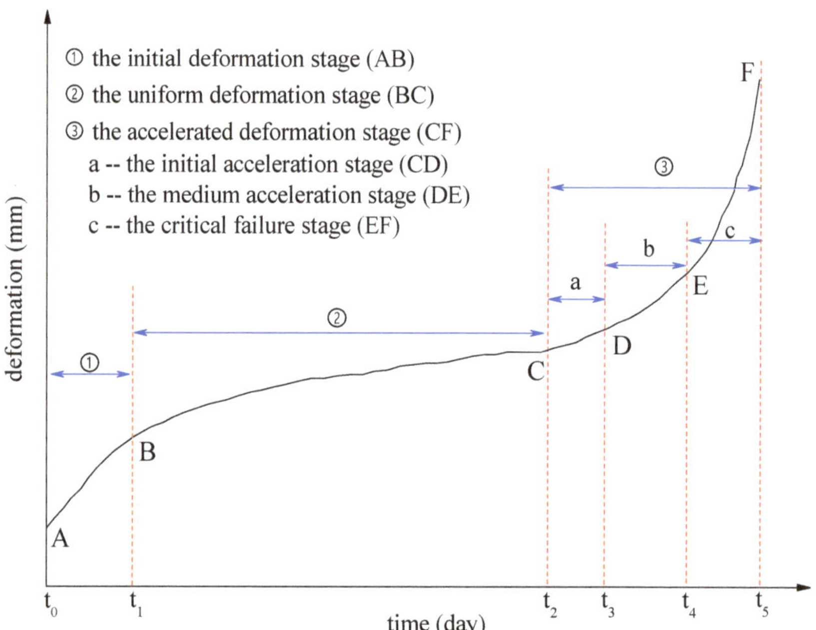


**Figure 1.** The deformation stage division of landslides.

The key to landslide prediction is to accurately estimate the dynamic variation and development trend of landslide stability. Generally, most scholars have used different types of functions for fitting, such as the Verhulst logistic function and Verhulst inverse function [37], which can be used to predict the displacement of landslides with time and divide the deformation stages. Meanwhile, random methods are the second most widely used type, such as the GM model [38–43], regression model [44], the BP network model [5,45,46], and the Markov Chain model [47–49]. These methods are applied to predict landslide displacement in a short time. The ARIMA model [50] and SVR model [51–54] are more suitable for short-term prediction. In addition, other scholars applied GIS technology to evaluate landslide susceptibility [55–57].

The prediction of landslides is mainly based on the relationship between the velocity change of displacement and time, but the research based on the acceleration and acceleration rate is limited. Considering the mechanism of landslide failure, the maximum acceleration is more consistent with the prediction criteria. At the same time, the landslide failure is mainly progressive failure, and the main feature of deformation is that the displacement acceleration is greater than zero and growing. However, scholars only deduce the formula of acceleration and acceleration maximum time for the Verhulst model, and no scholars have yet deduced and applied the formula of the acceleration and acceleration maximum time criterion for the inverse logic function curve, but they are not enough in the quantitative standard division of landslide evolution stage and the threshold of the critical sliding point.

In brief, although scholars divide the cumulative displacement time history curve of landslides into three, four, or five stages, a unified understanding is formed on its typical characteristics. Based on this unified understanding, this paper intends to adopt a new method, which is based on the inverse logistic function with a consideration of inverse distance weight, to predict landslide displacements, divide the landslide deformation stages, and realize the prediction of the critical sliding time. Lastly, four failed landslides are taken as examples to check the proposed method, and two landslides in deformation are predicted.

## 2. Displacement Prediction Model of Landslides

### 2.1. The Inverse Logistic Function Model of the Displacement-Time Curve

As shown in Figure 1, the cumulative displacement-time curve of a landslide is a typical inverse logistic function. The general expression of the logistic curve function can be written as follows:

$$y = \frac{\alpha}{1 + e^{-\beta(x-x_0)}} + \gamma \tag{1}$$

where $x_0$, $\alpha$ ($\alpha > 0$), $\beta$ ($\beta > 0$), and $\gamma$ are the unknown parameters to be determined. If $x$ and $y$ are replaced by the cumulative displacement of a landslide, $S(t)$ and $t$, respectively; then the inverse function of the logistic function can be obtained based on Equation (1),

$$S(t) = \frac{1}{\beta} ln \left( \frac{t - \gamma}{\alpha + \gamma - t} \right) + x_0 \tag{2}$$

where $t$ denotes the time length of monitoring or prediction, day, and $S(t)$ denote the cumulative displacements of the landslide, mm. Equation (2) is the general expression of the cumulative displacement-time curve.

The displacement monitoring work of a landslide could start at any time. Therefore, displacement occurring before monitoring could not be considered. However, the initial displacement value of the cumulative displacement-time function is usually set to zero at the monitoring start time, $t = 0$.

$$S(0) = 0 \tag{3}$$

Then, Equation (3) can be derived from Equation (2),

$$x_0 = \frac{1}{\beta} ln \frac{\alpha + \gamma}{-\gamma} \tag{4}$$

Then, Equation (2) can be transformed as follows:

$$S(t) = \frac{1}{\beta} ln \frac{\frac{\alpha + \gamma}{-\gamma} t + \alpha + \gamma}{\alpha + \gamma - t} \tag{5}$$

In order to simplify the above formula, we can assume $k = \frac{1}{\beta}, a = \frac{\alpha+\gamma}{-\gamma}, b = \alpha + \gamma$, then Equation (5) is simplified to as follows:

$$S(t) = kln\frac{at+b}{b-t} \tag{6}$$

where $k$ $(k > 0)$, $a$, and $b$ are all unknown parameters to be determined. Equation (6) is the general inverse logistic function expression of the cumulative displacement-time curve of a landslide. According to Equation (6), we know that the time, $t$, tends to the value of $b$ when the displacement value of the landslide tends toward positive infinity. That is, the right trend line of Equation (6) is $t = b > 0$.

In nature, it is a common rule that the smaller the distance is, the closer the attribute. Accordingly, when the unknown value of some attribute of one objective needs to be predicted, the measured value near the predicted time has a greater impact on the predicted value than that far from the predicted time. That is, the point nearer to the predicted position has a larger weight than the farther point, so this method is called inverse distance weighting (IDW). The simplest form of the IDW method is the reciprocal of the distance, which makes the weight sometimes very large or even infinite and then causes an error prediction result. For this reason, a constant is added to the distance when using the reciprocal mode,

$$w_i = \frac{1}{d_i + const} \tag{7}$$

where $w_i$ denotes the weight and $d_i$ denotes the distance of the point to the right.

Equation (7) assigns a large weight to the nearer points, and the weight decays quickly at a slightly farther distance. Although this is what we want, it makes the algorithm more sensitive to noise. The Gauss function is complex, but it overcomes the above shortcomings. Its form is as follows [58]:

$$w(t) = w_0 e^{-\frac{(t-b)^2}{2c^2}} \tag{8}$$

where $w(t)$ is the weight value at time $t$, $w_0$ is the basic weight, and $c$ denotes a constant to be determined. For the cumulative displacement-time curve of a landslide, $b$ denotes the time of the right trend line. Therefore, the weight range of the function is $(w_0 e^{-\frac{(b)^2}{2c^2}}, w_0)$. Here, let $w_0 = 1$ and $c = \frac{b}{\sqrt{2}}$. Then the Gauss inverse distance weight function is simplified as follows:

$$w(t) = e^{-(1-t/b)^2} \tag{9}$$

Equation (8) is the suggested formula to calculate the weight, which is in the range of $[e^{-1}, 1)$. Furthermore, the optimal parameters of the objective function, Equation (5), can be obtained according to the least square method of IDW, which is expressed as follows:

$$min\Delta = \sum_{i=1}^{n}\{w(t_i)[S'(t_i) - S(t_i)]\}^2 \tag{10}$$

where $S'(t_i)$ and $S(t_i)$ are the fit and monitoring displacements at time $t_i$, respectively.

### 2.2. The Division Standard of Landslide Deformation Stages

Equation (6) can be obtained from the landslide cumulative displacement-time monitoring curve, and then the function expressions of velocity, acceleration, acceleration rate with time can be obtained by the first derivative, second derivative and third derivative of Equation (5). Their formulas are expressed as follows,

$$\begin{cases} V(t) = \frac{kb(a+1)}{(at+b)(b-t)} \\ a_c(t) = \frac{-kb(a+1)(ab-b-2at)}{(at+b)^2(b-t)^2} \\ a_c'(t) = \frac{2kb(a+1)}{(at+b)^3(b-t)^3}[3a^2t^2 - 3ab(a-1)t + (a^2-a+1)b^2] \end{cases} \tag{11}$$

where $V(t)$, $a_c(t)$ and $a_c'(t)$ are in mm/day, mm/day$^2$ and mm/day$^3$, respectively.

In mathematics, curvature is the rate of change in the direction of a curve with respect to the distance along the curve, and its value equals the reciprocal of the radius of the circle that most closely conforms to the curve at a given point. Therefore, the larger the curvature is, the greater the change in direction of the curve. In a general curve function, $y = f(x)$, the curvature, $K$, can be calculated by the following formula:

$$K = \frac{|y''|}{\left(1 + y'^2\right)^{3/2}} \tag{12}$$

where $y'$ and $y''$ are the first and second derivatives, respectively, of $y$ with respect to $x$.

(1)  The curvature function of the cumulative displacement-time curve is

$$K_1 = \frac{|a_c(t)|}{(1 + V^2(t))^{3/2}} \tag{13}$$

There are two maximum curvature extreme points in the complete ideal displacement-time curve of landslides. The corresponding time of the first maximum curvature extreme point ($K_1^{\max 1}$) is the cut-off point of the initial deformation stage and the uniform deformation stage, and the second ($K_1^{\max 2}$) corresponds to the cut-off point of the uniform deformation stage and the accelerated deformation stage. However, in reality, the monitoring work may begin at any moment, which causes the monitored cumulative displacement-time curve to miss the first maximum curvature extreme point. If so, in most cases, the start moment of the monitor is in the uniform deformation stage of the landslide; then, only the second maximum curvature is encountered. Sometimes, the curve even misses two maximum curvature extreme points, which means the landslide has reached the acceleration deformation stage.

(2)  The curvature function of the velocity-time curve is

$$K_2 = \frac{|a_c'(t)|}{(1 + a_c^2(t))^{3/2}} \tag{14}$$

Generally, there is only one maximum curvature extreme point ($K_1^{\max 2}$) for the velocity-time curve of Equation (13), which corresponds to the cut-off point of the accelerated deformation stage and the critical sliding stage.

A whole cumulative displacement-time curve of landslides could be divided into four different stages: the initial deformation stage (the time before the time of $K_1^{\max 1}$), the uniform deformation stage (the time between $K_1^{\max 1}$ and $K_1^{\max 2}$), the acceleration deformation stage (the time between $K_1^{\max 2}$ and $K_2^{\max}$), and the critical sliding stage (the time after $K_2^{\max}$), as shown in Figure 2.

The above proposed method needs to be verified to determine whether it is available and accurate when used to predict the displacement of landslides. Here, four famous failed landslides in China, including the Xintan landslide, Wolongsi New landslide, Huangci landslide, and Saleshan landslide, are taken as typical examples for prediction by our proposed method. Finally, two ongoing landslides are predicted.

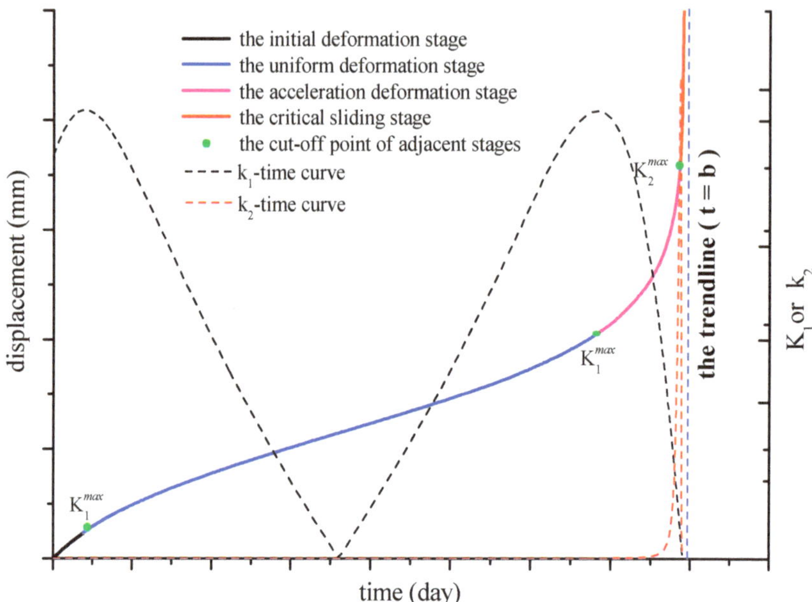


**Figure 2.** The typical deformation stages of landslides based on the inverse logistic function.

## 3. Verification with Typical Landslides

In order to certify the accuracy of the proposed method in this paper, four cases, Xintan Landslide, Wolongsi landslide, Huangci landslide and Saleshan's new landslide, were applied to divide the deformation stages and predict the sliding time. Furthermore, the prediction results were compared with other methods. Six other landslide cases were applied, and two ongoing landslides were predicted. Figure 3 shows the specific location of all the landslide cases.

### 3.1. Case 1: Xintan Landslide

The Xintan landslide is an accumulation landslide that failed on 12 June 1985. The area of the landslide is approximately $0.68 \times 10^6$ m$^2$, and its volume reaches $30.0 \times 10^6$ m$^3$, most of which slid into the Yangtze River. The landslide is approximately 2000 m in length from north to south and 450 m in width. The longitudinal average gradient of the landslide is 23 degrees. The height difference of the landslide is approximately 800 m. Permian limestone and Devonian quartz sandstone are exposed in the area of Guangjiaya at the back of Xintan landslide, while the landslide is located above the Silurian sandstone and sand shale, which are easily weathered, broken, and impermeable. Xintan landslide is the revival of ancient landslide deposits. The main reason is that there are often-produced collapse deposits from the Guangjiaya, which push the ancient landslide deposits moving forward. Furthermore, the surface land reclamation is digging and planting in disorder, and the surface drainage is not smooth.

Since 1978, the upper area of the Xintan landslide has experienced obvious deformation, and two displacement monitoring points (No. A3 and B3) were set up on the landslide in January 1978. The cumulative displacement data of point No. B3 on the Xintan landslide is shown in Table 1, which is used to predict through our proposed method, as shown in Figures 4 and 5.

**Figure 3.** The specific location of all landslide cases.

**Table 1.** The measured displacements of the Xintan landslide at point B3 (mm) [59].

| Year | Month | | | | | | | | | | | |
|------|-----|------|------|------|------|------|------|------|------|------|------|------|
| | Jan | Feb | Mar | Apr | May | June | July | Aug | Sept | Oct | Nov | Dec |
| 1978 | 13.3 | 16.7 | 20.5 | 24.8 | 28.8 | 43.4 | 60.5 | 66.5 | 75.0 | 77.6 | 77.8 | 87.8 |
| 1979 | 89.4 | 93.3 | 98.2 | 99.7 | 104.4 | 115.5 | 126.4 | 149.9 | 302.0 | 367.0 | 404.5 | 412.7 |
| 1980 | 427.9 | 437.8 | 442.4 | 451.4 | 456.4 | 471.4 | 520.4 | 588.4 | 688.4 | 690.1 | 721.8 | 723.9 |
| 1981 | 731.4 | 739.1 | 747.5 | 762.7 | 762.7 | 784.0 | 788.2 | 791.1 | 817.1 | 819.1 | 824.1 | 832.1 |
| 1982 | 837.1 | 840.6 | 848.1 | 926.1 | 926.1 | 1096.3 | 1117.8 | 1254.0 | 1504.8 | 1747.8 | 1823.6 | 1895.1 |
| 1983 | 1995.8 | 2024.5 | 2056.4 | 2087.6 | 2087.6 | 2148.5 | 2277.9 | 2430.0 | 2582.1 | 2734.2 | 2886.3 | 3038.1 |
| 1984 | 3190.5 | 3342.6 | 3494.7 | 3646.8 | 3646.8 | 3951.0 | 4188.7 | 4521.7 | 4673.5 | 5195.6 | 5467.3 | 5705.0 |
| 1985 | 5863.3 | 6012.7 | 6177.2 | 6304.5 | 6304.5 | - | - | - | - | - | - | - |

It is found that the annual cumulative displacement from 1978 to 1981 is hundreds of millimeters. However, from 1982 to 1985, cumulative displacement increased sharply to thousands of millimeters. The Xintan landslide continued to deform several years at an annual displacement of 1000 mm. Sun [60] considered that the Xintan landslide was in the initial creep stage before 1979, in the constant creep stage from August 1979 to July 1982, in the accelerated creep stage from July 1982 to 15 May 1985, entered the stage of catastrophic failure between 15 May 1985 and 12 June 1985, and then arrived in the critical sliding stage. The Rockfall and Landslide Research Institute of Hubei Province [61] thought that the Xintan landslide was in the potential deformation and creep stage before 1982, the revival stage from March 1982 to May 1983, the rapid deformation stage from June 1983 to June 1985, and the sliding stage from 9 to 12 June 1985. Furthermore, Table 2 lists the deformation stage division and prediction results of the Xintan landslide by the proposed method using the monitoring data during different durations of time.

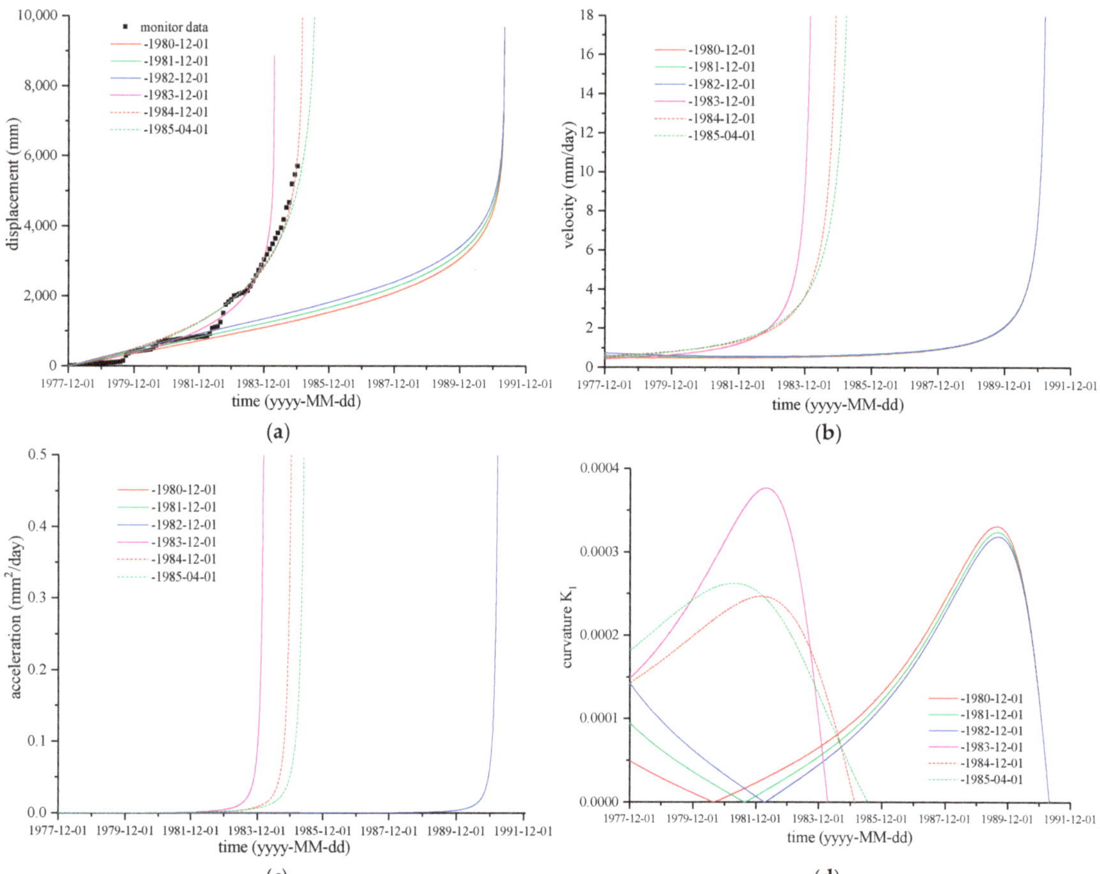

**Figure 4.** Prediction results of the Xintan landslide based on displacement monitoring data. (**a**) cumulative displacement-time curves; (**b**) velocity-time curves; (**c**) acceleration-time curves; (**d**) curvature $K_1$-time curves.

**Table 2.** The prediction results of the Xintan landslide based on different monitoring data.

| Parameters | | Deformation Stages | |
|---|---|---|---|
| | | The Start Point of the Acceleration Deformation Stage | The Start Point of the Critical Sliding Stage |
| ~1 April 1985 | time | 12 December 1980 | 18 June 1985 |
| k = 2935 | V(t) | 1.07 | 53.3 |
| a = −0.51 | a(t) | $8.75 \times 10^{-4}$ | 1.01 |
| b = 2810 | a′(t) | $1.24 \times 10^{-6}$ | 0.110 |

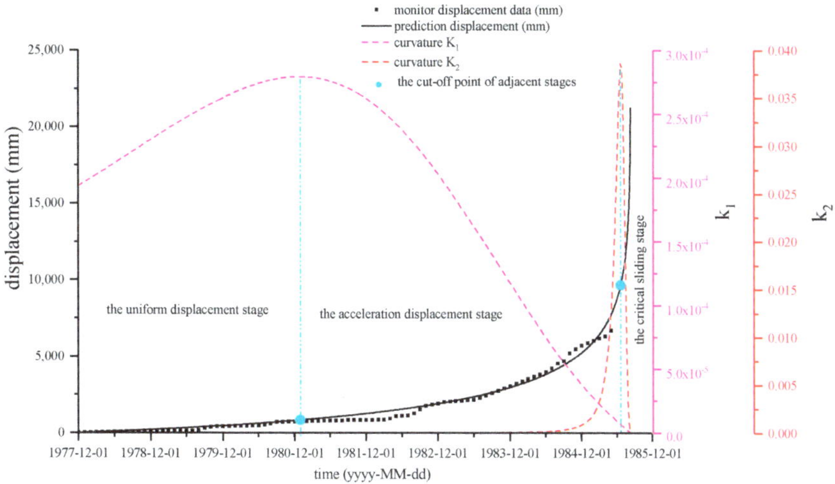



**Figure 5.** The critical sliding prediction of the Xintan landslide.

According to the inverse logic curve, the prediction time of the landslide is 18 June 1985, while the actual landslide is 12 June 1985. The prediction results of the landslide are compared with those of other different methods as shown in Table 3. It shows that the prediction accuracy of our method is much better than any other one.

**Table 3.** Prediction results of Xintan landslide.

| Model | Prediction Physical Parameter | Prediction Time | Δ/d * |
|---|---|---|---|
| Inverse Logic Function Model | Displacement acceleration | 18 June 1985 | 6 |
| Original synergetic model [59] | Displacement acceleration | 16 October 1987 | 856 |
| Improved synergetic model [59] | Displacement acceleration | 14 March 1985 | −90 |
| Displacement GM (1,1) Model [62] | Displacement velocity | 13 December 1981 | −1277 |
| Velocity GM (1,1) Model [62] | Displacement velocity | 1 July 1982 | −1077 |
| Original Verhulst model [63] | Displacement velocity | 24 July 1987 | 772 |
| Improved Verhulst model [63] | Displacement acceleration | 3 February 1986 | 236 |
| Improved Verhulst model [63] | Displacement accelerated acceleration | 7 February 1985 | 125 |

* where $\Delta$ is the absolute error, which refers to the difference between the predicted landslide occurrence time $t_y$ and the actual landslide occurrence time $t_s$, namely, $\Delta$ is $t_y$ minus $t_s$, and $\Delta$ unit is day. When the actual occurrence time of landslide is earlier than the prediction time, $\Delta > 0$. When the actual occurrence time of landslide is later than the prediction time, $\Delta < 0$.

### 3.2. Case 2: New Wolongsi Landslide

The New Wolongsi landslide is a loess landslide. The thickness of the landslide mass is more than 50 m to as much as 90 m, it is 645 m long and 650 m wide, and the volume scale is more than $20.0 \times 10^6$ m$^3$. It is a super-deep and super-large landslide. Damaged several times in the past, the failure range of each time is basically the same. It is the fourth terrace of the Weihe River at the back edge of the landslide. The upper part of the terrace is the loess of 97 m thickness, the lower part is river alluvium of 54 m thickness, and the lower part is Neogene mild clay and gravel. The gravel usually contains water. The rear sliding surface is arc-shaped with a steep slope, the middle front part is nearly horizontal, and the front edge is a reverse slope that overlaps the floodplain terrace. Since the crack was first discovered in March 1971, the No. 5 fissure on the New Wolongsi landslide was monitored with manual measurement for 66 days, and slip damage occurred on 5 May.

Taking the cumulative displacement-time monitoring data, which are shown in Table 4, of the No. 5 fracture as an example, the proposed method is used for prediction, and the results are shown in Figures 6 and 7 and Table 5.

**Table 4.** The measured displacements of the New Wolongsi landslide [59].

| Month | March | | | | | | | | | | | | | | | | |
|---|---|---|---|---|---|---|---|---|---|---|---|---|---|---|---|---|---|
| Date | 15 | 16 | 17 | 18 | 19 | 20 | 21 | 22 | 23 | 24 | 25 | 26 | 27 | 28 | 29 | 30 | 31 |
| displacement (mm) | 1.0 | 1.5 | 1.7 | 2.5 | 3.2 | 4.0 | 4.4 | 5.1 | 5.9 | 6.3 | 7.0 | 7.3 | 7.8 | 8.2 | 8.4 | 8.7 | 9.0 |
| Month | April | | | | | | | | | | | | | | | | |
| Date | 1 | 2 | 3 | 4 | 5 | 6 | 7 | 8 | 9 | 10 | 11 | 12 | 13 | 14 | 15 | 16 | 17 |
| displacement (mm) | 9.2 | 9.4 | 10.0 | 10.1 | 10.3 | 10.4 | 10.5 | 10.8 | 11.1 | 12.0 | 13.0 | 13.6 | 14.0 | 15.0 | 16.1 | 16.4 | 17.2 |
| Month | April | | | | | | | | | | | | May | | | | |
| Date | 18 | 19 | 20 | 21 | 22 | 23 | 24 | 25 | 26 | 27 | 28 | 29 | 30 | 1 | 2 | 3 | 4 | 5 |
| displacement (mm) | 17.0 | 18.0 | 19.0 | 19.0 | 20.0 | 23.0 | 24.0 | 25.2 | 26.0 | 27.0 | 28.2 | 30.0 | 31.0 | 32.0 | 33.0 | 42.0 | 47.0 | 61.0 |

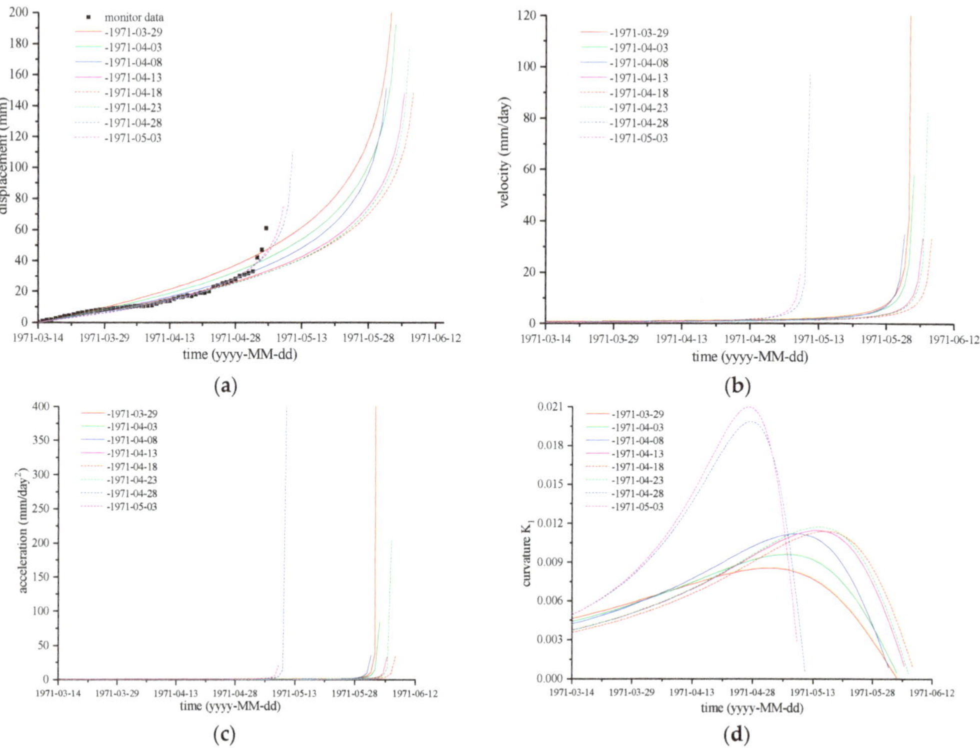

**Figure 6.** Prediction results of the New Wolongsi landslide based on displacement monitoring data. (**a**) cumulative displacement-time curves; (**b**) velocity-time curves; (**c**) acceleration-time curves; (**d**) curvature of $K_1$-time curves.

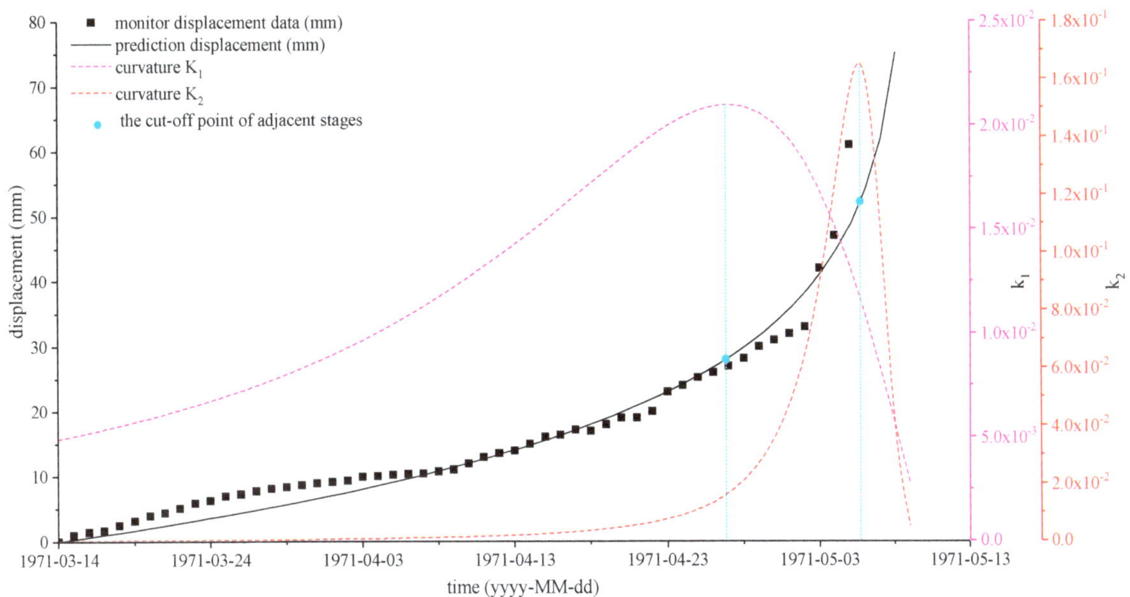


**Figure 7.** Deformation stage division and prediction of the New Wolongsi landslide.

**Table 5.** The prediction results of the New Wolongsi landslide based on different monitoring data.

| Parameters | | Deformation Stages | |
| --- | --- | --- | --- |
| | | The Start Point of the Acceleration Deformation Stage | The Start Point of the Critical Sliding Stage |
| ~3 May 1971 | time | 27 April 1971 | 3 May 1971 |
| k = 16.00 | V(t) | 1.42 | 4.03 |
| a = 0.10 | a(t) | 0.109 | 1.00 |
| b = 54.0 | a'(t) | 0.017 | 0.507 |

The prediction time of the landslide by the inverse logic curve is on 3 May 1971, which is basically consistent with the actual landslide time on 5 May 1971. The comparison with other methods is as shown in Table 6. It shows that most of the methods, including our proposed method, have good prediction accuracy.

**Table 6.** Prediction results of Wolongsi's new landslide.

| Model | Prediction Physical Parameter | Prediction Time | Δ/d |
| --- | --- | --- | --- |
| Inverse Logic Function Model | Displacement acceleration | 3 May 1971 | −2 |
| Original synergetic model [59] | Displacement acceleration | 26 May 1971 | 21 |
| Improved synergetic model [59] | Displacement acceleration | 5 May 1971 | 0 |
| Improved Verhulst model [63] | Displacement acceleration | 9 May 1971 | 4 |
| Improved Verhulst model [63] | Displacement accelerated acceleration | 28 April 1971 | −7 |
| Original Verhulst model [63] | Displacement velocity | 24 April 1971 | 19 |

### 3.3. Case 3: Huangci Landslide

The Huangci landslide is not entirely a new landslide; it has been damaged several times in its history. The landslide is close to the edge of the fourth terrace, and the settlement cracks, which evolved to the back-edge cracks of Huangci landslide caused by irrigation, have appeared for more than ten years. The sliding surface of the west bedding plane has also existed for several years. Huangci landslide is nearly 500 m wide at the rear and 300 m wide on the front; it is 370 m long in the north and south directions, and the volume of landslide is nearly $6.0 \times 10^4$ m$^3$. The thickness of the loess layer at Huangci landslide is 42.67 m, the thickness of light-yellow loess ($Q_3$) at the upper part is 36.40 m, and the thickness of brown-red/brown-yellow loess ($Q_2$) at the lower part is 6.27 m. The soft plastic layer with a thickness of 9 m is 31.73–40.90 m below the ground, and the plastic layer is 40.90–42.67 m below the ground. The lower part of the loess layer is an 8 m-thick gravel layer. Below the gravel is Cretaceous mudstone with argillaceous sandstone. The bedrock belongs to monoclinic structure; it inclines to Huangci village in front of the landslide.

A large amount of irrigation water is the main inducing factor of the Huangci landslide. The initial monitoring time of the Huangci landslide was on 1 August 1994, and the monitoring time interval was 15 days. The ground displacement monitoring data of No. A6 by GPS are shown in Table 7. At 2:30 a.m. on 30 January, 1995, the Huangci landslide slipped for 90 min before it completely stopped. The leading edge of the landslide went straight to Huangci Village. Our proposed method is also used on the Huangci landslide, and the results are shown in Figures 8 and 9 and Table 8.

**Table 7.** The measured displacements of the Huangci landslide at point A6 [59].

| Date | Year/Month | 1994/8 | | | 1994/9 | | 1994/10 | | 1994/11 | | 1994/12 | | 1995/1 | |
|---|---|---|---|---|---|---|---|---|---|---|---|---|---|---|
| | Date | 1 | 16 | 31 | 15 | 30 | 15 | 30 | 14 | 29 | 14 | 29 | 13 | 28 |
| displacement (mm) | | 1.0 | 3.3 | 6.1 | 8.3 | 10.5 | 15.3 | 18.1 | 21.6 | 27.5 | 33.6 | 39.0 | 48.7 | 64.3 |

**Table 8.** The prediction results of the Huangci landslide based on different monitoring data.

| Parameters | | Deformation Stages | |
|---|---|---|---|
| | | The Start Point of the Acceleration Deformation Stage | The Start Point of the Critical Sliding Stage |
| ~28 January 1995 | time | 29 January 1995 | 11 February 1995 |
| $k = 25.9$ | $V(t)$ | 1.44 | 5.19 |
| $a = 0.05$ | $a(t)$ | 0.080 | 1.04 |
| $b = 216$ | $a'(t)$ | $8.92 \times 10^{-3}$ | 0.416 |

According to the inverse logic curve, the prediction time of the landslide is February 11, 1995, and the actual landslide time is January 30, 1995. The comparison with other methods is shown in Table 9. It shows that the improved synergetic model has the best prediction accuracy, followed by our proposed method, and the other four methods have a larger prediction error.

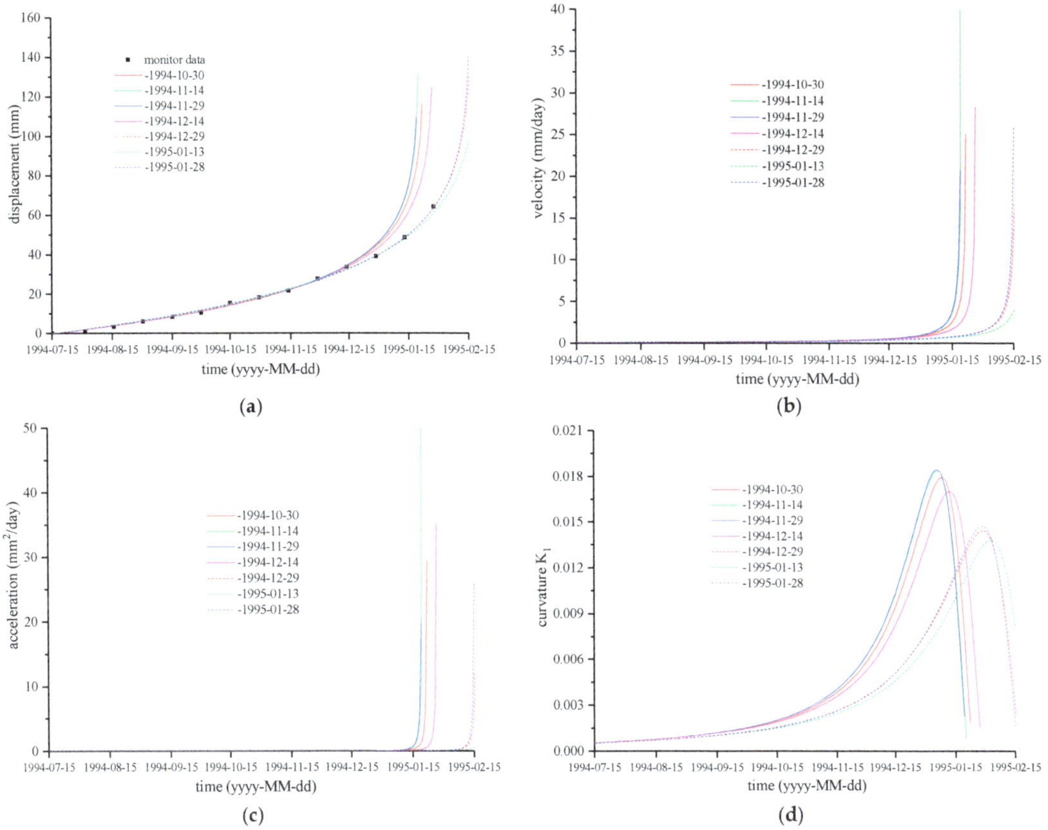

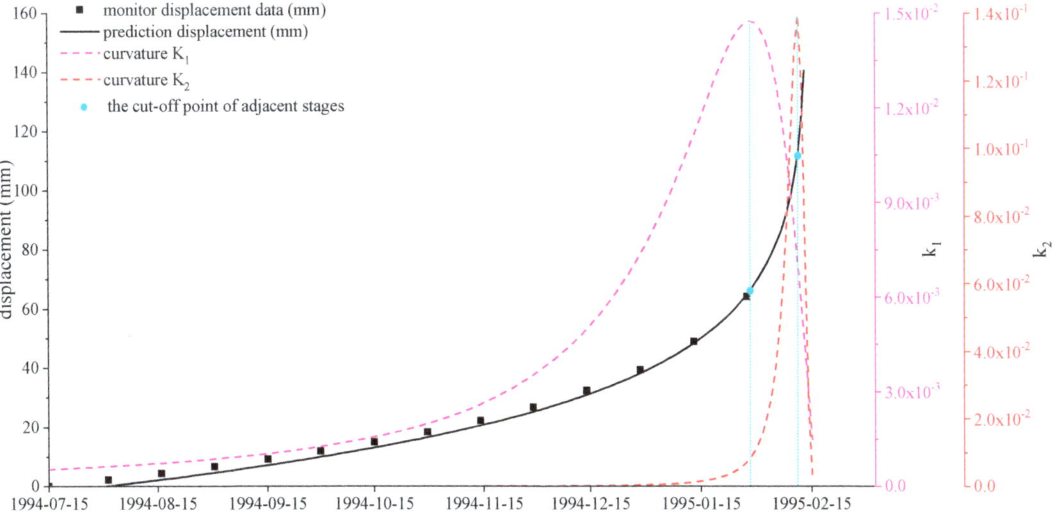

**Figure 8.** Prediction curves of the Huangci landslide based on displacement monitoring data. (**a**) cumulative displacement-time curves; (**b**) velocity-time curves; (**c**) acceleration-time curves; (**d**) curvature of $K_1$-time curves.

**Figure 9.** Deformation stage division and prediction of the Huangci landslide.

**Table 9.** Prediction results of Huangci landslide.

| Model | Prediction Physical Parameter | Prediction Time | Δ/d |
|---|---|---|---|
| Inverse Logic Function Model | Displacement acceleration | 11 February, 1995 | −13 |
| Original synergetic model [59] | Displacement acceleration | 26 March, 1995 | 55 |
| Improved synergetic model [59] | Displacement acceleration | 29 January, 1995 | −1 |
| Verhulst Model [62] | Displacement acceleration | 25 December, 1994 | −66 |
| Displacement GM (1,1) Model [62] | Displacement velocity | 17 December, 1994 | −74 |
| Velocity GM (1,1) Model [62] | Displacement velocity | 25 December, 1994 | −66 |

### 3.4. Case 4: Saleshan Landslide

The Saleshan landslide occurred in Dongxiang County, Gansu Province. Its failure first happened on 7 March 1983. The Saleshan landslide is mainly composed of red-layered clay rock of the Linxia Formation. The red layer of the Linxia Formation is covered with loess, in which some vertical discontinuities were developed. The sliding bed of the Saleshan landslide is mainly composed of mudstone.

On 25 March 1986, the Saleshan landslide failed again. Its volume was approximately $2.40 \times 10^6$ m$^3$, and the sliding distance was approximately 250 m. Displacement monitoring was carried out after it occurred the first time, and the displacement monitoring data are shown in Table 10. The prediction results using our proposed method are shown in Figures 10 and 11 and Table 11.

**Table 10.** The measured displacements of the Saleshan landslide [59].

| Year | | | | | | | | 1984 | | | | | | | | | | 1985 | | | 1986 | |
|---|---|---|---|---|---|---|---|---|---|---|---|---|---|---|---|---|---|---|---|---|---|---|
| Month | 6 | 7 | 8 | 9 | 10 | 11 | 12 | 1 | 2 | 3 | 4 | 5 | 6 | 7 | 8 | 9 | 10 | 11 | 12 | 1 | 2 |
| Displacement (mm) | 4 | 10 | 0 | 1.5 | 2.5 | 5.5 | 7.2 | 5.2 | 5.2 | 4.8 | 2.5 | 4.8 | 7 | 12 | 15 | 17 | 27 | 30 | 41 | 50 | 75 |

**Table 11.** Prediction results of the Saleshan landslide based on different monitoring data.

| | | Deformation Stages | |
|---|---|---|---|
| Parameters | | The Start Point of the Acceleration Deformation Stage | The Start Point of the Critical Sliding Stage |
| ~15 February 1986 | time | 10 February 1986 | 23 February 1986 |
| k = 25.4 | V(t) | 1.36 | 5.02 |
| a = −0.60 | a(t) | 0.078 | 1.02 |
| b = 654 | a′(t) | 0.042 | 2.15 |

The prediction time of the landslide by the inverse logic curve is 23 February 1986, while the actual time of the landslide is 25 March 1986. The comparison with other methods is shown in Table 12. It shows that the improved synergetic and Verhulst models have the best prediction accuracies, followed by our proposed method, and the other three methods have larger prediction errors.

**Table 12.** Prediction results of Saleshan's new landslide.

| Model | Prediction Physical Parameter | Prediction Time | Δ/d |
|---|---|---|---|
| Inverse Logic Function Model | Displacement acceleration | 23 February 1986 | −30 |
| Original synergetic model [58] | Displacement acceleration | 22 May 1986 | 58 |
| Improved synergetic model [58] | Displacement acceleration | 5 March 1986 | −20 |
| Original Verhulst model [62] | Displacement velocity | 4 July 1986 | 101 |
| Improved Verhulst model [62] | Displacement acceleration | 15 April 1986 | 21 |
| Improved Verhulst model [62] | Displacement accelerated acceleration | 16 February 1986 | −37 |

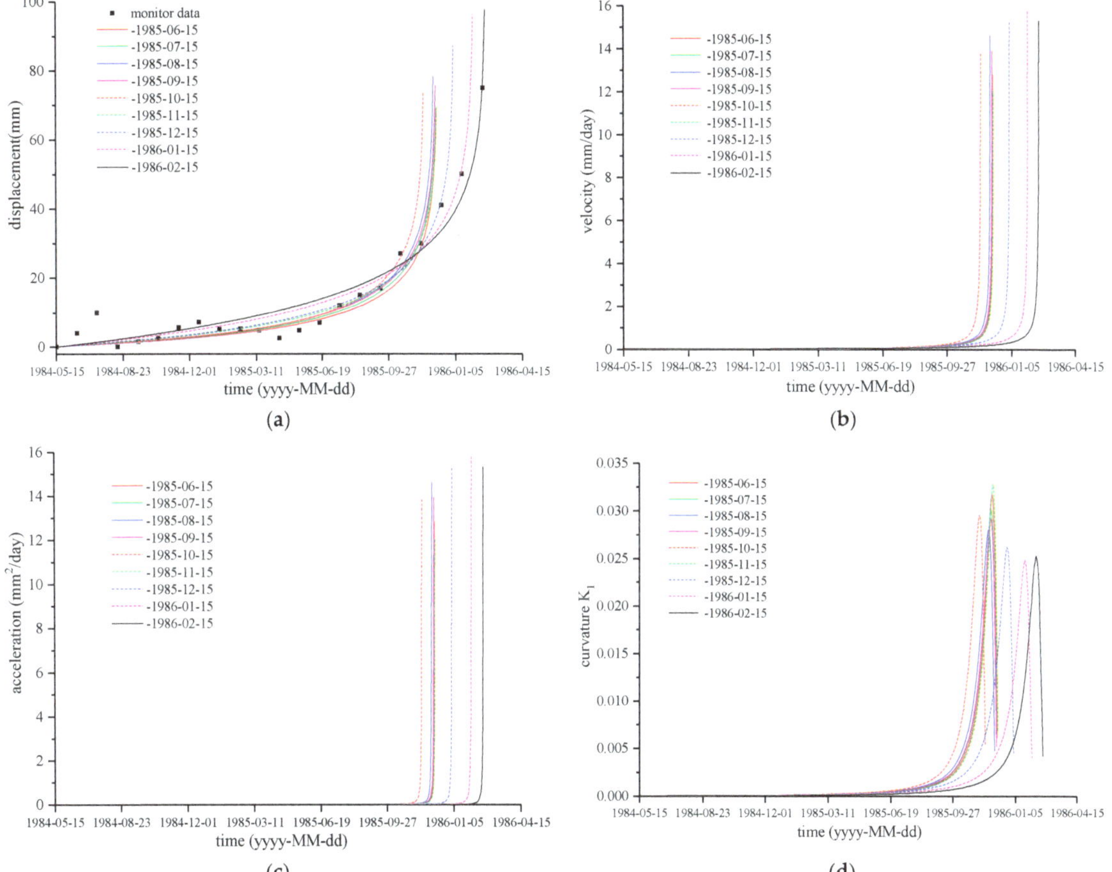

**Figure 10.** Prediction results of the Saleshan landslide based on displacement monitoring data. (**a**) cumulative displacement-time curves; (**b**) velocity-time curves; (**c**) acceleration-time curves; (**d**) curvature of $K_1$-time curves.

The factors affecting the prediction accuracy are as follows: (1) The more monitoring data that is accumulated, the higher the prediction accuracy will be. (2) The monitoring points could be representative. The deformation of different parts of the landslide is not synchronous. Therefore, the choice of the monitoring points applied to predict should be able to reflect the overall deformation trend of the landslide. For example, for the pull-type landslide, it is more appropriate to set monitoring points at the front edge of the landslide.

For the thrust-type landslide, it is more appropriate to set monitoring points at the back edge of landslide.

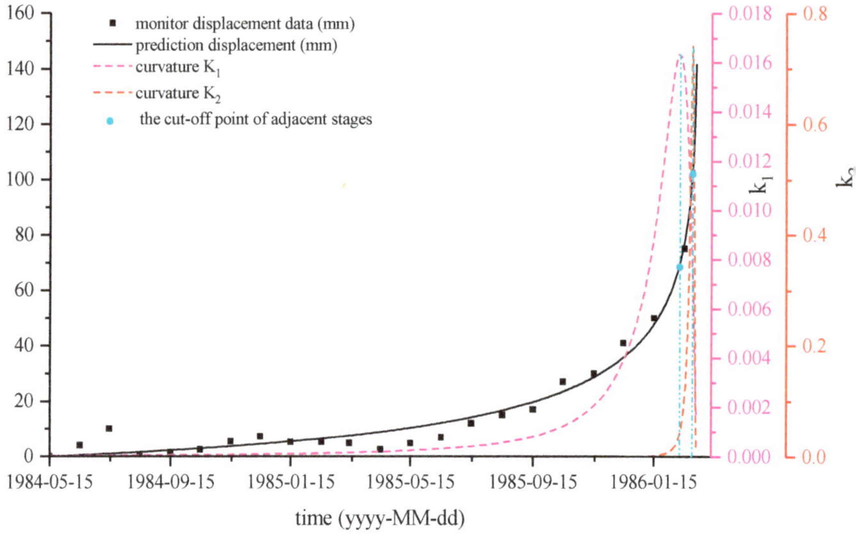

**Figure 11.** Deformation stage division and prediction of the Saleshan landslide.

## 4. Landslide Predictions

The long-term displacement predictions of the Liangshuijing landslide and Gapa landslide were carried out by the proposed method, and the deformation evolution stage of each landslide was studied and divided; then, the critical sliding time and the change of each parameter were predicted.

### 4.1. Case 1: Long Displacement Prediction of Liangshuijing Landslide

Liangshuijing landslide is located on the right bank of the Yangtze River with a slope angle of 30 to 35 degree. The height of the leading edge of the landslide is approximately 100 m, the height of the trailing edge is approximately 319.5 m, the relative height difference is approximately 219.5 m, the plane longitudinal length is approximately 434 m, the lateral width is 358 m, the area is approximately $11.82 \times 10^4$ m$^2$, the average thickness of the slide body is approximately 34.5 m, and the total volume is approximately $4.08 \times 10^6$ m$^3$. The professional monitoring was just carried out after the landslide experienced a large deformation in March 2009.

Manual horizontal displacement monitoring began on 5 April 2009 and stopped on 22 April 2009. Automatic horizontal displacement monitoring began on 20 April 2009, and the surface displacement monitoring point ZJC22 on the right side of the central part of the landslide was selected for prediction, as shown in Table 13. The monitoring data were maintained until October 2009.

The cumulative variation curve of horizontal displacement at the monitoring points is shown in Figure 12, and the monitoring results are shown in Table 14.

**Table 13.** The measured displacements of Liangshuijing landslide.

| Time yy/mm/dd | Displacement /mm | Time yy/mm/dd | Displacement /mm | Time yy/mm/dd | Displacement /mm |
|---|---|---|---|---|---|
| 2009/04/05 | 0.00 | 2009/06/14 | 215.90 | 2009/08/23 | 269.87 |
| 2009/04/10 | 18.83 | 2009/06/19 | 229.71 | 2009/08/28 | 266.11 |
| 2009/04/15 | 59.00 | 2009/06/24 | 242.26 | 2009/09/02 | 273.64 |
| 2009/04/20 | 96.65 | 2009/06/29 | 241.00 | 2009/09/07 | 273.64 |
| 2009/04/25 | 105.44 | 2009/07/04 | 243.52 | 2009/09/12 | 266.11 |
| 2009/04/30 | 112.97 | 2009/07/09 | 243.52 | 2009/09/17 | 269.87 |
| 2009/05/05 | 115.48 | 2009/07/14 | 252.30 | 2009/09/22 | 273.64 |
| 2009/05/10 | 133.05 | 2009/07/19 | 251.05 | 2009/09/27 | 272.39 |
| 2009/05/15 | 134.31 | 2009/07/24 | 259.83 | 2009/10/02 | 278.66 |
| 2009/05/20 | 156.90 | 2009/07/29 | 259.83 | 2009/10/07 | 273.64 |
| 2009/05/25 | 166.95 | 2009/08/03 | 254.81 | 2009/10/12 | 278.66 |
| 2009/05/30 | 184.52 | 2009/08/08 | 258.58 | 2009/10/17 | 277.41 |
| 2009/06/04 | 197.07 | 2009/08/13 | 264.85 | 2009/10/22 | 281.17 |
| 2009/06/09 | 207.11 | 2009/08/18 | 264.85 | 2009/10/27 | 287.45 |

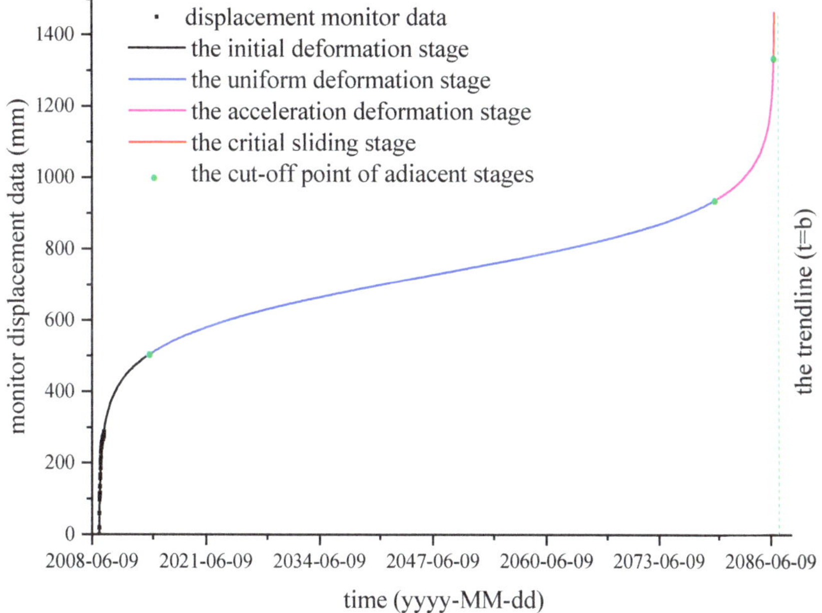


**Figure 12.** Deformation stage division and prediction of the Liangshuijing landslide.

**Table 14.** The prediction results of the Liangshuijing landslide based on different monitoring data.

| Parameters | | Deformation Stages | | |
|---|---|---|---|---|
| | | The Start Point of the Uniform Deformation Stage | The Start Point of the Acceleration Deformation Stage | The Start Point of the Critical Sliding Stage |
| ~15 November 2009 | time | 30 May 2009 | 3 September 2086 | 26 October 2086 |
| k = 89.51 | V(t) | 1.42 | 1.43 | 9.14 |
| a = 3520.0 | a(t) | −0.023 | 0.023 | 0.932 |
| b = 28,337.8 | a'(t) | $7.14 \times 10^{-4}$ | $2.17 \times 10^{-3}$ | 0.571 |

### 4.2. Case 2: Long-Term Displacement Prediction of the Gapa Landslide

The Gapa landslide is located on the right bank of the Yalong River. This area is approximately 11 km upstream of the Jinping first-class hydropower station, and the volume of the Gapa landslide is approximately $13.0 \times 10^6$ m$^3$, The landslide has a length of 980 m, a width of 320 to 400 m, an area of 0.28 km$^2$.

There are seven surface displacement monitoring points, denoted as GNSS1–GNSS6, and GNSS9 near the landslide area. GNSS1 and GNSS2 are located in the number one sliding body, which is the most likely failure area (see Figure 13). Therefore, GNSS1 and GNSS2 displacement monitoring data are selected for prediction, and the monitoring results are shown in Table 15.

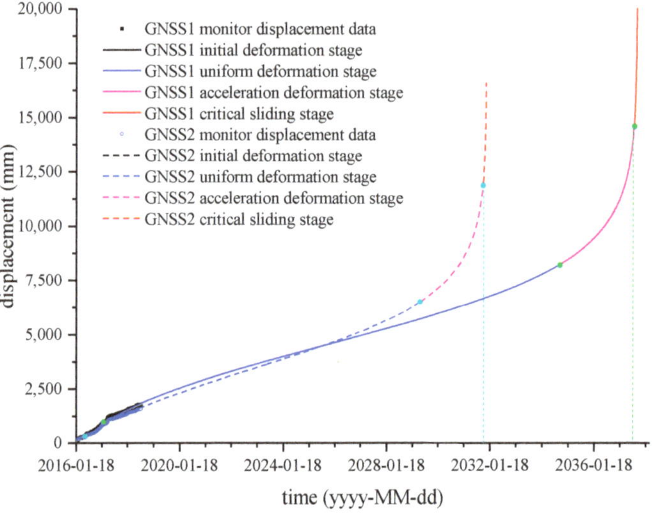


**Figure 13.** Deformation stage division and prediction of the Gapa landslide.

The prediction results show that the deformation is not synchronized in different parts of the Gapa landslide. In the number one sliding body, the former part of the Gapa landslide is destroyed before the trailing part, which illustrates that the Gapa landslide is a retrogressive type of landslide. Furthermore, the Gapa landslide will not fail in the next 10 years.

**Table 15.** The prediction results of the Gapa landslide based on different monitoring data.

| Parameters | | Deformation Stages | | |
|---|---|---|---|---|
| | | The Start Point of The Uniform Deformation Stage | The Start Point of the Acceleration Deformation Stage | The Start Point of the Critical Sliding Stage |
| GNSS1 | time | 10 February 2017 | 25 September 2034 | 26 July 2037 |
| k = 1877 | V(t) | 1.98 | 1.98 | 40.20 |
| a = 11.41 | a(t) | $-1.57 \times 10^{-3}$ | $1.57 \times 10^{-3}$ | 0.85 |
| b = 7907 | a'(t) | $2.99 \times 10^{-6}$ | $8.56 \times 10^{-6}$ | 0.10 |
| GNSS2 | time | 6 June 2016 | 19 May 2029 | 16 October 2031 |
| k = 1706 | V(t) | 2.15 | 2.15 | 42.9 |
| a = 7.40 | a(t) | $-1.95 \times 10^{-3}$ | $1.96 \times 10^{-3}$ | 1.07 |
| b = 5793 | a'(t) | $4.37 \times 10^{-6}$ | $1.20 \times 10^{-5}$ | 0.14 |

## 5. Warning Threshold of Critical Sliding

In order to find out one or some parameters as an effective tool to predict the critical sliding point of landslides, the accumulated displacements, velocities and accelerations of above four landslides and some other landslide cases from references at the moment of the critical sliding are listed in Table 16.

**Table 16.** Velocity and acceleration characters at the critical sliding moment of different landslides.

| Landslides | Parameters | | | | | | | | | |
|---|---|---|---|---|---|---|---|---|---|---|
| | Degree of Slope Surface (°) | Degree of Sliding Surface (°) | Depth (m) | Different Elevator (m) | Width (m) | Length (m) | Volume ($10^6 \times m^3$) | V (mm/day) | a (mm/day$^2$) | a' (mm/day$^3$) |
| Xintan landslide | 25–28 | 25–28 | 40–50 | 800 | 270–700 | 2000 | 30 | 53.3 | 1.01 | 0.110 |
| Wolongsi landslide | 0–25 | 0–45 | 50–90 | 200 | 645 | 650 | 20 | 4.65 | 1.00 | 0.507 |
| Huangci landslide | 15–35 | 15–20 | 40 | 100 | 300–500 | 370 | 6 | 5.19 | 1.04 | 0.416 |
| Saleshan landslide | 15–45 | 25–35 | 30–40 | 425 | 275 | 250 | 2.4 | 5.02 | 1.02 | 2.150 |
| Liangshuijing landslide | 30–35 | 25–50 | 34.5 | 221.5 | 358 | 434 | 4.08 | 9.14 | 0.93 | 0.571 |
| Baishi Landslide [64] | 40 | 38 | 25 | 270 | 260 | 300 | 2.0 | 4.00 | 1.07 | 0.569 |
| Tianhuangping [65] | 35–50 | 40.3 | 40 | 70 | 60 | 100 | 0.062 | 6.45 | 1.02 | 0.329 |
| Xikou Landslide [47] | 65 | 37 | 24 | 108 | 80–90 | 80 | 0.168 | 8.89 | 0.99 | 0.220 |
| Jianshanbeibang landslide [66] | / | / | 5–15 | 240 | 240 | 105 | 0.252 | 15.26 | 1.02 | 0.710 |
| Jimingshi Landslide [67] | 45 | 30 | 15 | 230 | 150 | 250–300 | 0.6 | 4.04 | 1.00 | 1.49 |
| Shuiwenzhan Landslide | 28–33 20–21 | 28–33 | 64–120 (80) | 500 | 570 | 1000 | 15 | 28.75 | 0.85 | 0.138 |
| Gapa landslide | 30–34 | 23–34 | 60 | 470 | 360 | 980 | 13 | 42.90 | 1.07 | 0.140 |

According to Table 16, Figure 14 is obtained. Obviously, it is found that the values of acceleration at the critical sliding moment are very close to all landslides; their values are at a range of between 0.85 and 1.07 mm/day$^2$. Therefore, the acceleration can be thought as a key index to predict the critical sliding point, and the threshold of displacement acceleration judging that the critical sliding point is 0.90 mm/day$^2$.

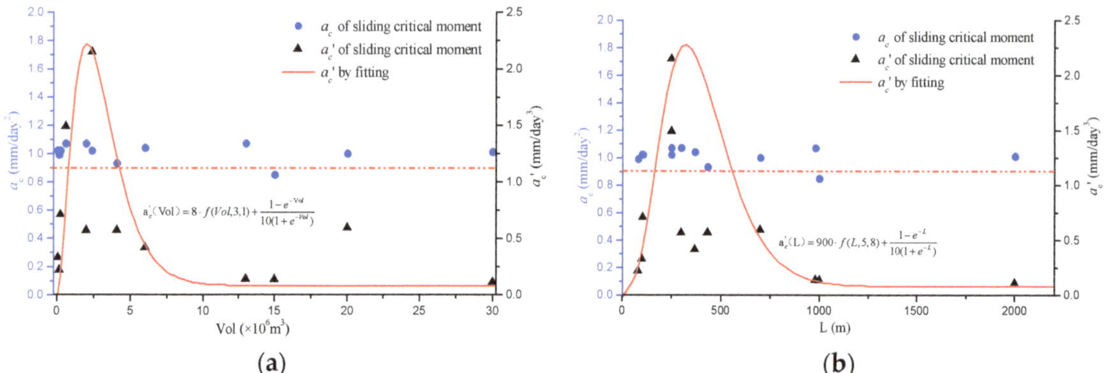

**Figure 14.** The acceleration and its rate of landslides at the critical sliding moment. (**a**) the relationship of acceleration and its rate with the volume of landslides; (**b**) the relationship of acceleration and its rate with the length of landslides

Furthermore, the relationship of the acceleration rate with the geometric parameters of landslides has skewed distribution characteristics, as shown in Figure 14, while Gamma function is one of the commonly used skewed distribution functions. Gamma function is shown as the following formula:

$$\begin{cases} f(x, \alpha, \beta) = \frac{1}{\beta^\alpha \Gamma(\alpha)} x^{\alpha-1} e^{-\frac{x}{\beta}}, \ x > 0 \\ \Gamma(\alpha) = \int_0^{+\infty} e^{-t} t^{\alpha-1} dt \end{cases} \tag{15}$$

where $\alpha$ and $\beta$ are shape parameter and scale parameter. Then, we can obtain the shape and scale parameters of Gamma function, which fits the relationship of the acceleration rate $(mm/day^3)$, $a_c'$, with the volume of landslide $(\times 10^6 \ m^3)$, $Vol$, and the length of landslide (m), $L$. Based on the data in Table 16, the critical threshold function can be fitted as shown in Equation (15).

$$\begin{cases} a_c'(L) = 900 f(L, 5, 8) + \frac{1-e^{-L}}{10\left(1+e^{-L}\right)} \\ a_c'(Vol) = 8 f(Vol, 3, 1) + \frac{1-e^{-Vol}}{10\left(1+e^{-Vol}\right)} \end{cases} \tag{16}$$

In the above formula, the function, $f$, is Gamma function as Equation (15), and the second part on the right side is an additive term, which is for improving fitting precision.

Sometimes, the geometric parameters are evaluated; they are not so accurate, and will reduce the prediction, effect shown in Equation (16). The displacement acceleration is almost a constant when a landslide enters a sliding state. Therefore, the value of displacement acceleration, 0.9 mm/day$^2$, is suggested as the first early warning standard of sliding, and Equation (16) can be used as the secondary critical threshold function of landslide failure.

## 6. Conclusions

The prediction model and early warning criteria are the key to landslide prediction, most of which are judged by the characteristics of the cumulative displacement. However, due to the different sizes of landslides and the complexity, nonhomogeneity, and uncertainty of their attributes, the cumulative displacement before failure is different by a few centimeters to dozens of meters, and the deformation velocity of landslides at the moment of critical sliding can also vary from a few millimeters per day to dozens of centimeters per day. Therefore, the general threshold values of displacement and velocity are difficult to predict for different landslides. Moreover, there is no quantitative standard for dividing the landslide deformation stages.

Based on the analysis and study of the variations in the cumulative displacement, deformation rate, displacement-time curvature, and velocity-time curvature of typical landslides, a suitable method is built to predict the displacement of landslides that can calculate the critical sliding time of landslides. This proposed method can predict the landslide displacement, divide the landslide deformation stages, and gain the threshold values of the critical sliding point by displacement acceleration and acceleration rate. Some meaningful conclusions and understandings are obtained:

(1) The cumulative displacement-time curves of landslides are in good agreement with the inverse logistic function. A general expression of the cumulative displacement-time function of landslides based on the logistic inverse function is proposed. Furthermore, the least squares formula of the inverse logistic function prediction with the IDW method is suggested to fit the inverse logistic function based on the displacement monitoring data of landslides.

(2) Based on the prediction model of the inverse logistic function with the IDW method, a new standard to divide the landslide deformation stages is proposed. This method was applied in some typical landslides and verified as effective and accurate. Lastly, this method was also used to predict deformation in two landslides—the Gapa and Liangshuijing landslides—and the critical sliding times were calculated. It should be noted that these two landslides remain in the initial deformation stage or just enter into the uniform deformation stage according to the monitoring displacement data. The prediction results are suggested to be continuously renewed when more monitoring displacement data are obtained.

(3) The prediction accuracy is mainly decided by the cumulative monitoring time and the distance far from the critical sliding time, and the critical sliding prediction time will be more accurate when the monitoring data are closer to the failure times of the landslides; otherwise, the prediction time may be farther from the real failure time. Therefore, we should predict the critical sliding time continually with the increase in monitoring data, which will improve the prediction accuracy.

(4) The displacement acceleration is recommended as a key index to predict the critical sliding of landslide, and its threshold value is suggested as $0.90 \text{ mm/day}^2$. Furthermore, the supplementary index of the critical sliding moment is the displacement acceleration rate, and the critical threshold function is suggested as Equation (15).

**Author Contributions:** G.Z. and X.H. designed the framework of the whole thesis; L.B., G.Z. and S.W. proposed the predict model, analyzed the data and drew the figures; G.Z. and L.B. helped with writing the paper. All authors have read and agreed to the published version of the manuscript.

**Funding:** This research was funded by the Natural Key R&D Program of China (2017YFC1501302), the National Natural Science Foundation of China (41877263), and the Fundamental Research Funds for the Central Universities, China University of Geosciences (Wuhan) (CUGCJ1802).

**Institutional Review Board Statement:** Not applicable.

**Informed Consent Statement:** Not applicable.

**Data Availability Statement:** Data is contained with the article.

**Conflicts of Interest:** The authors declare no conflict of interest.

# References


1. Saito, M. Forecasting the time of occurrence of a slope failure. In Proceedings of the 6th International Conference on Soil Me-chanics and Foundation Engineering, Montreal, QC, Canada, 8–15 September 1965; pp. 537–541.
2. Gokceoglu, C.; Sezer, E.A. A statistical assessment on international landslide literature (1945–2008). *Landslides* **2009**, *6*, 345–351. [CrossRef]
3. Bai, S.-B.; Wang, J.; Lü, G.-N.; Zhou, P.-G.; Hou, S.-S.; Xu, S.-N. GIS-based logistic regression for landslide susceptibility mapping of the Zhongxian segment in the Three Gorges area, China. *Geomorphology* **2010**, *115*, 23–31. [CrossRef]
4. Du, J.; Yin, K.; Lacasse, S. Displacement prediction in colluvial landslides, Three Gorges Reservoir, China. *Landslides* **2013**, *10*, 203–218. [CrossRef]
5. Peng, L.; Niu, R.; Huang, B.; Wu, X.; Zhao, Y.; Ye, R. Landslide susceptibility mapping based on rough set theory and support vector machines: A case of the Three Gorges area, China. *Geomorphology* **2014**, *204*, 287–301. [CrossRef]

6.    Ma, J.; Tang, H.; Hu, X.; Bobet, A.; Zhang, M.; Zhu, T.; Song, Y.; Eldin, M.A.M.E. Identification of causal factors for the Majiagou landslide using modern data mining methods. *Landslides* **2017**, *14*, 311–322. [CrossRef]
7.    Ma, J.; Tang, H.; Liu, X.; Hu, X.; Sun, M.; Song, Y. Establishment of a deformation forecasting model for a step-like landslide based on decision tree C5.0 and two-step cluster algorithms: A case study in the Three Gorges Reservoir area, China. *Landslides* **2017**, *14*, 1275–1281. [CrossRef]
8.    Qin, S.Q.; Jiao, J.J.; Wang, S.J. The predictable time scale of landslides. *Bull. Int. Assoc. Eng. Geol.* **2001**, *59*, 307–312. [CrossRef]
9.    Saito, M. Forecasting time of slope falure by trtiary creep. In *Proceedings of the 7th International Conference on Soil Mechanics and Foundation Engineering, Mexico City*; Oyo Chishitsu Co., Itd.: Tokyo, Japan, 1969; Volume 2, pp. 677–683.
10.   Fukuzono, T. A new method for predicting the failure time of a slope. In Proceedings of the 4th International Conference on Landslides, Tokyo, Japan, 23–31 August 1985; pp. 145–150. [CrossRef]
11.   Carlà, T.; Intrieri, E.; Di Traglia, F.; Casagli, N. A statistical-based approach for determining the intensity of unrest phases at Stromboli volcano (Southern Italy) using one-step-ahead forecasts of displacement time series. *Nat. Hazards* **2016**, *84*, 669–683. [CrossRef]
12.   Azimi., C.; Biarez., J.; Oesvarreux, P.; Eime, F. Forecasting time of failure for a rockslide in gypsum. In Proceedings of the 5th International Symposium on Landslides, Lausanne, Switzerland, 10–15 July 1988; Volume 1, pp. 531–536. (In French).
13.   Hayashi, S.; Park, B.-W.; Komamura, F.; Yamamori, T. On the Forecast of Time to Failure of Slope (II). *Landslides* **1988**, *25*, 11–16. [CrossRef]
14.   Voight, B. A method for prediction of volcanic eruptions. *Nat. Cell Biol.* **1988**, *332*, 125–130. [CrossRef]
15.   Crosta, G.B.; Agliardi, F. Failure forecast for large rock slides by surface displacement measurements. *Can. Geotech. J.* **2003**, *40*, 176–191. [CrossRef]
16.   Corominas, J.; Moya, J.; Ledesma, A.; Lloret, A.; Gili, J.A. Prediction of ground displacements and velocities from groundwater level changes at the Vallcebre landslide (Eastern Pyrenees, Spain). *Landslides* **2005**, *2*, 83–96. [CrossRef]
17.   Busslinger, M. *Landslide Time-Forecast Methods—A Literature Review Towards Reliable Prediction of Time to Failure*; HSR University of Applied Science: Rapperswil, Switzerland, 2009.
18.   Mufundirwa, A.; Fujii, Y.; Kodama, J. A new practical method for prediction of geomechanical failure-time. *Int. J. Rock Mech. Min. Sci.* **2010**, *47*, 1079–1090. [CrossRef]
19.   Wang, R.; Nie, L. Landslide prediction in Fushun west open pit mine area with quadratic curve exponential smoothing method. In Proceedings of the 2010 18th International Conference on Geoinformatics, Beijing, China, 18–20 June 2010; pp. 1–6.
20.   Crosta, G.; Agliardi, F. How to obtain alert velocity thresholds for large rockslides. *Phys. Chem. Earth Parts A/B/C* **2002**, *27*, 1557–1565. [CrossRef]
21.   Li, X.; Kong, J.; Wang, Z. Landslide displacement prediction based on combining method with optimal weight. *Nat. Hazards* **2012**, *61*, 635–646. [CrossRef]
22.   Bozzano, F.; Cipriani, I.; Mazzanti, P.; Prestininzi, A. A field experiment for calibrating landslide time-of-failure prediction functions. *Int. J. Rock Mech. Min. Sci.* **2014**, *67*, 69–77. [CrossRef]
23.   Lian, C.; Zeng, Z.; Yao, W.; Tang, H. Multiple neural networks switched prediction for landslide displacement. *Eng. Geol.* **2015**, *186*, 91–99. [CrossRef]
24.   Liao, K.; Wu, Y.; Miao, F.; Li, L.; Xue, Y. Using a kernel extreme learning machine with grey wolf optimization to predict the displacement of step-like landslide. *Bull. Int. Assoc. Eng. Geol.* **2020**, *79*, 673–685. [CrossRef]
25.   Krkač, M.; Špoljarić, D.; Bernat, S.; Arbanas, S.M. Method for prediction of landslide movements based on random forests. *Landslides* **2016**, *14*, 947–960. [CrossRef]
26.   Meng, M.; Chen, Z.Q.; Huang, D.; Zeng, B.; Chen, C.J. Displacement prediction of landslide in Three Gorges Reservoir area based on H-P filter, ARIMA and VAR models. *Rock Soil Mech.* **2016**, *37*, 552–560. [CrossRef]
27.   Tang, H.; Wasowski, J.; Juang, C.H. Geohazards in the three Gorges Reservoir Area, China—Lessons learned from decades of research. *Eng. Geol.* **2019**, *261*, 105267. [CrossRef]
28.   Terzaghi, K. Mechanism of landslides. In *Application of Geology to Engineering Practice (Berkeley Volume)*; Geological Society of America: Washington, DC, USA, 1950; pp. 83–123.
29.   Ter-Stepanian, G. Creep on natural slopes and cutting. In Proceedings of the 3rd International Symposium on Landslides, New Delhi, India, 7–11 April 1980; Volume 2, pp. 95–108.
30.   Tavenas, F.; Leroueil, S. Creep and failure of slopes in clays. *Can. Geotech. J.* **1981**, *18*, 106–120. [CrossRef]
31.   Cruden, D.M.; Masoumzadeh, S. Accelerating creep of the slopes of a coal mine. *Rock Mech. Rock Eng.* **1987**, *20*, 123–135. [CrossRef]
32.   Intrieri, E.; Gigli, G.; Mugnai, F.; Fanti, R.; Casagli, N. Design and implementation of a landslide early warning system. *Eng. Geol.* **2012**, *147*, 124–136. [CrossRef]
33.   Xu, Q.; Yuan, Y.; Zeng, Y.; Hack, H. Some new pre-warning criteria for creep slope failure. *Sci. China Ser. E Technol. Sci.* **2011**, *54*, 210–220. [CrossRef]
34.   Ma, J.W.; Tang, H.M.; Hu, X.L.; Yong, R.; Xia, H.; Song, Y.J. Application of 3D laser scanning technology to landslide physical model test. *Rock Soil Mech.* **2014**, *35*, 1495–1505.
35.   Qin, S.Q.; Wang, S.J. Advances in research on nonlinear evolutionary mechanisms and process of in stabilization of planar-slip slope. *Earth Environ.* **2005**, *33*, 75–82. (In Chinese) [CrossRef]

36. Tang, H.; Hu, X.; Xu, C.; Li, C.; Yong, R.; Wang, L. A novel approach for determining landslide pushing force based on landslide-pile interactions. *Eng. Geol.* **2014**, *182*, 15–24. [CrossRef]
37. Li, T.B.; Chen, M.D. Time prediction of landslide using verhulst inverse function-model. *J. Geol. Hazards Environ. Preserv.* **1996**, *3*, 13–17. (In Chinese)
38. Lu, P.; Rosenbaum, M.S. Artificial Neural Networks and Grey Systems for the Prediction of Slope Stability. *Nat. Hazards* **2003**, *30*, 383–398. [CrossRef]
39. Liu, S.; Lin, Y. Introduction to Grey Systems Theory. *Underst. Complex Syst.* **2010**, *1*, 1–18. [CrossRef]
40. Deng, J.L. Grey modeling resource theory and GM (1, 1, bk). *J. Grey Syst.* **2005**, *17*, 201.
41. Kayacan, E.; Ulutas, B.; Kaynak, O. Grey system theory-based models in time series prediction. *Expert Syst. Appl.* **2010**, *37*, 1784–1789. [CrossRef]
42. Jin, X.-G.; Zeng, J.; Liu, X.-R. Application of GM (1, 1) Optimized Model in Prediction of Landslide. In Proceedings of the Third International Conference on Natural Computation (ICNC 2007), Haikou, China, 24–27 August 2007; Volume 5, pp. 735–739.
43. Gao, W.; Feng, X. Study on displacement predication of landslide based on grey system and evolutionary neural network. *J. Rock Soil Mech.* **2004**, *25*, 514–517.
44. Jibson, R.W. Regression models for estimating coseismic landslide displacement. *Eng. Geol.* **2007**, *91*, 209–218. [CrossRef]
45. Aleshin, Y.; Torgoev, I. Landslide Prediction Based on Neural Network Modelling. In *Landslide Science and Practice*; Springer: Berlin/Heidelberg, Germany, 2013; pp. 311–317.
46. Neaupane, K.; Achet, S. Use of backpropagation neural network for landslide monitoring: A case study in the higher Himalaya. *Eng. Geol.* **2004**, *74*, 213–226. [CrossRef]
47. Zhao, Y.N.; Niu, R.Q.; Li, J.; Peng, L.; Wang, Y. Prediction of Landslide Displacement Based on Kernel Principal Component Analysis and Neural Network-Markov Chain. *Adv. Mater. Res.* **2013**, *726*, 1512–1520. [CrossRef]
48. Chuang, C.-W.; Lin, C.-Y.; Chien, C.-H.; Chou, W.-C. Application of Markov-chain model for vegetation restoration assessment at landslide areas caused by a catastrophic earthquake in Central Taiwan. *Ecol. Model.* **2011**, *222*, 835–845. [CrossRef]
49. Victorov, A. Probabilistic Model of Landslide Processes Based on Markov Chains. In Proceedings of the 15th International Multidisciplinary Scientific GeoConference SGEM2015, Ecology, Economics, Education and Legislation, Albena, Bulgaria, 18–24 June 2015; Volume 2, p. 579. [CrossRef]
50. Zhang, G. Time series forecasting using a hybrid ARIMA and neural network model. *Neurocomputing* **2003**, *50*, 159–175. [CrossRef]
51. Pradhan, B.; Lee, S. Landslide susceptibility assessment and factor effect analysis: Backpropagation artificial neural networks and their comparison with frequency ratio and bivariate logistic regression modelling. *Environ. Model. Softw.* **2010**, *25*, 747–759. [CrossRef]
52. Deng, Y.-F.; Jin, X.; Zhong, Y.-X. Ensemble SVR for prediction of time series. In Proceedings of the 2005 International Conference on Machine Learning and Cybernetics, Guangzhou, China, 18–21 August 2005; Volume 6, p. 3528.
53. Huang, C.-L.; Tsai, C.-Y. A hybrid SOFM-SVR with a filter-based feature selection for stock market forecasting. *Expert Syst. Appl.* **2009**, *36*, 1529–1539. [CrossRef]
54. Squarzoni, C.; Delacourt, C.; Allemand, P. Nine years of spatial and temporal evolution of the La Valette landslide observed by SAR interferometry. *Eng. Geol.* **2003**, *68*, 53–66. [CrossRef]
55. An, K.; Kim, S.; Chae, T.; Park, D. Developing an accessible landslide susceptibility model using open-source. *Sustainability* **2018**, *10*, 293. [CrossRef]
56. Moresi, F.V.; Maesano, M.; Collalti, A.; Sidle, R.C.; Matteucci, G.; Mugnozza, G.S. Mapping Landslide Prediction through a GIS-Based Model: A Case Study in a Catchment in Southern Italy. *Geoscience* **2020**, *10*, 309. [CrossRef]
57. Pal, S.C.; Chowdhuri, I. GIS-based spatial prediction of landslide susceptibility using frequency ratio model of Lachung River basin, North Sikkim, India. *SN Appl. Sci.* **2019**, *1*, 416. [CrossRef]
58. Kuo, B.C.; Yang, J.M.; Sheu, T.W.; Yang, S.-W. Kernel-Based KNN and Gaussian Classifiers for Hyperspectral Image Classifica-tion. In Proceedings of the Geoscience and Remote Sensing Symposium, IGARSS 2008, Boston, MA, USA, 7–11 July 2008.
59. He, X.H.; Wang, S.J.; Xiao, R.H.; Rao, X.Y.; Luo, B. Improvement and application of synergetic forecast model for landslides. *Chin. J. Geotech. Eng.* **2013**, *35*, 1839–1848. (In Chinese)
60. Sun, G.Z. Significance of Successful Prediction of Xintan Landslide and Deformation Monitoring (Sequence). *J. Geol. Hazard Control* **1996**, 1–4. [CrossRef]
61. Lu, Y.H. Signs of Xintan Landslide and its successful monitoring and prediction. *J. Water Soil Conserv. Not.* **1985**, *5*, 1–9. (In Chinese)
62. Hu, H.; Xie, J.H. GM (1, 1) Model of Landslide Time Prediction Based on Velocity Parameters. *J. Yangtze River Acad. Sci.* **2018**, *35*, 70–76, 87.
63. He, X.H.; Wang, S.J.; Xiao, R.H.; Rao, X.Y.; Luo, B. *Improvement of Verhulst Forecast Model of Landslide and Its Application*; 2013 Annual (13th) Academic Paper Collection of Institute of Geology and Geophysics; Chinese Academy of Sciences, Engineering Geology and Water Resources Research Office: Beijing, China, 2014.
64. Tang, R.; Deng, R.; An, S.Z. Deformation Monitoring and Failure Mechanism Analysis of Baishi Landslide in Beichuan County. *J. Eng. Geol.* **2015**, *23*, 760–768.
65. Mei, Q.Y. Forming conditions and sliding mechanism of switch yard slope at Tianhuangping power station. *Chin. J. Rock Mech. Eng.* **2001**, *20*, 25–28.

66. Xu, Q.; Zeng, Y.P. Research on acceleration variation characteristics of creep landslide and early-warning prediction indicator of critical sliding. *Chin. J. Rock Mech. Eng.* **2009**, *28*, 1099–1106.
67. Lu, G.F. Formation and Monitoring and Forecast of Jimingsi Landslide. *Chin. J. Geol. Hazard Control* **1994**, *5*, 376–383.

*Article*

# Research on the Evolution and Damage Mechanism of Normal Fault Based on Physical Simulation Experiments and Particle Image Velocimetry Technique

Xianfeng Peng [1], Hucheng Deng [1,2,*], Jianhua He [1,3,*], Hongde Chen [4] and Yeyu Zhang [5]

1    College of Energy Resources, Chengdu University of Technology, Chengdu 610059, China; pengxianfeng@stu.cdut.edu.cn

2    State Key Laboratory of Oil and Gas Reservoir Geology and Exploitation, Chengdu University of Technology, Chengdu 610059, China

3    Key Laboratory of Deep Earth Science and Engineering, Ministry of Education, Sichuan University, Chengdu 610065, China

4    Institute of Sedimentary Geology, Chengdu University of Technology, Chengdu 610059, China; chd@cdut.edu.cn

5    Shale Gas Evaluation and Exploitation Key Laboratory of Sichuan Province, Chengdu 610051, China; zhangyy.shale@foxmail.com

*    Correspondence: denghucheng@cdut.cn (H.D.); hejianhua19@cdut.cn (J.H.)



**Citation:** Peng, X.; Deng, H.; He, J.; Chen, H.; Zhang, Y. Research on the Evolution and Damage Mechanism of Normal Fault Based on Physical Simulation Experiments and Particle Image Velocimetry Technique. *Energies* **2021**, *14*, 2825. https://doi.org/10.3390/en14102825

Academic Editor: Emanuele Tondi

Received: 5 April 2021
Accepted: 12 May 2021
Published: 14 May 2021





**Abstract:** The formation and evolution of (normal) fault affect the formation and preservation of some reservoirs, such as fault-block reservoirs and faulted reservoirs. Strain energy is one of the parameters describing the strength of tectonic activity. Thus, the formation and evolution of normal fault can be studied by analyzing the variation of strain energy in strata. In this work, we used physical simulation to study the formation and evolution of normal fault from a strain energy perspective. Based on the similarity principle, we designed and conducted three repeated physical simulation experiments according to the normal fault in the Yanchang Formation of Jinhe oilfield, Ordos Basin, China, and obtained dip angle, fault displacement, and strain energy via the velocity profile recorded by high-resolution Particle Image Velocimetry (PIV). As a result, the strain energy is mainly released in the normal fault line zone, and can thus serve as channels for oil/gas migration and escape routes connecting to the earth's surface, destroying the already formed oil/gas reservoirs. One might need to avoid drilling near the fault line. Besides, a significant amount of strain energy remaining in the hanging wall is the reason why the normal fault continues to evolve after the normal fault formation until the antithetic fault forms. Our findings provide important insights into the formation and evolution of normal fault from a strain energy perspective, which plays an important role in the oil/gas exploration, prediction of the shallow-source earthquake, and post-disaster reconstruction site selection.

**Keywords:** normal fault; physical simulation experiment; Particle Image Velocimetry (PIV); strain energy; strain energy release rate


---

## 1. Introduction

In geology, a fault is a planar fracture or discontinuity in a volume of rock. If the rock mass above an inclined fault moves down, then the fault is termed normal, whereas, if the rock above the fault moves up, the fault is termed a reverse fault [1]. During conventional oil/gas exploration, the formation and evolution of normal fault are mainly studied using outcrop research, geostatistics, and finite element numerical simulation [2,3]. While these approaches can provide important information regarding the formation and evolution of normal fault, they bear certain drawbacks. For example, while strata have complex internal structures and mechanical properties, these conventional methods generally assume homogeneous rocks. Previous studies have shown that the formation and evolution of normal fault are due to the instability of energy-driven local strata, including micro-crack closure, elastic deformation, micro-defect expansion, and catastrophic failure [4,5]. During these processes, strata rocks continuously exchange energy with the adjacent ones, transforming external mechanical energy into strain energy, which produces a negligible amount of heat. The strain energy is then released in the form of electromagnetic radiation, acoustic emission, and kinetic energy. Such strain energy release causes the fault localizes and forms, leading to the formation and evolution of faults [4–7]. As a result, strain energy evolution relates to the formation and evolution of faults [8]. Thus, the knowledge about strata strain energy evolution is of significant importance for understanding the underlying mechanisms of the formation and evolution of normal fault.

Extensive studies were conducted to investigate the formation and evolution of normal fault from the strain energy perspective. Some works measured the accumulation and release of strain energy using theoretical approaches. They reported that the fault localizes and forms are due to not only the deformation exceeding fracture pressure of the strata, but also the release of strain energy [9–12]. Using the time and frequency domain methods, Akiyama et al. found that the sudden release of strain energy is the primary source for normal fault formation [13]. The seismic moment tensor from GPS data has been used to measure strata strain energy in geophysics. However, this approach is strictly limited due to the lack of seismic stations for recording the GPS data [14–16]. Recently, using acoustic emission technology and rock compression tests, previous studies found that the formation of faults is a rapid transitioning process from the triaxial stress state to a unidirectional stress one, in which it is difficult to measure the strain energy release in laboratory conditions [17–19]. Although these studies provide valuable insights for the understanding of the formation and evolution of normal fault, the underlying mechanism of strain energy release and its relation to the formation and evolution of normal fault are still unclear.

A physical simulation experiment is a powerful and effective tool for studying the formation and evolution of normal fault [20–22]. It can simulate structural deformation under various conditions, such as different boundaries and physical parameters. It can also simulate the basin and structures through direct visualization. The traditional physical simulation experiment uses an interval photographing method to record experimental data, while normal fault formation and evolution is a continuous process. In this work, we use the high-resolution particle imaging velocimetry (PIV), which can continuously monitor the experimental procedures. We use three layers of horizontally and homogeneously laid dry quartz sands in a simulation chamber to model the formation and evolution of normal fault and monitor the bending of the colored marker beds that will be incorporated into the model. The strain energy is calculated based on the velocity field of quartz sand particles, which is recorded by PIV. By applying the prototype similarity criterion, our physical simulation can monitor the formation and evolution of normal fault from the strain energy perspective to simulate real conditions.

We studied the damage mechanism of normal fault based on physical simulation experiments and particle image velocimetry technique. A case study conducted in Yanchang Formation is presented to demonstrate the efficiency of the proposed method. The validation results prove that this approach could effectively and accurately explain the tectonic deformation of normal faults. More importantly, this paper tries to give a new perspective to study the formation and evolution of normal faults, and it quantifies the formation and evolution process of normal faults with strain energy.

## 2. Experimental Section

### 2.1. Materials

We use the extensional structural model that was proposed by Cloos to study the formation and evolution of normal fault [23]. The basement and boundary of the normal fault model are both assumed to be rigid and simulated with transparent tempered glass. The internal friction angle is an important indicator of the rock shear strength. Previous studies have shown that the internal friction angle of dry quartz sand is approximately the same as that of strata [24,25]. Therefore, dry quartz sand has been widely used in the physical simulation experiments to simulate structural deformation [26–28]. The main material is white dry quartz sand (China ISO Standard Sand Co., Ltd. No. 45, Yanghe Road, Xinyang Industrial Zone, Haicang District, Xiamen City, Fujian Province, China) with a $SiO_2$ mass fraction of more than 99.5%. The density of dry quartz sand is $1.35 \times 10^3$ kg/m$^3$, with particle sizes between 0.30 mm and 0.45 mm and internal friction angle between $29°$ and $31°$. Silicone resin has Newtonian fluid properties at low strain rates. Silicone resin is often used to simulate the plastic deformation of the upper crust [25]. Therefore, the detachment layers in the structural model were simulated with canvas that was coated with 15 mm thick silicone resin. The density of silicone resin is $0.92 \times 10^3$ kg/m$^3$, and the elongation is 600% to 700%.

### 2.2. Experimental Setup

Figure 1 shows the experimental apparatus, which is a comprehensively experimental platform for the physical simulation of geological structural deformation, which is provided by the State Key Laboratory of Oil and Gas Reservoir Geology and Exploitation (Chengdu University of Technology). It includes a control system, a simulation chamber, a hydraulic system, a camera system, and a thermostat (air conditioner, with a temperature of 293.15 K).

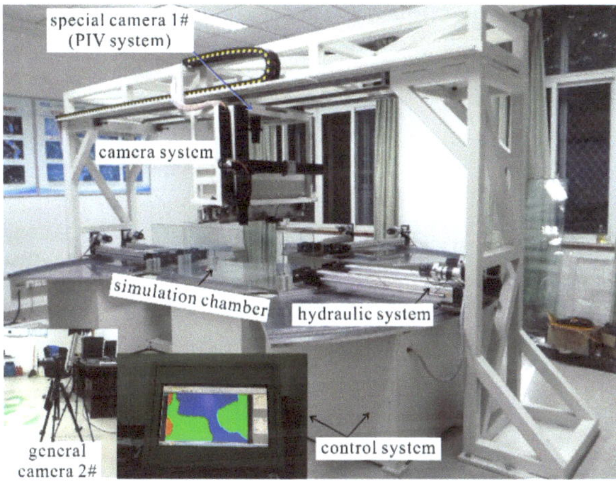


**Figure 1.** The comprehensively experimental platform for physical simulation of geological structural deformation.

The control system is used to control each system (including a simulation chamber, a hydraulic system, and a camera system) by computer (Figure 1). The simulation chamber is used to place dry quartz sands that simulates normal faults. Its size can be customized according to the experimental requirements, but it cannot be larger than 200 cm × 200 cm × 100 cm (length × width × height). The hydraulic system is used to provide tension or extrusion force with a maximum load of 10 kN and loading speed range of 0.0001 mm/s to 1 mm/s. The camera system includes a special camera 1# (PIV system, made by Beijing Cubetiandi Science and Technology Development Co., Ltd., Beijing, China) and a general camera 2# installed before and above the simulation chamber (Figure 2). The special camera 1# measures the displacement of the particles by processing the photo taking at 30-ms intervals. These photos are preprocessed using the *Micro Vec 3* that comes with the PIV system and saved as digital format files (*.dat*); the digital format files can be opened using *Tecplot. 360. 2010.* The general camera 2# takes pictures of the experiment process and saves them.

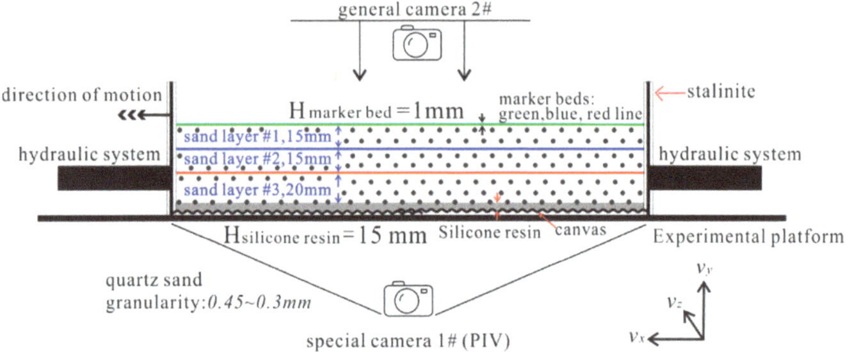



**Figure 2.** The schematic representation of the simulation chamber.

We place the dry quartz sands in the simulation chamber, as shown in Figure 2. The hydraulic system provides tensions from the two ends of the simulation chamber. The recording system (the camera system) includes a special camera 1# (PIV system) and a general camera 2# installed before and above the simulation chamber, which is used to monitor the experiment continuously, as shown in Figure 2. The entire physical simulation is conducted at ambient conditions of 293.15 K, with the system temperature being maintained using a thermostat.

### 2.3. Experiment Procedure

The prototype of the normal fault in this study is from the Yanchang Formation of Jinhe oilfield, Ordos Basin, China, which has 289.31 million tons of oil-in-place and it is a vital development zone of the Sinopec North China Branch. The Jinhe Oilfield is located between Qingyang and Zhengning, where the area is 20 km × 17.5 km, as shown in Figure 3. Three Wells, JH 17, JH 62, and JH 64, have been drilled into this normal fault. The seismic interpretation and drilling indicate that this normal fault passes through the Yan'an, Yanchang and Ermaying Formation, and the thicknesses are 750 m, 750 m, and 1000 m, respectively; the Yanchang Formation sandstone is the main oil-producing layer in Jinhe Oilfield; the dip angle is in the range of 60°–65°; and, the fault displacement is falling in 30 m–45 m.

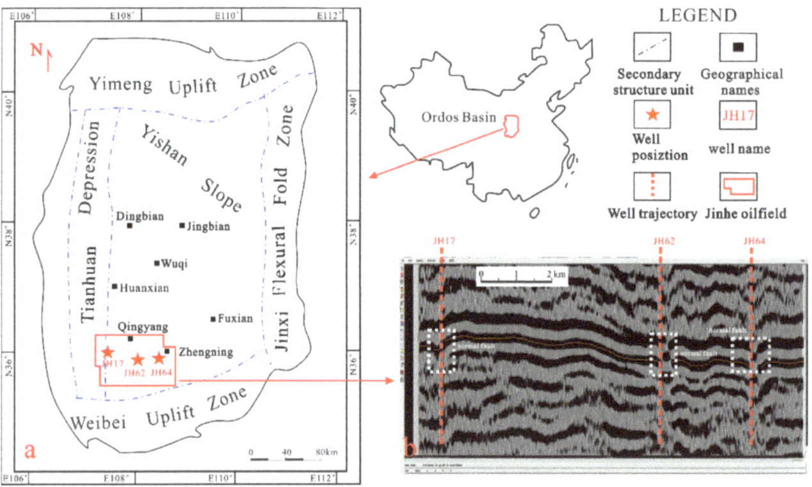



**Figure 3.** Prototype of physical simulation of geological structural deformation experiment. (**a**) The geographical position; (**b**) The seismic interpretation section of Jinhe Oilfield.

We set the experimental parameters following the similarity criterion between the model and prototype, including geometric, dynamic, and motion similarities [28–30]. Various similarity criteria are given, as follows:

(1) Geometric similarity. The area of the simulation chamber is 40 cm $\times$ 35 cm, while the area of the simulated strata is 20.0 km $\times$ 17.5 km with the geometric similarity ratio of $L^* = 2 \times 10^{-5}$. The thicknesses of Yan'an, Yanchang, and Ermaying Formation in Jinghe Oilfield are 750 m, 750 m, and 1000 m, respectively. Therefore, we set the thicknesses of quartz sands in the simulation chamber as 15 mm, 15 mm, and 20 mm, respectively. Because the physical properties of the dry quartz sands do not change throughout the simulations, to facilitate the observation of structural deformation in the simulation chamber, millimeter-thick colored quartz sands are laid on each layer surface as a marker bed, as shown in Figure 3.

(2) Dynamic similarity. We use the density similarity ratio of $\rho^* = 0.5$ and acceleration of gravity ratio of $g^* = 1$, as presented in Table 1. Consequently, the dynamic similarity ratio of $\sigma^*$ and kinematic similarity ratio $v^*$ are $1 \times 10^{-5}$ and $2 \times 10^5$, respectively.

(3) Kinematic similarity. Previous studies have shown that the detachment layers in the prototype slipped and deformed at a strain rate of 0.23 mm/year [31], so the uniform velocity of $v = 1.5$ μm/s is taken as the strain rate in our experiments based on the kinematic similarity ratio, as presented in Table 1.

**Table 1.** The experimental parameters between the model and prototype.

| Parameters | Units | Model | Prototype | Similarity Ratio |
|---|---|---|---|---|
| geometric similarity ($L^*$) | cm | 40 (length)<br>35 (width) | $20.0 \times 10^5$<br>$17.5 \times 10^5$ | $2 \times 10^{-5}$<br>$2 \times 10^{-5}$ |
| Density ($\rho^*$) | kg/m$^3$ | $1.35 \times 10^3$ | $2.7 \times 10^3$ | 0.5 |
| acceleration of gravity ($g^*$) | m/s$^2$ | 9.8 | 9.8 | 1 |
| dynamic similarity ($\sigma^*$) | | $\sigma^* = \rho^* \times g^* \times L^* = 0.5 \times 1 \times 2 \times 10^{-5}$ | | $1 \times 10^{-5}$ |
| coefficient of viscosity ($\eta^*$) | Pa·s | $1 \times 10^4$ (★) | $1 \times 10^{19}$ (★) | $1 \times 10^{-15}$ |
| Kinematic similarity ($v^*$) | | $v^* = \sigma^*/\eta^* \times L^* = (1 \times 10^{-5})/(1 \times 10^{-15}) \times (2 \times 10^{-5})$ | | $2 \times 10^5$ |

(★) The viscosity coefficient of the silicone resin used to simulate the detachment layer is set as $1 \times 10^4$ Pa·s in this experiment; the slip coefficient of sedimentary stratum is set as $1 \times 10^{19}$ Pa·s [30,32,33].

We first layer-fill dry quartz sands in the simulation chamber, as shown in Figure 2. Subsequently, we fix the hydraulic system at the right end of the simulation chamber, while open the left end to stretch the canvas at the bottom of sand layers at a uniform velocity ($v = 1.5$ μm/s) until the occurrence of antithetic faults. The experimental phenomena are monitored by the general camera 2# and the special camera 1# (PIV system), which records data at a time interval of 30 ms. The entire experimental procedures are repeated three times for replicability.

*2.4. Data Analysis*

2.4.1. Data Collection by PIV Technique

The high-resolution Particle Image Velocimetry (PIV) is a technique, in which a series of images are obtained by a high-resolution special camera, and the velocity vectors of points are obtained by cross-correlation algorithms [34,35]. The PIV measurement can observe the whole dynamic process without disturbing the test object, and then obtain the instantaneous velocity profiles. The cross-correlation algorithms allow for the calculation of displacement vectors in the sand with sub-pixel accuracy (<0.1 pixels). With a given optical vector accuracy of better than 0.1 pixels, the absolute accuracy of the displacement vectors (dx) depends on the image scaling and correction [36].

In this study, the PIV system is from Beijing Cubetiandi Science and Technology Development Co., Ltd. (http://www.piv.com.cn, accessed on 1 April 2021) and has an optical resolution of 16 megapixels (4 k $\times$ 4 k) and a given optical vector accuracy of better than 0.5 pixels. In this experiment, with an optical resolution of 16 megapixels (4 k $\times$ 4 k) and an experiment with 40 cm width, the absolute accuracy of the length of displacement vectors is 0.1 mm. PIV is used to measure the displacement of the particles by processing the photos taken every 30 ms in this paper, as shown in Figure 4. *Micro Vec 3* and *Tecplot. 360. 2010* are the software used by the authors.

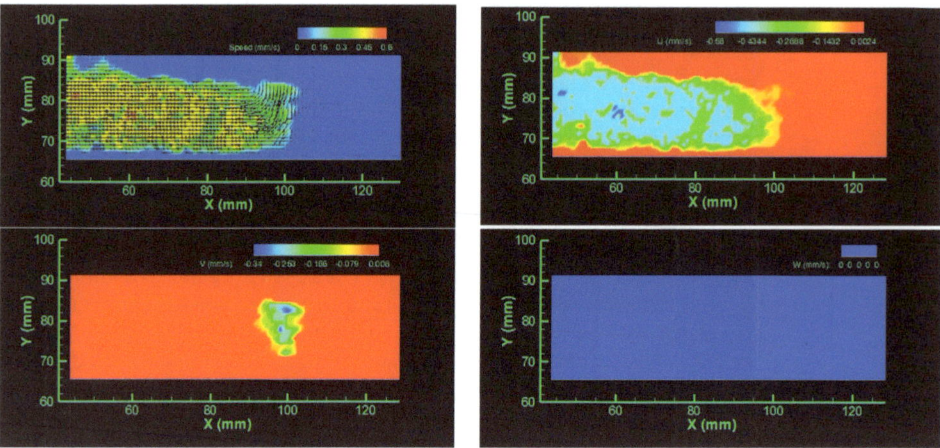

**Figure 4.** The instantaneous velocity profiles of the particles measured by the PIV technique (U, V, and W are the velocities in x-, y- and z-coordinates, respectively).

However, the inevitable micro-vibration of tempered glass baffles during the experiment and air disturbance on the top surface of sand layers in the simulated chamber would affect the experimental precision. In the actual analysis, the data within 2 cm near the glass baffle and within 1 cm of the top surface of sand layers in the simulated chamber are discarded. The data used for the analysis are between 85 mm and 110 mm in the x-coordinate and 70 mm and 90 mm in the y-coordinate. In addition, the hydraulic system takes about 30 min. to start and balance, so the initial recording time is from 2040 s to eliminate the systematic errors.

According to current metrological recommendations of GUM [37–39], every correctly performed measurement requires that its result be supplemented with a qualitative parameter characterizing this measurement, i.e., with the value of error or uncertainty.

To verify the accuracy of the measurement method and the measurement results by PIV, the velocity measurements at the center point (97.5 mm, 80 mm) of the simulation chamber were analyzed, as shown in Figure 4. Table 2 presents the information of standard deviation, standard error (SE) of mean, lower 95% confidence interval (CI) of mean, upper 95% CI of mean and mean absolute deviation values of 34 groups data of measurements for total velocity, U velocity, and V velocity by PIV. The result presented in Figure 2 shows that the measurement method and measurement results by the PIV technique can be compared with each other.

**Table 2.** The value of measurement uncertainty of PIV measured quantity.

| Velocity [mm/s] | Mean | Standard Deviation | SE of Mean | Lower 95% CI of Mean | Upper 95% CI of Mean | Mean Absolute Deviation |
|---|---|---|---|---|---|---|
| total | 0.439 | 0.03927 | 0.00674 | 0.42611 | 0.45341 | 0.03211 |
| U | 0.413 | 0.03171 | 0.00544 | 0.40247 | 0.42459 | 0.02685 |
| V | 0.145 | 0.03671 | 0.00629 | 0.13249 | 0.15811 | 0.03035 |

### 2.4.2. Dip Angle and Fault Displacement

The dip angle ($\alpha$) in the unit of degree (°) is the angle between the normal fault line and horizontal projection line [40], which can be obtained as,

$$\alpha = \arctan\frac{v_y}{v_x} \tag{1}$$

where $v_x$ and $v_y$ are velocities in x- and y- coordinates in the unit of mm/s, respectively, where x- and z- are used as the directions in easting and northing, and y- would be the depth, as shown in Figure 2.

The sand layers in the simulation chamber are designed to be horizontally stressed, i.e., $v_y = 0$ mm/s, so that the dip angle is $0°$ or no dip angle is seen before the breakage of sand layers. $\alpha \neq 0°$ indicates the fracture of sand layers. It is important to note that this is only true in this sandbox experiment.

In geology, a fault is a planar fracture or discontinuity in a volume of rock, across which there has been significant displacement as a result of rock-mass movement. This significant displacement is called fault displacement (L) [41], which can be calculated by,

$$L = \sum_{t=0}^{t=m} L_{i,j,t} = \sum_{t=0}^{t=m} (s_{i+1,j,t} - s_{i,j,t}) = \sum_{t=0}^{t=m} (v_{i+1,j} - v_{i,j})t \tag{2}$$

where $L$ is the fault displacement in the unit of m, $S_{i+1,j,t}$ and $S_{i,j,t}$ are the displacements of two adjacent points located on the same layer in the unit of m, $v_{i+1,j}$ and $v_{i,j}$ are the speeds of two adjacent points in the unit of m/s, and $t$ is the total time in the unit of s.

### 2.4.3. Strain Energy and Strain Energy Density

Strain energy is the potential energy that is stored in strata in the form of strain and stress. Differential equations of motion in a single-degree-of-freedom system can only be transformed into energy balance equations under the horizontal tensile force [12]. Given a viscously damped single-degree-of-freedom system subjected to a horizontal earthquake ground motion, the equation of motion can be written as,

$$m\ddot{x}_t + c\dot{x} + f(x) = 0 \tag{3}$$

$$x_t = x + x_g \tag{4}$$

where $m$ represents mass, $c$ is the viscous damping coefficient, $f(x)$ is the restoring force, $x_t$ is the total displacement of the mass with respect to the ground, $x$ is the displacement of the simulation chamber, and $x_g$ is the relative displacement of the mass with respect to the simulation chamber, as shown in Figure 5.

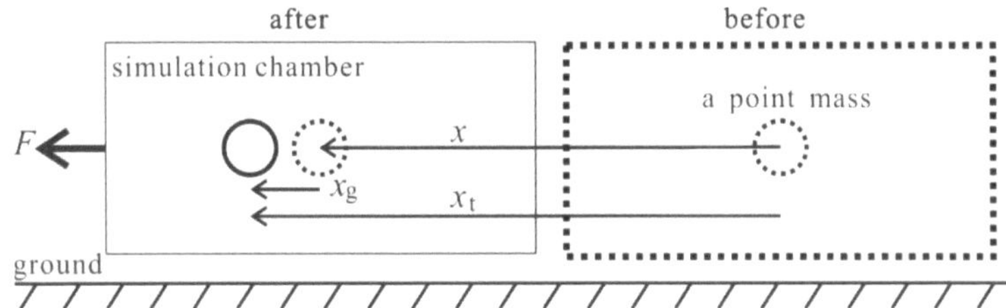


**Figure 5.** The relationship between $x$, $x_t$, and $x_g$.

Integrating Equation (3) with respect to $x$ results in,

$$\int m\ddot{x}_t dx + \int c\dot{x}dx + \int f(x)dx = 0 \tag{5}$$

Replacing $x = x_t - x_g$, in the first term of Equation (5), then

$$\int m\ddot{x}_t dx = \int m\ddot{x}_t (dx_t - dx_g) = \int m\frac{\dot{x}_t}{dt}dx_t - \int m\ddot{x}_t dx_g = m\frac{(\dot{x}_t)^2}{2} - \int m\ddot{x}_t dx_g \quad (6)$$

Substituting Equation (6) into Equation (5) yields,

$$m\frac{(\dot{x}_t)^2}{2} + \int c\dot{x}dx + \int f(x)dx = \int m\ddot{x}_t dx_g \quad (7)$$

where $E_k = \frac{m(\dot{x}_t)^2}{2}$ is the kinetic energy, $E_\xi = \int c\dot{x}dx$ is the damping energy, $E_a = \int f(x)dx$ is the absorbed energy, and $E_i = \int m\ddot{x}_t dx_g$ is the input energy.

We assume that the unit volume of rock has no heat exchange with the outside environment during the deformation process under external force, based on the first law of thermodynamics. Based on the first law of thermodynamics, we have

$$E_i = E_d + E_e \quad (8)$$

where: $E_d$ is the unit dissipation, forming internal damage and shaping deformation, i.e., the change of internal state conforms to the trend of entropy increase; $E_e$ is the unit released strain energy, in which the elastic strain energy can be released after unloading in strata [42].

Because the earth's crust is an isothermal layer, we conduct the experimental data at a constant temperature without heat exchange. In other words, $E_d = 0$, so we have $E_e$ as

$$E_e = E_i = \int m\ddot{x}_t dx_g \quad (9)$$

The strain energy accumulated in the unit volume of strata is called the strain energy density [18], and it can be calculated as

$$\rho_e = \int \frac{m}{V}\ddot{x}_t dx_g \quad (10)$$

where $\rho_e$ is the unit released strain energy density and $V$ is the unit volume.

The strain energy release rate (SERR) is an effective parameter to characterize normal fault formation and evolution [43]. In this study, the ratio of strain energy reduction to the total strain energy during the strain energy release process ($E_{eb} > E_{ea}$) is defined as the strain energy release rate that is given as [44],

$$SERR = \frac{E_{eb} - E_{ea}}{E_{eb}} \times 100\% \quad (11)$$

where $E_{eb}$ is the total strain energy before the release and $E_{ea}$ is the total strain energy after the release.

## 3. Results and Discussion

### 3.1. Identifying the Formation and Evolution of Normal Fault

In Figure 6, we present the digital image of the experimental procedures. The dry colored quartz sands laid on each layer surface comprise the marker bed, while track reflects the progress of normal fault tectonic deformation. The marker bed in the simulation chamber remains intact, and there is no sign of fracture from t = 2040 s to t = 3830 s, as shown in Figure 6. The marker bed is gradually bent to form the first depression and the vertical depression braking distance on the cross-section increases from t = 3830 s to t = 7426 s. Subsequently, a second depression forms from t = 7426 s to t = 8670 s. During the first depression period, the first micro-crack (the micro-crack refers to the fault line that can be seen on the top surface as shown in Figure 6) on the horizontal plane forms at t = 5118 s and the second at t = 7182 s. The length of the first micro-crack gradually increases until a distinct fault line forms on the horizontal plane from t = 5118 s to t = 7182 s and the second from t = 7182 s to t = 7426 s.

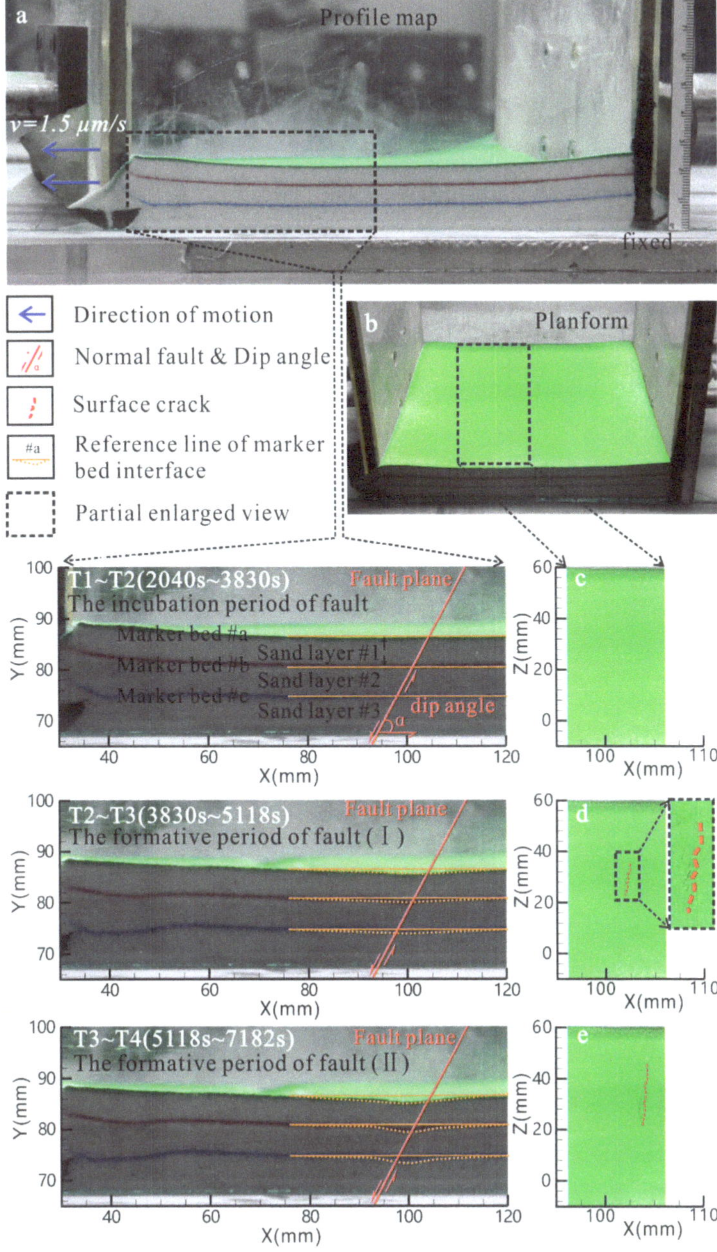


**Figure 6.** *Cont.*

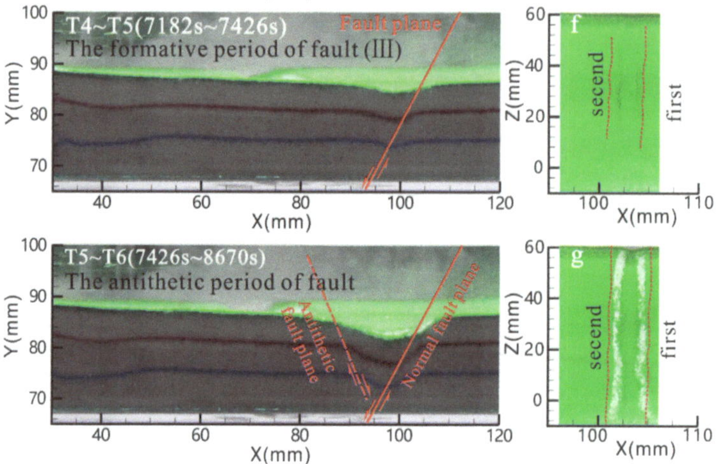



**Figure 6.** Experimental phenomenon of a physical simulation experiment.

We depict the experimental phenomenon in detail and the calculated PIV, as shown in Figure 7. The triangular facet, which is the outcrop characteristic of the normal fault [45] can be observed in the simulation chamber. Generally, a fault line is a line that is commonly plotted on geologic maps to represent a fault and a fault plane where the fault can be seen or mapped on the surface. However, the fault plane is a plane that represents the fracture surface of a fault. In this study, we define the fault line as the localized shear zone, and the fault zone as the more distributed deformation around the localized shear zone. It is well known that the fault line is the central region of the normal fault that best reflects its formation and evolution. Four equally spaced recording points on the fault line are selected to analyze the dip angle and fault displacement change over time in this study.

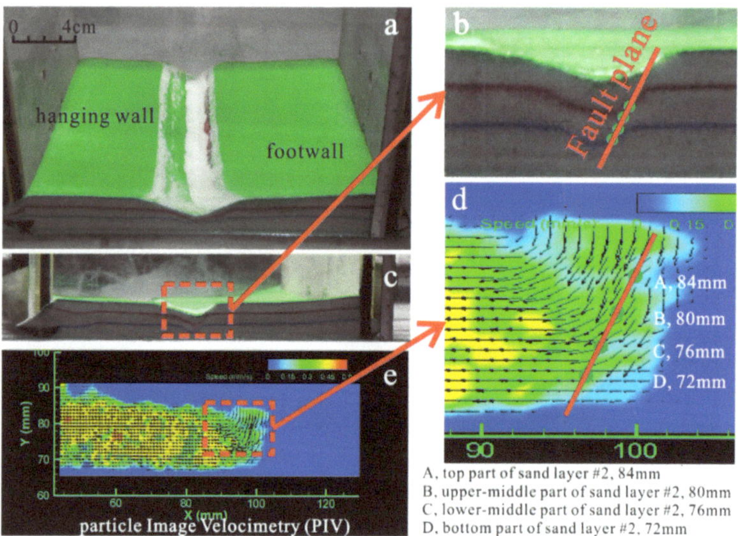



**Figure 7.** Experimental phenomena, including top (**a**) and front (**c**) views and partial enlargement (**b**). PIV results (**e**) and partial enlargement (**d**), as well as the four recording points, are depicted on the fault line.

According to the kinematic similarity ($v^* = 2 \times 10^5$) seen in Table 1, the strain energy of normal fault in the experiment is converted into the energy of the real geological prototype, which is convenient to compare with the energy released when a real normal fault is formed. Figure 8 presents the strain energy that is obtained by Equation (9) between 85 mm and 110 mm in the x-coordinate and 70 mm and 90 mm in the y-coordinate from Figure 7d during the experiment.

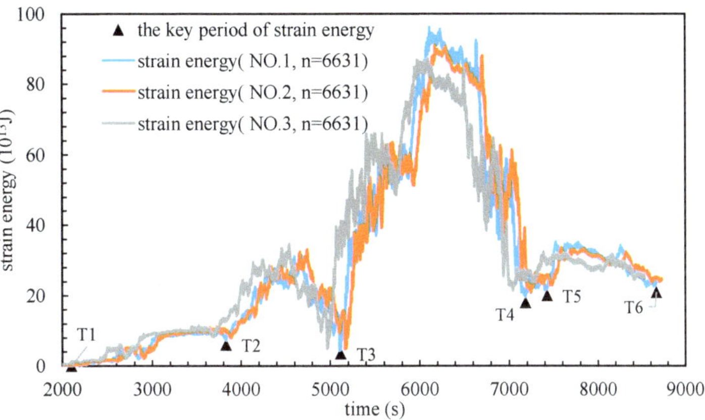


**Figure 8.** Strain energy over time of three repeated experiments, and the key period of strain energy.

The three repeated experiments show similar strain energy curves, as shown in Figure 9. As a result, we use the mean data of three experiments to illustrate the strain energy evolution, as shown in Figures 10 and 11.

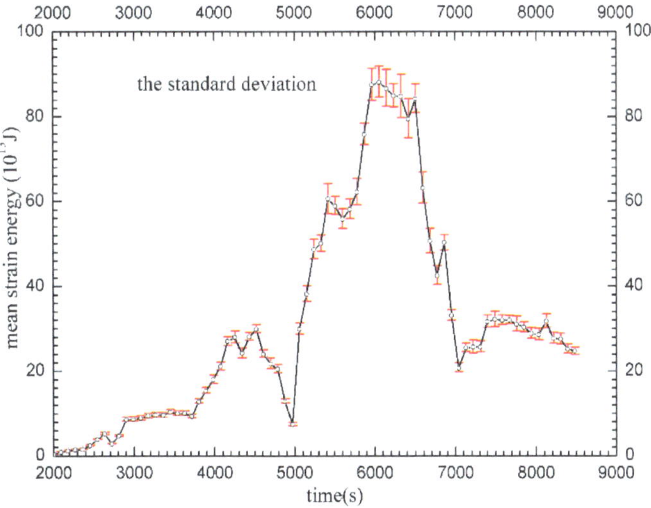


**Figure 9.** The mean and standard deviation values of strain energy over time of three repeated experiments.

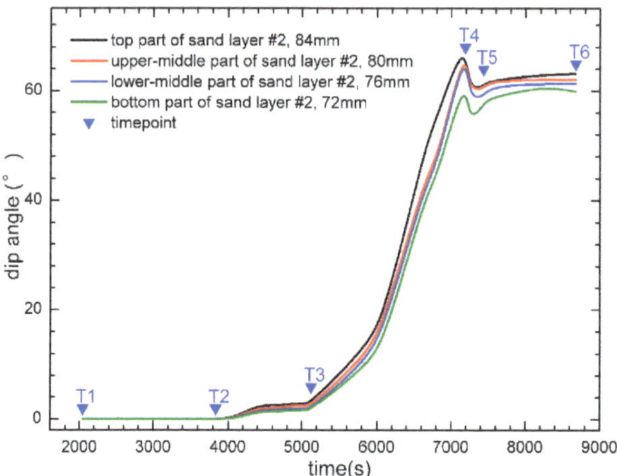

**Figure 10.** The dip angle of the normal fault over time. The data from Figure 6d, between 85 mm and 110 mm in the x-coordinate and 70 mm and 90 mm in the y-coordinate.

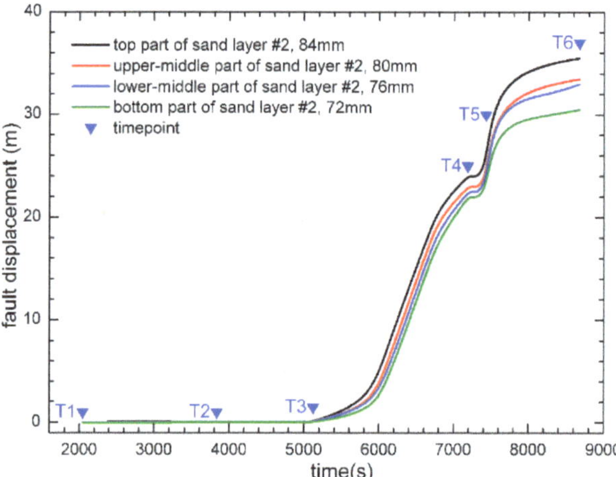

**Figure 11.** The fault displacement of the normal fault over time. The data from Figure 6d, between 85 mm and 110 mm in the x-coordinate and 70 mm and 90 mm in the y-coordinate.

It is observed that the strain energy has a periodic variation of accumulation and release, conforming to the elastic rebound theory [46], which claims that the total strain energy that can be accumulated in the strata has a fixed upper limit, as shown in Figure 9. When such a limit is exceeded, the strata fracture will form normal fault and release strain energy. When the total strain energy stored in the strata declines to a certain low level, it begins to accumulate in a new cycle. Based on such strain energy evolution, we classify six different time intervals, as depicted in Table 3. The first time interval, T1, is set at t = 2040 s, when the experiment starts; the second time interval, T2, is set at t = 3830 s, when the marker bed is bent to form the first depression; the third time interval, T3, is set at t = 5118 s, when the first micro-crack forms on the horizontal plane; the fourth time interval, T4, is set at t = 7182 s, when the second micro-crack forms on the horizontal plane; the fifth time interval, T5, is set at t = 7426 s, when the second depression forms; and, the sixth time interval, T6, is set at t = 8670 s, when the experiment ends. Overall, the formation and evolution of normal fault have distinct stages, which should be studied separately.

**Table 3.** Time intervals during the formation and evolution of normal fault.

| Timepoint | Time, s | Strain Energy, $\times 10^{13}$ J |
|---|---|---|
| T1 | 2040 | 0.0161 |
| T2 | 3830 | 7.9011 |
| T3 | 5118 | 4.7146 |
| T4 | 7182 | 19.2356 |
| T5 | 7426 | 22.7641 |
| T6 | 8670 | 23.2352 |

We depict the dip angle evolution during the third experiment in Figure 10. It shows that the dip angle is almost 0° and the sand layers do not have a fracture between T1 and T2. Between T2 and T3, the dip angle is between 0° and 3° and fracture occurs. Moreover, the dip angle increases sharply and can be up to 70° between T3 and T4. However, the dip angle shows a decreasing trend between T4 and T5, indicating that the normal fault activity begins to weaken, and it enters a stable stage. Between T5 and T6, the antithetic fault occurs with normal fault evolution, which is called the antithetic fault formation period. Additionally, we can find that the tendency of $\alpha$ decreases from the top part of sand layer #2 to bottom part of sand layer #2, indicating that, the shallower the buried depth of the sand layers, the more active normal fault.

Figure 11 presents the fault displacement of the normal fault. According to the geometric similarity ($L^* = 2 \times 10^{-5}$) of Table 1, the fault displacement of the normal fault in the experiment is converted into the fault displacement of the real geological prototype, which is convenient to compare with the actual fault displacement. When the fault displacement is 0 mm, the sand layers do not fracture between T1 and T3, as shown in Figure 11. The fault displacement increases rapidly between T3 and T4, which is the main period of normal fault formation. The fault displacement tends to be stable between T4 and T5, indicating that the occurrence of antithetic fault weakens the normal fault activity. The fault displacement continues to increase during between T5 and T6, because the formation of the antithetic fault increases the total displacement. We can also find that the tendency of the fault displacement decreases from the top part of sand layer #2 to the bottom part of sand layer #2. The trend of the dip angle and fault displacement with buried depth is the same, showing that the activity of normal fault gradually increases as the buried depth increases.

According to the previous works [47,48], the dip angle, fault displacement, and strain energy characteristics, the formation and evolution of normal fault are divided into three distinct periods in this work: the incubation period, formative period, and antithetic fault period, as shown in Table 4.

**Table 4.** The dividing basis of the formation and evolution of normal fault periods.

| Items | Incubation Period | Formative Period | | | Antithetic Faults Period |
|---|---|---|---|---|---|
| | | Elementary Stage | Unstable Stage | Stable Stage | |
| time point (s) | T1~T2 (2040~3830) | T2~T3 (3830~5118) | T3~T4 (5118~7182) | T4~T5 (7182~7426) | T5~T6 (7426~8670) |
| dip angle ($\alpha$, °) | almost 0 | 0~3 | close to 70 | decline | stable |
| fault displacement ($L$, m) | 0 | 0 | increase rapidly | stable | increase rapidly |

### 3.2. Strain Energy Density Variation during the Formation and Evolution of Normal Fault

In this section, we illustrate the variation of strain energy density in different periods of normal fault formation and evolution. The Surfer 12 software draws the contour maps of strain energy density.

### 3.2.1. The Incubation Period (T1~T2)

During this period, the strain energy density increases gradually and slowly under the horizontal tectonic stress, as shown in Figure 12. The average strain energy density is less than $5 \times 10^6$ KJ/m$^3$, and the strata do not break to form a normal fault. The first region of strain energy accumulation forms and then expands below the fault line. As the structural stress continues, the strain energy accumulation occurs over the fault line. Finally, they merge to form the normal fault nucleation.

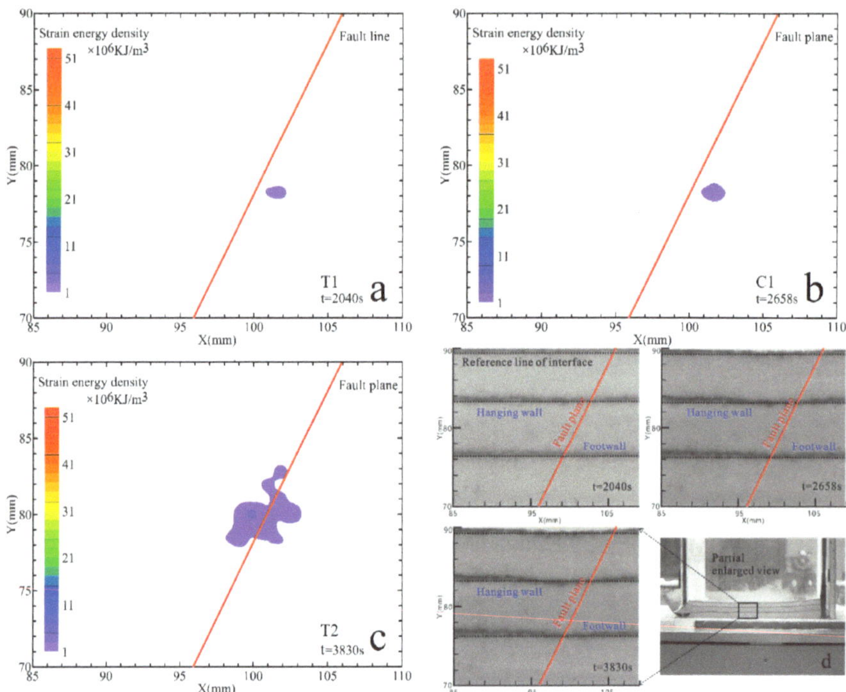


**Figure 12.** Strain energy density distribution at (**a**) T1, t = 2040 s; (**b**) C1, t = 2658 s; (**c**) T2, t = 3830 s. (**d**) the normal fault evolution observed from the simulation chamber. C refers to the time point between key periods.

### 3.2.2. The Formative Period (T2~T5)

1.  The elementary stage (T2~T3)

During this stage, the strain energy density first increases and then decreases, as shown in Figure 13, and the average strain energy density is $10 \times 10^6$ KJ/m$^3$. The strain energy density of the fault line region increases fastest, reaches the limit of strata failure value E = $15 \times 10^6$ KJ/m$^3$, and then decreases first. Moreover, after that, the strain energy density of the hanging wall and footwall regions increases and forms two new strain energy accumulation regions, as shown in Figure 13c, and the strain energy density of the footwall reaches the limit of strata failure value E and is then released. The strain energy density of the hanging wall in the region (98 mm, 75 mm) is completely released at T3, as shown in Figure 13f,g. However, the sand layers in the simulation chamber remain stable, and the fault line does not appear, as shown in Figure 6d. We can conclude that the multi-region strain energy release is the precursor of normal fault formation.

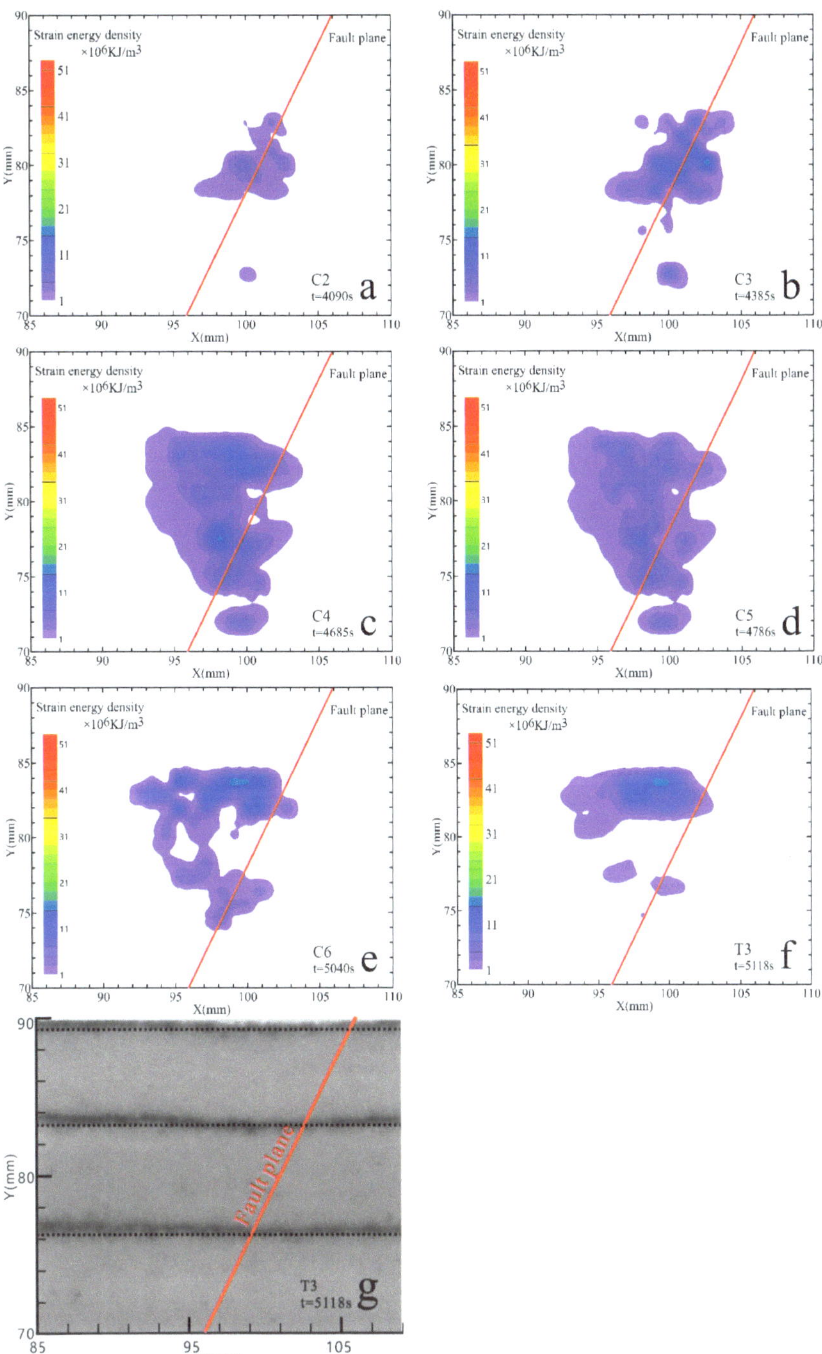

**Figure 13.** Strain energy density distribution at (**a**) C2, t = 4090 s; (**b**) C3, t = 4385 s; (**c**) C4, t = 4685 s; (**d**) C5, t = 4786 s; (**e**) C6, t = 5040 s; (**f**) T3, t = 5118 s; and, (**g**) T3, t = 5118 s, the normal fault evolution observed from the simulation chamber.

2. The unstable stage (T3~T4)

The strain energy is also accumulated before the release at this stage, as shown in Figure 14, but the average strain energy density is $20 \times 10^6$ KJ/m$^3$, which is quadruple that during the incubation stage. The strain energy of the hanging wall in the region (98 mm, 75 mm) is accumulated again, in a large area. A small area of strain energy accumulation region forms below the fault line and then merges into the large-scale one, which results in unstable strain energy density distribution in the layers. Figure 15 shows that the integrity of the fault zone is formed, and the fault line penetrates the strata. The main strain energy release occurs in the fault line zone, which is the reason for the fault line stability loss.

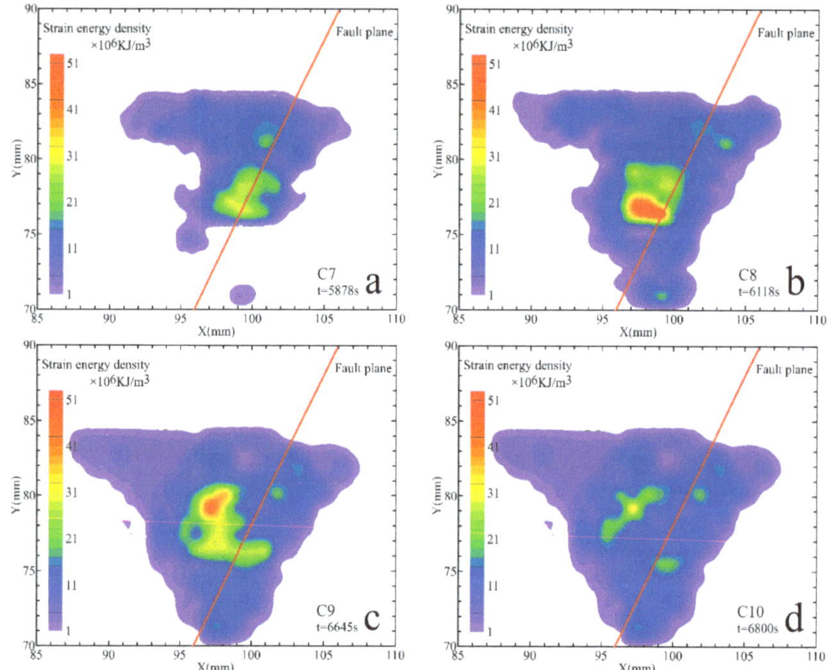

**Figure 14.** Strain energy density distribution at (**a**) C7, t = 5878 s; (**b**) C8, t = 6118 s; (**c**) C9, t = 6645 s; and, (**d**) C10, t = 6800 s. Strain energy bullseye at the top left is the "incubation stage" of the antithetic fault.

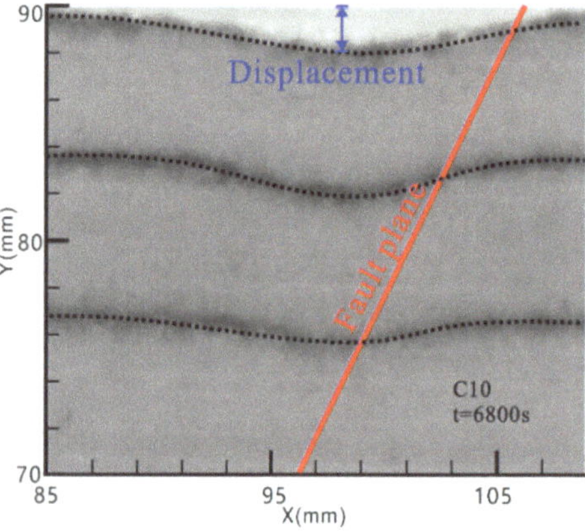


**Figure 15.** The cross-sectional image from the simulation chamber at C10, t = 6800 s.

3. The stable stage (T4~T5)

A typical normal fault fracture zone structure is formed with the initiation of a fault line, a fault has formed. It then just continues to grow/propagate downwards at T5, as shown in Figure 16. The hanging wall moves past the footwall on the fault, which generates heat and dissipates strain energy. The amount of strain energy increase is equivalent to that of the released. The total strain energy tends to be stable, and the normal fault is in a stable period.

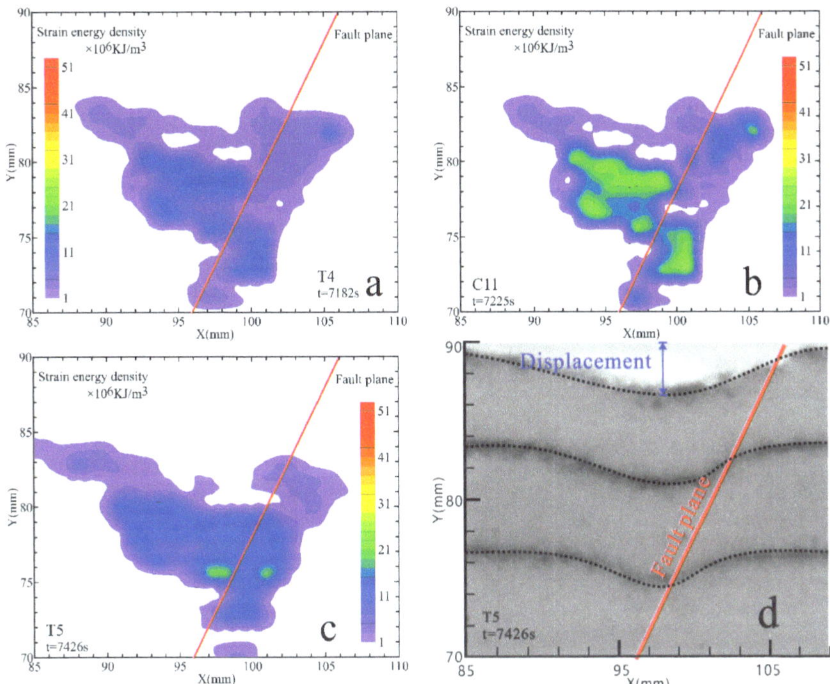


**Figure 16.** Strain energy density distribution at (**a**) T4, t = 7182 s; (**b**) C11, t = 7225 s; (**c**) T5, t = 7426 s; and, (**d**) the normal fault from the simulation chamber at T5, t = 7426 s. The inception stage of the antithetic goes into the start of the elementary stage for the antithetic fault.

### 3.2.3. The Antithetic Fault Period (T5~T6)

Figure 17 shows that the fault lines cut off the connection between the hanging wall and footwall layers after flexure. The footwall is no longer subjected to the horizontal tectonic stress, and its strain energy gradually becomes stable. However, the hanging wall has been under structural stress, and its strain energy increases again after a short period of stability. As the strain energy of the hanging wall reaches the limit of strata failure value, the antithetic faults form in a region that is far from the fault line on the hanging wall. Moreover, the residual strain energy after the formation of normal fault is mainly located on the hanging wall, which is also the cause of the instability of the hanging wall.

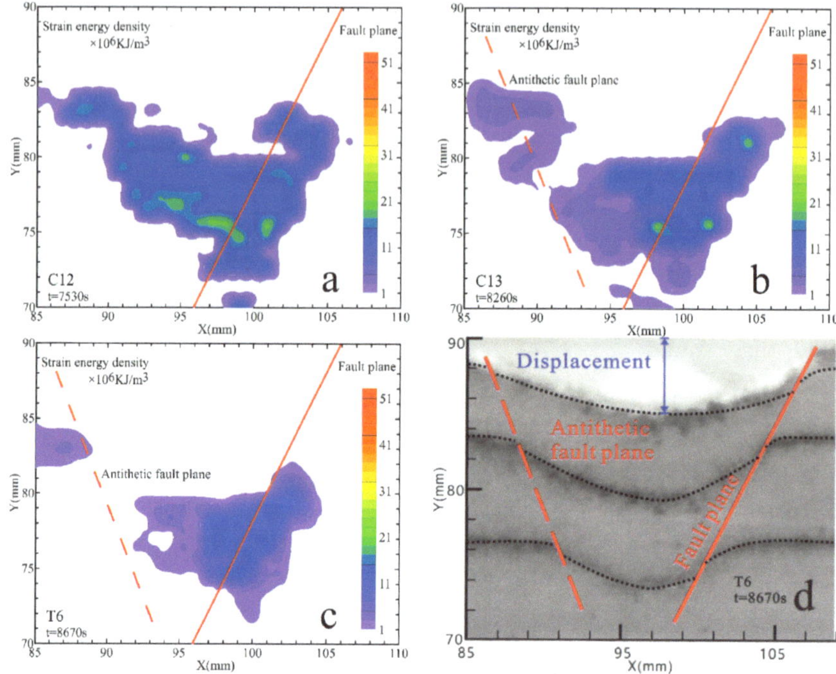

**Figure 17.** Strain energy density distribution at (**a**) C 12, t = 7530 s; (**b**) C 13, t = 8260 s; (**c**) T6, t = 8670 s; and, (**d**) the normal fault from the simulation chamber at T6, t = 8670 s. The inception stage of the antithetic going into the start of the elementary stage for the antithetic fault.

### 3.2.4. The Strain Energy of Each Period

The average strain energy density at different periods has the following characteristics. The average strain energy density of the elementary stage is twice as big as the one of the incubation periods. The one at the stable stage is larger than the one at the antithetic fault period. The one at the unstable stage is $20 \times 10^6$ KJ/m$^3$, which is the largest among them. Therefore, the average strain energy density can be used as an indicator of normal fault formation and evolution.

Figure 18 presents that the different regions of strata have different strain energy density characteristics. The strain energy density in the hanging wall is larger than that of the footwall, which is only 75% of the hanging wall. The strain energy is mainly located in region B of the hanging wall, the C region of the footwall, and D region, where the antithetic fault is formed. The strain energy in the A region near the fault line is the biggest. The strain energy density has the characteristic that, the farther away from the fault line, the smaller it is in the footwall and the upper 10 mm of the hanging wall.

### 3.3. Strain Energy Characteristics of Normal Fault

### 3.3.1. Distribution of Strain Energy Density

The strain energy distributions in the hanging wall and footwall are different, as shown in Figure 19. Thus, in this study, the normal fault zone is divided into five regions, namely the fault cores, damage zone in the hanging wall, damage zone in the footwall [49], surrounding rocks in the hanging wall, and surrounding rocks in the footwall, as shown in Figure 19. According to the previous achievements of the normal fault zone of Yanchang Formation in the Jinghe Oilfield, the distribution range of each region is known according to the geometric similarity ratio L*, as shown in Table 5.

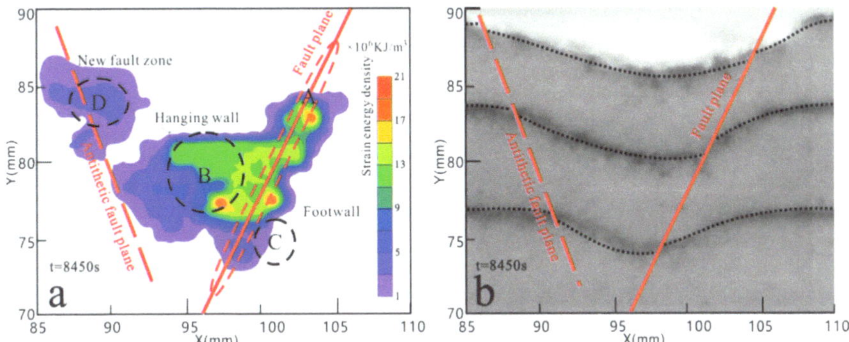

**Figure 18.** Strain energy density distribution characteristics of different regions. (**a**) The strain energy density distribution mode, A is fault plane zone; B is hanging wall zone; C is footwall zone; D is new fault zone; (**b**) the normal fault from the simulation chamber at t = 8450 s.

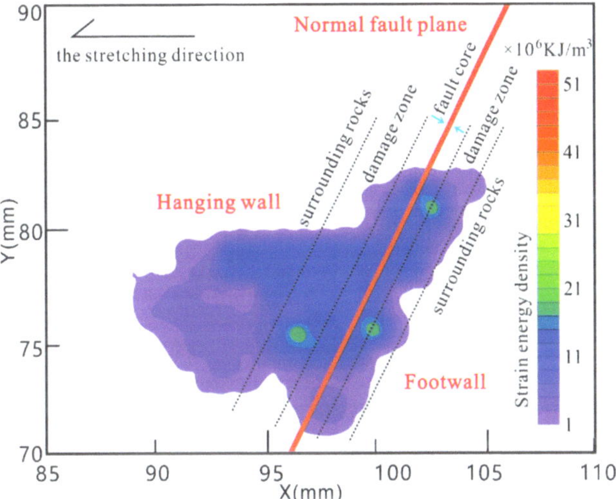


**Figure 19.** Schematic diagram of the five regions of a normal fault.

**Table 5.** Range conversion table for each region of the normal fault in the prototype and experiment ($L^* = 2 \times 10^{-5}$).

| Normal Fault Zones | Range in the Prototype, m | Range in the Model, mm |
| --- | --- | --- |
| Fault core | (0, 50] | (0, 1] |
| damage zone in the hanging wall | (50, 140] | (1, 2.8] |
| damage zone in the footwall | (50, 125] | (1, 2.5] |
| Surrounding rocks in the hanging wall | (140, ∞) | (2.8, ∞) |
| Surrounding rocks in the footwall | (125, ∞) | (2.5, ∞) |

Figure 20 shows that the five regions have similar strain energy characteristics. The strain energy of each region gradually increases during the incubation period. The strain energy of each region has a similar feature to increasing first and then decreasing. Among them, the normal fault core has the largest strain energy, and the rupture zone in the hanging wall is second during the elementary stage. The strain energy increases significantly after a brief decrease in all five regions, and the strain energy increases at the fastest rate in the region of the rupture zone in the hanging wall. The strain energy is reduced in all five regions during the stable stage. The strain energy begins to decrease after a brief increase in the surrounding rock in the hanging wall during the antithetic fault period.

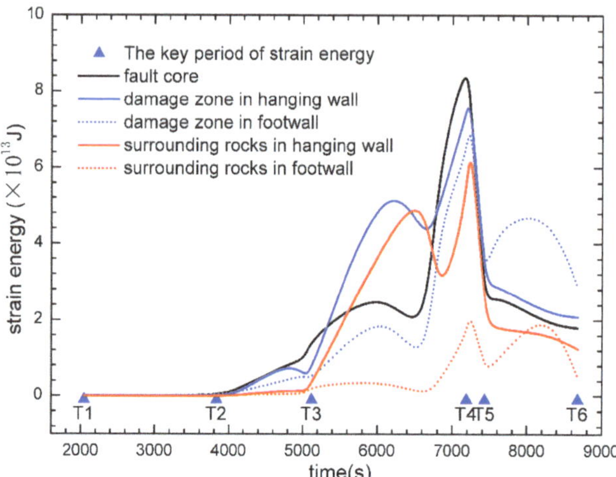



**Figure 20.** The distribution of strain energy at different locations in the normal fault over time.

### 3.3.2. The Central Position of Strain Energy

In general, the energy source causing the tectonic deformation of the strata cannot be assumed as a point source. The size of structural deformation does not significantly affect the energy attenuation direction of normal fault. Therefore, the energy source of a normal fault can be assumed to be a point source [49]. Calculating the position of the energy source that causes the formation to generate a normal fault based on the material point method, also known as the central position of strain energy (CPSE). In this work, the CPSE is calculated by Equation (A8) in Appendix A and shown in Figure 21. The area of the dot represents the magnitude of the strain energy at each time point (19 in total), and the position of the dot represents the position of the strain energy at each time point, as shown in Figure 21. The CPSE is mainly located in the fault line and the hanging wall, and its position and magnitude change with time, as following: the CPSE is located at the bottom of the footwall and moving to the fault line during the incubation period, while it is located at the fault line during the elementary stage. As the strain energy reaches the limit of strata value and is released, the CPSE jumps from the fault line to the hanging wall. The strain energy release results in an unstable state of the hanging wall, forming a new CPSE. The CPSE moves to the depth of the hanging wall during the stable stage, while it moves to the bottom position of the fault line, and the strain energy first increases and then decreases during the antithetic fault period.

The CPSE change with time indicates that the location of the fracture is not fixed during normal fault formation and evolution. The fracture mainly occurs in the hanging wall, especially at the bottom of the hanging wall. The fracture at the bottom creates space, and the sands at the top of the strata collapse under gravity to replenish the missing space. The constant transformation of the CPSE is the reason why the associated fracture of the normal fault has both tension and shear cracks. The Anderson model of fault analysis from the perspective of triaxial force cannot explain this phenomenon. It is necessary to study the normal fault formation and evolution from the perspective of strain energy.

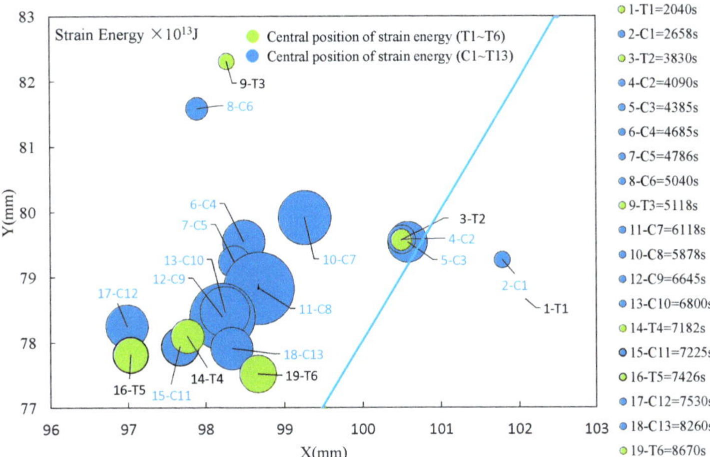



**Figure 21.** The central position of strain energy in each period.

### 3.3.3. Strain Energy Release Rate

Figure 22 shows the strain energy release rate during normal fault formation. There is always strain energy release during the normal fault formation, but the release rate of strain energy is low, with a maximum value of 26.1% and an average value of 1.7%. The moment of strain energy release rate higher than average corresponds to the critical period of a normal fault (refers to T1–T6, as shown in Table 3). Therefore, the strain energy release rate is an indicator of the further evolution of normal fault. During the formation of a normal fault, the strain energy release rate has no relationship with the fracturing activity in the layers. On the one hand, the strain energy release rate during the normal fault formation period does not increase significantly, so the low strain energy release rate will not dictate that the layer fracture activity is weakened. On the other hand, the strain energy release rate increases significantly during the period before the stabilization period, which indicates that the sustained increase of the strain energy release rate is an indicator that normal fault is about to enter a stable period.

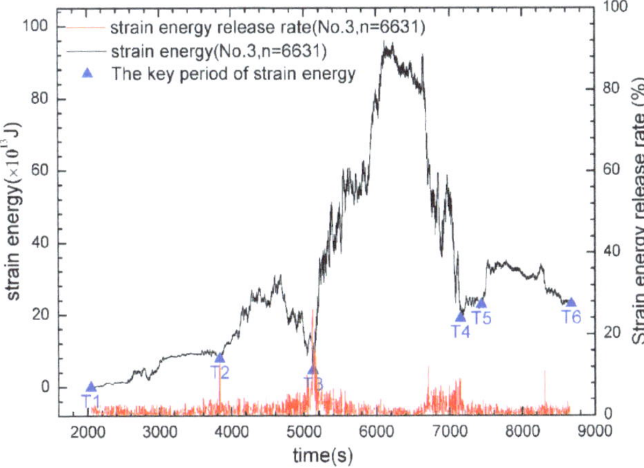



**Figure 22.** Strain energy release rate during the formation and evolution of normal fault.

## 4. Discussion

Strain energy is one of the parameters that describes the strength of tectonic activity. Thus, the formation and evolution of (normal) fault can be studied by analyzing the variation of strain energy in stratas. Physical simulation experiments can only simulate the structural deformation under various conditions, but they cannot simulate basins and structures through direct visualization [20–22]. Therefore, we cannot study the formation and evolution of normal faults by this approach. In this paper, we use high-resolution particle imaging velocimetry (PIV), which can continuously monitor the experimental procedures, and the strain energy is calculated based on the velocity field of quartz sand particles, which is recorded by PIV. A case study in Yanchang Formation is presented to demonstrate the efficiency of the proposed method. The validation results prove that the approach in this paper could effectively and accurately explain the tectonic deformation of normal faults. As a result, the strain energy is mainly released in the normal fault line zone and, thus, can serve as channels for oil/gas migration and escape conduits connecting to the earth's surface, destroying the already formed oil/gas reservoirs. Besides, a large amount of strain energy remaining in the hanging wall is the reason why the normal fault continues to evolve after the normal fault formation until the antithetic fault forms. One might need to avoid drilling near the fault line. More importantly, this paper may give a new perspective to study the formation and evolution of normal faults.

## 5. Conclusions

In this study, we used physical simulation experiments with PIV technology to study the formation and evolution of normal fault from the strain energy perspective. The following conclusions are drawn.

(1) The formation and evolution of normal faults are phased. We identified three phases, including the include incubation period, the formative period, and the antithetic fault period. During the incubation period, strain energy accumulates along the fault line, while it is generally small, and the strata are not broken. During the formative period, strain energy increases significantly, and it reaches the limit of strata and that in the fault line region is released, and normal fault tends to be in a stable stage. When the fault line cuts off the connection between the footwall and hanging wall before the antithetic fault period, the footwall is in a stable period, and the hanging wall would continue to move to form the antithetic fault period.

(2) Overall, the strain energy before normal fault formation is continuously accumulated. The normal fault is formed when the strain energy reaches the limit of strata. A low level of strain energy release randomly distributed cannot form normal faults. The main strain energy release occurs in the fault line zone. Consequently, one might need to avoid drilling near the fault line, but on the normal fault zone. The strain energy is released during the whole process of normal fault formation, and the maximum release rate is 26.1%. However, the low rate of strain energy release does not mean that the fault activity is weakened, and the sustained increase of strain energy release rate is a sign that the normal fault is about to enter a stable period.

(3) The average strain energy density is an indicator of the formation and evolution of normal fault. The farther from the fault line, the smaller the strain energy density. The influence range of the strain energy in the hanging wall is greater than that of the footwall. A large amount of strain energy remaining in the hanging wall is the reason why the normal fault continues to evolve after the normal fault formation until the antithetic fault is formed.

**Author Contributions:** Methodology, H.D.; software, X.P.; validation, Y.Z.; formal analysis, X.P.; investigation, X.P.; resources, H.D.; visualization; H.C.; data curation, H.D.; writing—original draft preparation, X.P. and J.H.; writing—review and editing, X.P., J.H.; supervision, H.D. and J.H.; funding acquisition, J.H. All authors have read and agreed to the published version of the manuscript.

**Funding:** This research was funded by the Key Laboratory of Deep Earth Science and Engineering (Sichuan University), grant number DESEYU 202102, National Natural Science Foundation of China, grant number 41672133, and the National Strategic Research Program, grant number 2016ZX05048-001-04-LH.

**Institutional Review Board Statement:** Not applicable.

**Informed Consent Statement:** Not applicable.

**Data Availability Statement:** The study did not report any data.

**Acknowledgments:** This work was performed at State Key Laboratory of Oil and Gas Reservoir Geology and Exploitation of the Chengdu University of Technology, Chengdu. We are grateful to Xinjian Wang for him a hearty welcome to the laboratory, for their numerous suggestions and their patients. The authors greatly acknowledge the financial support from the research by the Key Laboratory of Deep Earth Science and Engineering (Sichuan University) (DESEYU 202102), National Natural Science Foundation of China (Grant No. 41672133) and the National Strategic Research Program (Grant No. 2016ZX05048-001-04-LH) are gratefully acknowledged.

**Conflicts of Interest:** The authors declare no conflict of interest.

## Appendix A. Derivation of the Central Position of Strain Energy (CPSE)

The calculation method of the central position of strain energy is as follows:

$$\vec{P} = \sum m_i \vec{v}_i == m\vec{v}_c \tag{A1}$$

$$\vec{v}_c = \frac{\sum m_i \vec{v}_i}{m} = \frac{\sum m_i \frac{\vec{dr_i}}{dt}}{m} = \frac{\sum m_i d\vec{r}_i}{mdt} \tag{A2}$$

Moreover, because, $\vec{v}_c = \frac{\vec{dr_c}}{dt}$, so

$$\frac{\vec{dr_c}}{dt} = \frac{\sum m_i d\vec{r}_i}{mdt} \tag{A3}$$

Integration against time (t)

$$d\vec{r}_c = \frac{\sum m_i d\vec{r}_i}{m} \tag{A4}$$

Integral over displacement (r)

$$\vec{r}_c = \frac{\sum m_i \vec{r}_i}{\sum m_i} \tag{A5}$$

Base on the work-energy Principle

$$E = \frac{1}{2} m \vec{v}_c^2, \sum E_i = \sum \frac{1}{2} m_i \vec{v}_c^2, E_i = \frac{1}{2} m_i \vec{v}_c^2, m_i = \frac{2E_i}{\vec{v}_c^2} \tag{A6}$$

Combining (A5) and (A6),

$$\vec{r}_c = \frac{\sum m_i \vec{r}_i}{\sum m_i} = \frac{\sum \frac{2E_i}{v_c^2} \vec{r}_i}{\sum \frac{2E_i}{v_c^2}} = \frac{\sum E_i \vec{r}_i}{\sum E_i} \tag{A7}$$

Rectangular coordinate system, expression of each component:

$$x_c = \frac{\sum E_i x_i}{\sum E_i}, y_c = \frac{\sum E_i y_i}{\sum E_i}, z_c = \frac{\sum E_i z_i}{\sum E_i} \tag{A8}$$

where:
$\vec{P}$ = Total momentum, kg·m/s;
$m_i$ = mass of the $i$-th object, kg;
$m$ = The total mass of the objects, kg;
$\vec{v}_i$ = Speed of the $i$-th object, m/s;

$\vec{v_c}$ = Speed of particles with equivalent effects, m/s;

$t$ = time, s;

$\vec{r}_i$ = Displacement of the $i$-th object in polar coordinates, m;

$\vec{r}_c$ = Displacement of particles with equivalent effects in polar coordinates, m;

$E$ = Total kinetic energy of the object, J;

$E_i$ = Total kinetic energy of the $i$-th object, J;

$x_i$ = $x$-coordinate of the $i$-th object in Cartesian coordinates, m;

$y_i$ = $y$-coordinate of the $i$-th object in Cartesian coordinates, m;

$z_i$ = $z$-coordinate of the $i$-th object in Cartesian coordinates, m;

$x_c$ = $x$-coordinate of the particles with equivalent effects in Cartesian coordinates, m;

$y_c$ = $y$-coordinate of the particles with equivalent effects in Cartesian coordinates, m;

and,

$z_c$ = $z$-coordinate of the particles with equivalent effects in Cartesian coordinates, m.

## References


1. Wernicke, B.; Axen, G.J. On the role of isostasy in the evolution of normal fault systems. *Geology* **1988**, *16*, 848–851. [CrossRef]
2. Martinsen, O.J.; Pulham, A.J.; Haughton, P.D.; Sullivan, M.D. *Outcrops Revitalized: Tools, Techniques and Applications*; SEPM Tulsa: Tulsa, OK, USA, 2011.
3. Sibson, R. Fault rocks and fault mechanisms. *J. Geol. Soc.* **1977**, *133*, 191–213. [CrossRef]
4. Bernabé, Y.; Revil, A. Pore-scale heterogeneity, energy dissipation and the transport properties of rocks. *Geophys. Res. Lett.* **1995**, *22*, 1529–1532. [CrossRef]
5. Sujatha, V.; Kishen, J.C. Energy release rate due to friction at bimaterial interface in dams. *J. Eng. Mech.* **2003**, *129*, 793–800. [CrossRef]
6. Zhao, Z. *Research on Rock Deformation and Failure Based on Energy Dissipation and Energy Release*; Sichuan University: Chengdu, China, 2007.
7. Zhao, Z.; Lu, R.; Zhang, G. Analysis on Energy Transformation for Rock in the Whole Process of Deformation and Fracture. *Min. Res. Dev.* **2006**, *26*, 8–11.
8. Faulkner, D.; Jackson, C.; Lunn, R.; Schlische, R.; Shipton, Z.; Wibberley, C.; Withjack, M. A review of recent developments concerning the structure, mechanics and fluid flow properties of fault zones. *J. Struct. Geol.* **2010**, *32*, 1557–1575. [CrossRef]
9. Fajfar, P.; Vidic, T.; Fischinger, M. Seismic demand in medium and long period structures. *Earthq. Eng. Struct. Dyn.* **1989**, *18*, 1133–1144. [CrossRef]
10. Sucuoglu, H. Effect of connection rigidity on seismic response of precast concrete frames. *PCI J.* **1995**, *40*, 94–103. [CrossRef]
11. Teran-Gilmore, A. *Performance-Based Earthquake-Resistant Design of Framed Buildings Using Energy Concepts*; University of California: Berkeley, CA, USA, 1996.
12. Uang, C.M.; Bertero, V.V. Evaluation of seismic energy in structures. *Earthq. Eng. Struct. Dyn.* **1990**, *19*, 77–90. [CrossRef]
13. Akiyama, H. *Earthquake Resistant Limit State Design for Buildings*; University of Tokyo Press: Tokyo, Japan, 1985.
14. Holt, W.; Chamot Rooke, N.; Le Pichon, X.; Haines, A.; Shen Tu, B.; Ren, J. Velocity field in Asia inferred from Quaternary fault slip rates and Global Positioning System observations. *J. Geophys. Res. Solid Earth* **2000**, *105*, 19185–19209. [CrossRef]
15. Iinuma, T.; Ohzono, M.; Ohta, Y.; Miura, S. Coseismic slip distribution of the 2011 off the Pacific coast of Tohoku Earthquake (M 9.0) estimated based on GPS data—Was the asperity in Miyagi-oki ruptured? *EarthPlanets Space* **2011**, *63*, 24. [CrossRef]
16. Jing, Y.; Li, H.; Xiong, Y.Z.; Fan, L.L.; Zhang, S.Z.; Sun, Q.-W.; Dong, J.Y.; Liu, F.-Q.; Wang, H.-Z. Use Seismic Moment Tensor and GPS Data to Analyse the Recent CrustalM ovement Energy Distribution Characteristics of China Continent. *Geol. J. China Univ.* **2009**, *1*, 011.
17. He, M.; Miao, J.; Feng, J. Rock burst process of limestone and its acoustic emission characteristics under true-triaxial unloading conditions. *Int. J. Rock Mech. Min. Sci.* **2010**, *47*, 286–298. [CrossRef]
18. Sih, G.C. Strain energy density factor applied to mixed mode crack problems. *Int. J. Fract.* **1974**, *10*, 305–321. [CrossRef]
19. Young, W.C.; Budnyas, R.G. *Roark's Formulas for Stress and Strain*; McGraw-Hill: New York, NY, USA, 2017.
20. Gutscher, M.A.; Klaeschen, D.; Flueh, E.; Malavieille, J. Non-Coulomb wedges, wrong-way thrusting, and natural hazards in Cascadia. *Geology* **2001**, *29*, 379–382. [CrossRef]
21. Lohrmann, J.; Kukowski, N.; Adam, J.; Oncken, O. The impact of analogue material properties on the geometry, kinematics, and dynamics of convergent sand wedges. *J. Struct. Geol.* **2003**, *25*, 1691–1711. [CrossRef]
22. Waltham, D. Folding and faulting in coulomb materials. *Basin Res.* **2002**, *14*, 319–328. [CrossRef]
23. Cloos, E. Experimental analysis of Gulf Coast fracture patterns. *AAPG Bull.* **1968**, *52*, 420–444.
24. McClay, K. Extensional fault systems in sedimentary basins: A review of analogue model studies. *Mar. Pet. Geol.* **1990**, *7*, 206–233. [CrossRef]
25. De Pater, C.; Cleary, M.; Quinn, T.; Barr, D.; Johnson, D.; Weijers, L. Experimental verification of dimensional analysis for hydraulic fracturing. *SPE Prod. Facil.* **1994**, *9*, 230–238. [CrossRef]
26. Bonini, M.; Corti, G.; Ventisette, C.D.; Manetti, P.; Mulugeta, G.; Sokoutis, D. Modelling the lithospheric rheology control on the Cretaceous rifting in West Antarctica. *Terra Nova* **2007**, *19*, 360–366. [CrossRef]

27. Costa, E.; Vendeville, B. Experimental insights on the geometry and kinematics of fold-and-thrust belts above weak, viscous evaporitic décollement. *J. Struct. Geol.* **2002**, *24*, 1729–1739. [CrossRef]
28. Hubbert, M.K. Theory of scale models as applied to the study of geologic structures. *Bull. Geol. Soc. Am.* **1937**, *48*, 1459–1520. [CrossRef]
29. Bonini, M. Deformation patterns and structural vergence in brittle–ductile thrust wedges: An additional analogue modelling perspective. *J. Struct. Geol.* **2007**, *29*, 141–158. [CrossRef]
30. Cotton, J.T.; Koyi, H.A. Modeling of thrust fronts above ductile and frictional detachments: Application to structures in the Salt Range and Potwar Plateau, Pakistan. *Geol. Soc. Am. Bull.* **2000**, *112*, 351–363. [CrossRef]
31. Xu, S.; Fukuyama, E.; Yamashita, F.; Mizoguchi, K.; Takizawa, S.; Kawakata, H. Strain rate effect on fault slip and rupture evolution: Insight from meter-scale rock friction experiments. *Tectonophysics* **2018**, *733*, 209–231. [CrossRef]
32. Couzens-Schultz, B.A.; Vendeville, B.C.; Wiltschko, D.V. Duplex style and triangle zone formation: Insights from physical modeling. *J. Struct. Geol.* **2003**, *25*, 1623–1644. [CrossRef]
33. Leturmy, P.; Mugnier, J.; Vinour, P.; Baby, P.; Colletta, B.; Chabron, E. Piggyback basin development above a thin-skinned thrust belt with two detachment levels as a function of interactions between tectonic and superficial mass transfer: The case of the Subandean Zone (Bolivia). *Tectonophysics* **2000**, *320*, 45–67. [CrossRef]
34. Adam, J.; Urai, J.; Wieneke, B.; Oncken, O.; Pfeiffer, K.; Kukowski, N.; Lohrmann, J.; Hoth, S.; Van Der Zee, W.; Schmatz, J. Shear localisation and strain distribution during tectonic faulting—New insights from granular-flow experiments and high-resolution optical image correlation techniques. *J. Struct. Geol.* **2005**, *27*, 283–301. [CrossRef]
35. White, D.; Take, W.; Bolton, M. Soil deformation measurement using particle image velocimetry (PIV) and photogrammetry. *Geotechnique* **2003**, *53*, 619–631. [CrossRef]
36. White, D.J.; Take, W.; Bolton, M. Measuring soil deformation in geotechnical models using digital images and PIV analysis. In Proceedings of the International Conference on Computer Methods and Advances in Geomechanics, Tucson, AZ, USA, 7–12 January 2001; pp. 997–1002.
37. Guide to the Expression of Uncertainty in Measurement, Joint Committee for Guides in Metrology (JCGM) 100:2008. Available online: https://www.bipm.org/documents/20126/2071204/JCGM_100_2008_E.pdf/cb0ef43f-baa5-11cf-3f85-4dcd86f77bd6 (accessed on 5 September 2008).
38. Golijanek-Jędrzejczyk, A.; Mrowiec, A.; Hanus, R.; Zych, M.; Świsulski, D. Uncertainty of mass flow measurement using centric and eccentric orifice for Reynolds number in the range $10{,}000 \leq \mathrm{Re} \leq 20{,}000$. *Measurement* **2020**, *160*, 107851. [CrossRef]
39. Golijanek-Jędrzejczyk, A.; Mrowiec, A.; Hanus, R.; Zych, M.; Świsulski, D. Determination of the uncertainty of mass flow measurement using the orifice for different values of the Reynolds number. *Eur. Phys. J. Conf.* **2019**, *213*, 02022. [CrossRef]
40. Lahee, F.H. *Field Geology*; McGraw-Hill Book Company, Incorporated: New York, NY, USA, 1923.
41. Newmark, N.M.; Hall, W.J. Pipeline design to resist large fault displacement. In Proceedings of the US National Conference on Earthquake Engineering, Ann Arbor, MI, USA, 18–20 June 1975; pp. 416–425.
42. Solecki, R.; Conant, R.J. *Advanced Mechanics of Materials*; Oxford University Press: New York, NY, USA, 2003; Volume 198.
43. Ding, Y.; Zhu, X. Probe into Input Energy for the Characterization of Earthquake Ground Motions. *J. Beijing Jiaotong Univ.* **2007**, *31*, 49–51.
44. Hussain, M.; Pu, S.; Underwood, J. Strain energy release rate for a crack under combined mode I and mode II. In *Fracture Analysis: Proceedings of the 1973 National Symposium on Fracture Mechanics*; Part II; ASTM International: West Conshohocken, PA, USA, 1974.
45. Ye, Q.; Yuan, M. *Explanations of Common Terms in Oil and Gas Field Development*; Petroleum Industry Press: Conshohocken, PA, USA, 1996.
46. Reid, H.F. The mechanics of the earthquake. In *The California Earthquake of April 18, 1906, Report of the State Earthquake Investigation Commission*; The Caknegie Institution of Washington: Washington, DC, USA, 1910.
47. Caine, J.S.; Evans, J.P.; Forster, C.B. Fault zone architecture and permeability structure. *Geology* **1996**, *24*, 1025–1028. [CrossRef]
48. Doughty, P.T. Clay smear seals and fault sealing potential of an exhumed growth fault, Rio Grande rift, New Mexico. *AAPG Bull.* **2003**, *87*, 427–444. [CrossRef]
49. Zhang, L.; Lv, Y.; Peng, Y.; Xie, Z. Analysis of Seismological Energy Density of Minor-moderate Earthquakes. *Earthq. Res. China* **2008**, *24*, 407–414.

Article

# Mechanical Behaviors of Granite after Thermal Shock with Different Cooling Rates

Peng Xiao [1,2], Jun Zheng [1,2,*], Bin Dou [1,2,*], Hong Tian [1,2], Guodong Cui [1,2] and Muhammad Kashif [3]

[1]  Faculty of Engineering, China University of Geosciences, Wuhan 430074, China; xiaopeng805@cug.edu.cn (P.X.); htian@cug.edu.cn (H.T.); cuiguodong@cug.edu.cn (G.C.)
[2]  National Center for International Research on Deep Earth Drilling and Resources Development, Wuhan 430074, China
[3]  Department of Earth Sciences, University of Sargodha, Sargodha 40100, Pakistan; Muhammad.kashif@uos.edu.pk
*  Correspondence: junzheng@cug.edu.cn (J.Z.); doubin@cug.edu.cn (B.D.); Tel.: +86-180-7174-8712 (J.Z.); +86-189-8615-3360 (B.D.)

**Abstract:** During the construction of nuclear waste storage facilities, deep drilling, and geothermal energy development, high-temperature rocks are inevitably subjected to thermal shock. The physical and mechanical behaviors of granite treated with different thermal shocks were analyzed by non-destructive (P-wave velocity test) and destructive tests (uniaxial compression test and Brazil splitting test). The results show that the P-wave velocity ($V_P$), uniaxial compressive strength ($UCS$), elastic modulus ($E$), and tensile strength ($s_t$) of specimens all decrease with the treatment temperature. Compared with air cooling, water cooling causes greater damage to the mechanical properties of granite. Thermal shock induces thermal stress inside the rock due to inhomogeneous expansion of mineral particles and further causes the initiation and propagation of microcracks which alter the mechanical behaviors of granite. Rapid cooling aggravates the damage degree of specimens. The failure pattern gradually transforms from longitudinal fracture to shear failure with temperature. In addition, there is a good fitting relationship between P-wave velocity and mechanical parameters of granite after different temperature treatments, which indicates P-wave velocity can be used to evaluate rock damage and predict rock mechanical parameters. The research results can provide guidance for high-temperature rock engineering.

**Keywords:** high temperature; granite; cooling rates; P-wave velocity; mechanical properties; failure patterns

**Citation:** Xiao, P.; Zheng, J.; Dou, B.; Tian, H.; Cui, G.; Kashif, M. Mechanical Behaviors of Granite after Thermal Shock with Different Cooling Rates. *Energies* **2021**, *14*, 3721. https://doi.org/10.3390/en14133721

Academic Editor: Joel Sarout

Received: 25 May 2021
Accepted: 16 June 2021
Published: 22 June 2021

**Publisher's Note:** MDPI stays neutral with regard to jurisdictional claims in published maps and institutional affiliations.

## 1. Introduction

The utilization of fossil energy has several worrisome environmental implications, such as the greenhouse effect and acid rain. In these circumstances, replacing gas and oil combustion with renewable energy resources can significantly reduce $CO_2$ emissions and environmental pollution [1–3]. Therefore, clean renewable energy is the future direction of energy development and is attracting more and more attention [2,4]. Geothermal energy is a pollution-free and renewable energy source that has achieved rapid development [4–6]. Thus, many countries have focused on the exploitation of deep geothermal energy, including subsidizing deep geothermal exploitation and addressing challenges in geothermal energy production [2,4,7]. During deep geothermal energy exploitation, it is important to circularly pump a working fluid into an underground fracture network to realize heat extraction. Hot dry rocks (HDR) with a temperature up to 400 °C in a deep geothermal reservoir may be directly exposed to cool water and suffer great temperature gradients [8,9]. As a result, great temperature gradients will cause the propagation of microcracks and impact on the mechanical properties of HDR [8,9]. The effects of these changes are twofold. On the one hand, the stability of the shaft wall rock is reduced, which may lead to shaft wall collapse. On the other hand, the permeability of reservoir rock will increase, which is conducive to the improvement of the thermal recovery rate [10,11]. The effect of thermal shock on the physical and mechanical behaviors of deep rocks is also involved in other high-temperature deep rock engineering projects, such as the construction of nuclear waste storage facilities and deep hydraulic fracturing [4,12,13]. Geological exploration data have shown that deep crystalline rock (granite, biotite gneiss, etc.) reservoirs have sufficient geothermal energy to serve as geothermal reservoirs [14,15]. So, a deep understanding of the mechanical behaviors of crystalline rocks suffering drastic temperature changes is of great importance for engineering applications involving high-temperature rock mechanics [4,16]. It is significant to find out the change law and mechanism of mechanical properties of granite after suffering a thermal shock [4].

Past studies have shown that there is no doubt that high-temperature treatment and rapid cooling have a thermal shock on the mechanical properties of crystalline rocks. Scholars have done extensive research on the physical, mechanical, and microstructure performances of granites exposed to thermal shock [4,5,9,10,14–24]. Martyushev et al. confirmed that effective pressure can alter reservoir properties [25,26]. David et al. [27] found that rock microstructure has strong control of physical properties. Griffiths [28] and Fredrich [29] et al. established a link between the microstructural parameters and the mechanical behavior of rock by micromechanical models. With a thermal shock, the microstructure and the physical and mechanical properties of specimens may be changed. In general, the thermal treatment weakens the mechanical behavior and the dynamic elasticity properties of crystalline rock [30–39]. Zhao et al. [35] found that thermal shock reduces the mechanical properties of granite, especially when the temperature is over 400 °C, and the mechanical behaviors of granite deteriorated sharply. A high cooling rate or high-temperature gradients means intensive thermal shock, which will make the mechanical properties of rock deteriorate more seriously. Kumari et al. [23] studied the effect mechanism of cooling rate on the physical and mechanical properties of granite and found that thermal shock will weaken the mechanical behaviors of rock, which is mainly achieved by forming microcracks in rock. Zhu et al. studied experimentally the influence of thermal shock on the physical and mechanical properties of granite and the cracking characteristics of granite under the thermal shock [5,14,18,40,41]. The results show that the high temperature treatment causes lots of microcracks in the specimens. The existence of microcracks increases the volume of the rock, reduces the density and $V_P$ of the rock, and weakens it's mechanical strength (uniaxial compressive strength, elastic modulus, etc.) [42]. Acoustic emission monitoring and ultrasonic velocity measurements can be used to study the influence of temperature on the microstructure of granite [43]. In addition, Zhu et al. [21] also explored the relationship between the P-wave velocity and the mechanical parameters of granite that underwent different thermal shocks. Shao et al. [44] investigated the fracture toughness of granite specimens after elevated temperature treatment and liquid nitrogen cooling. The scholars also considered the influence of heating/cooling rate on thermal cracking in granite, and their results confirmed that a rapid cooling rate caused more microcracks in granite [10,45–47]. However, Heap et al. confirmed that the heating/cooling rate does not alter mechanical behaviors of some rock, such as andesitic dome rock from Volcan de Colima (Mexico) [48,49]. In addition, Yang et al. [50] investigated the relationship between temperature treatment and thermal-mechanical properties by a fully coupled thermal-mechanical model.

In the above-mentioned previous studies, only some of the literature has reported the failure patterns or mechanical behaviors of granite exposed to different thermal shock. Therefore, the aim of this paper is to experimentally reveal the influence of exposure temperature (25 to 600 °C) and different cooling rates (cooling with water or air) on the mechanical behavior of granite through a uniaxial compression experiment and the Brazil splitting test. This study is expected to make a clear understanding of the mechanical behavior of high-temperature granite to provide a contribution to reservoir construction and well borehole stability during the development of geothermal energy.

## 2. Materials and Methods

### 2.1. Sample Preparation

The granite specimens chosen in this research work were collected from Xuzhou, Jiangsu Province, China. All the specimens were cored from an identical granite block with a depth of about 200 m to reduce the scatter of the experimental results. Then, the specimens were processed into cylinders with φ 50 mm × H 100 mm (for Uniaxial compression test) or φ 50 mm × H 50 mm (for Brazilian split test). In order to ensure the reliability of the experimental achievement, the non-parallelism error of the two end faces and the diameter error along the height of the specimens were less than 0.05 and 0.3 mm respectively, and the end faces were perpendicular to the axis of the test piece with the deviation angle less than 0.25°, as shown in Figure 1. The main mineral compositions of the granite, which was determined by X-ray diffraction analysis, were albite 53.44%, microcline 30.07%, quartz 9.04%, and muscovite 7.45%. In order for the specimens have better uniformity and consistency which can ensure the accuracy of the experimental results, the density and $V_P$ of the original specimens were taken as measurements, and the abnormal rock specimens were eliminated. The average density and $V_P$ of the tested specimens were 2.65 g/cm$^3$ and 5365.67 m/s, respectively.

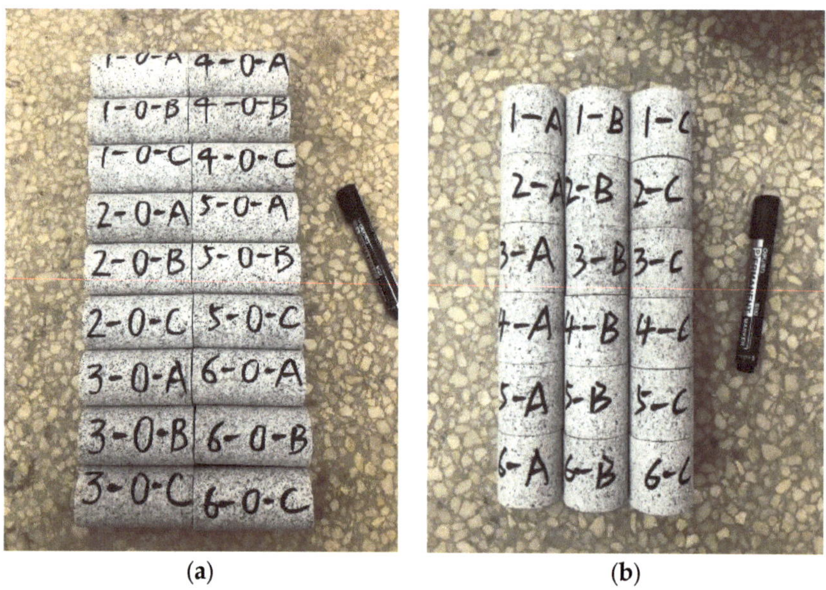

(a)          (b)

**Figure 1.** Granite specimens used in the test. (**a**) Specimens for the uniaxial compression test; (**b**) specimens for the Brazilian split test.

### 2.2. Test Equipment

As shown in Figure 2, the test equipment used in this study includes an SG-XL 1200 high-temperature furnace, DC-1020 constant-temperature bath, RSM-SY5 wave velocity detector, Universal strength testing machine, and RFP-03 Brazilian split testing machine. The SG-XL 1200 high-temperature furnace applied to heat rock specimens to different target temperatures, can reach a maximum temperature of 1200 °C with a constant heating rate and maintenance of the content temperature with a temperature control accuracy of ±3 °C. The DC-1020 constant temperature bath, which was used to cool the high-temperature rock specimens, maintained a constant temperature from −10 to 99 °C with a temperature error less than 0.3%. The maximum axial loading of the Universal strength testing machine was 2000 kN, and the axial loading can be performed as displacement-controlled with a stress resolution of 0.1 kN or stress-controlled conditions with a displacement resolution of 0.001 mm.

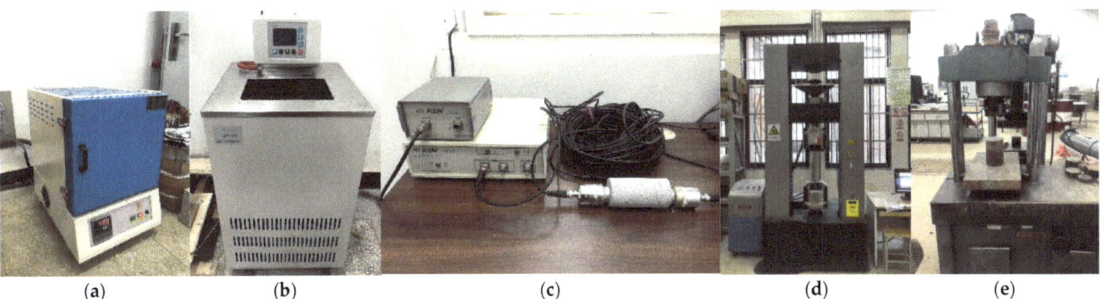



**Figure 2.** Test equipment used in this study. (**a**) SG-XL 1200 high-temperature furnace; (**b**) DC-1020 constant temperature bath; (**c**) RSM-SY5 wave velocity detector; (**d**) universal strength testing machine; (**e**) RFP-03 Brazilian split testing machine.

### 2.3. Experimental Procedure

The test flow chart is displayed in Figure 3. The treatment temperatures of specimens were 200, 300, 400, 500, and 600 °C, respectively. According to the different aim treating temperatures, the rock specimens were put into the high-temperature furnace in batches, and the specimens were heated to each aim temperature with the heating rate of 5 °C/min. After realizing the target temperature, the rock specimens were maintained in the furnace at the target temperature for 2 h to ensure the uniform temperature distribution of the rock specimens (Figure 4). The rock specimens exposed to a high temperature were cooled with two cooling shocks. Half of the specimens were taken out from the high-temperature furnace and put into a DC-1020 constant temperature bath filled with room-temperature water to cool them to room temperature, and then put into a drying vessel, which is a closed glass container with a desiccant used to absorb water from granite samples. For the remaining specimens, they were taken out from the high-temperature furnace and directly put into the drying vessel to cool to room temperature. After the rock specimens were completely dry, the physical and mechanical tests of the rock specimens were carried out.

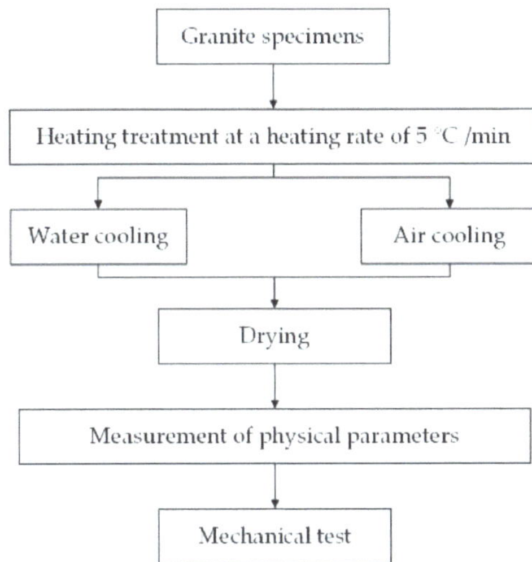



**Figure 3.** Experimental flow chart.

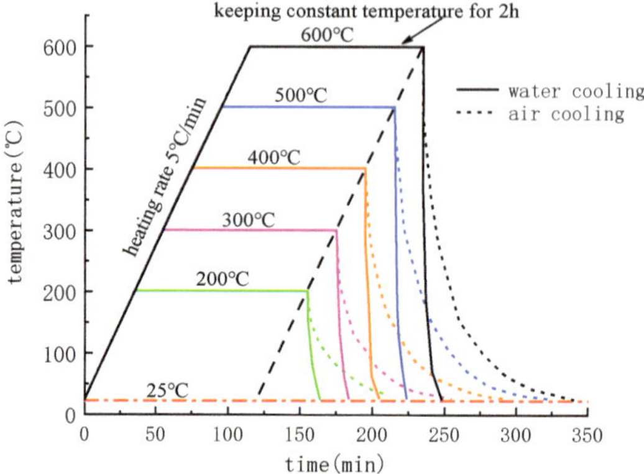



**Figure 4.** Testing scheme of heating and cooling processes.

For the uniaxial compression test, the treated granite specimens were put in the universal strength testing machine, and then the axial load was used at a loading rate of 0.03 mm/min until specimens failed to determine the uniaxial compression mechanical behaviors of specimens. The stress–strain curve is automatically recorded through the universal strength testing machine during the test. For the Brazilian split test, the treated rock specimens were put into the Brazilian split testing machine, and then the radial load was applied at a loading rate of 0.5 kN/s under radial stress control until the specimens failed.

## 3. Results

### 3.1. P-Wave Velocity

The variations in $V_P$ of granite specimens with temperature under different cooling methods are shown in Figure 5. High-temperature treatment decreased the $V_P$ of rock specimens, and the decrease degree of $V_P$ of water-cooling rock specimens was more dramatic than that of air-cooling specimens. When the heating temperature increases from 25 to 600 °C, the values of $V_P$ of the rock specimens cooled with water decreased by 5%, 19%, 39%, 52%, 56%, and 78%, respectively, while these of air-cooling specimens decreased by 6%, 15%, 33%, 44%, 46%, and 71%, respectively. In different temperature ranges, the decreasing amplitude of the P-wave velocity presents a different trend. The P-wave velocity of specimens decreased more rapidly in stage C than in stages A and B.

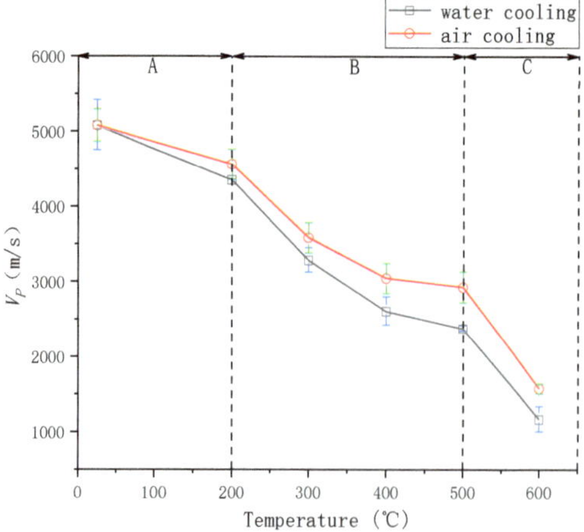



**Figure 5.** Variation of $V_P$ of rock specimens with temperature under different cooling routes.

### 3.2. Stress–Strain Relations

The uniaxial compression stress–strain relations of specimens treated with different temperature routes are shown in Figure 6. The stress–strain relations are very similar and can be divided into four stages according to the changing characteristics: compaction, elastic deformation, yield, and failure.

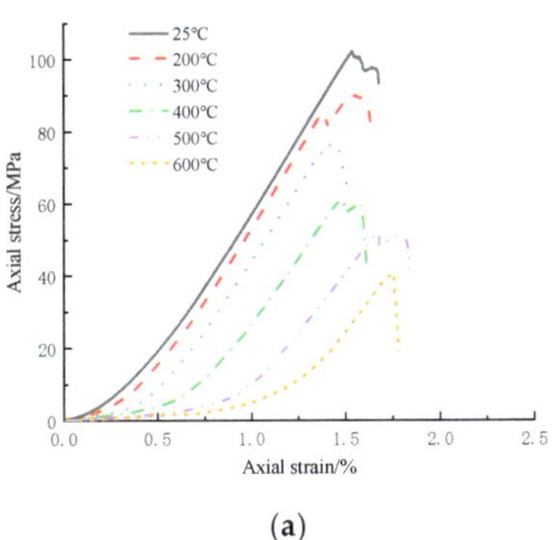


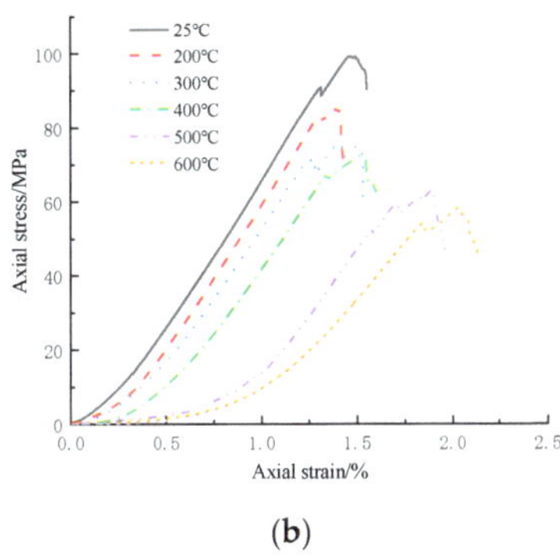


(**a**)    (**b**)

**Figure 6.** Stress–strain relations of rock specimens with different temperature treating routes. (**a**) water cooling; (**b**) air cooling.

As shown in Figure 6, heating temperature and cooling routes both have significant influences on the characteristics of different stages in the stress–strain curve. Under the same cooling condition, the compaction stage (the initial nonlinear part) showed a more obvious elongation tendency when the treating temperature increased from 25 to 600 °C. Meanwhile, granite specimens cooled by water have a more obvious compaction stage than those cooled by air. The compaction stage reflects the closure of microcracks, and both high temperature and cooling treatment induced the generation and propagation of microcracks in the granite specimens [30,31,51,52]. Therefore, the variation of stress–strain curve characteristics during the compression stage indirectly indicates the quantity of augmentation of microcracks [53]. The elastic deformation stage in the stress–strain curves is close to a straight line. As can be seen from Figure 6, under the same cooling rate, the slope of the line representing the elastic deformation stage shows a downward trend with the rise in heating temperature, which indicates that the elastic modulus of the granite specimens decreases with the rise in temperature. Under the same treatment temperature, the elastic modulus of air-cooling granite specimens is higher than that of water-cooling granite specimens. The obvious yield platform stage can be observed in the rock deformation's yield stages, and the characteristics of the yield stage of granite specimens are not consistent under various experimental conditions. The peak strain of the rock specimen increases with the treatment temperature. Meanwhile, the peak strain of the water-cooling rock specimen is less than that of the air-cooling rock specimen at the same treatment temperature. The stress decreases significantly after the specimens reach the peak strength. In a word, the increase in temperature and the cooling treatment both decrease the elasticity of specimens and increase the plasticity of specimens. Cooling with water also provides the same effect, comparing to air cooling.

### 3.3. Mechanics Parameters

The *UCS* of granite is derived from the peak stress of the stress–strain curve. The E is defined by the tangent slope of the elastic deformation phase of the stress–strain curve from the uniaxial compression test. The splitting peak strength of granite specimens can be used to characterize the $\sigma_t$ of specimens, which can be obtained from the Brazilian splitting test. The experimental results are shown in Table 1.

**Table 1.** Mechanical parameters of granite specimens after two different thermal shocks.

| Cooling Route | | Water Cooling | | | | | | Air Cooling | | | | | |
|---|---|---|---|---|---|---|---|---|---|---|---|---|---|
| Temperature/°C | | 25 | 200 | 300 | 400 | 500 | 600 | 25 | 200 | 300 | 400 | 500 | 600 |
| $UCS$/MPa | No. 1 | 94.38 | 89.93 | 71.06 | 60.72 | 47.55 | 41.20 | 102.99 | 91.37 | 73.73 | 72.26 | 69.09 | 57.02 |
| | No. 2 | 102.00 | 81.30 | 76.32 | 59.44 | 51.89 | 37.33 | 102.96 | 94.50 | 82.61 | 72.09 | 61.74 | 61.71 |
| | No. 3 | 95.91 | 81.02 | 71.81 | 61.98 | 56.64 | 40.34 | 99.13 | 85.11 | 76.07 | 73.17 | 62.96 | 57.80 |
| | Ave. | 97.43 | 84.08 | 73.06 | 60.71 | 52.03 | 39.62 | 101.70 | 90.33 | 77.47 | 72.51 | 64.6 | 58.84 |
| $E$/GPa | No. 1 | 7.99 | 7.91 | 7.38 | 6.83 | 5.99 | 5.00 | 8.09 | 7.71 | 7.84 | 7.64 | 7.27 | 5.97 |
| | No. 2 | 8.09 | 7.85 | 6.92 | 6.65 | 6.16 | 5.03 | 8.25 | 8.12 | 7.85 | 7.36 | 6.85 | 6.19 |
| | No. 3 | 8.01 | 7.8 | 7.15 | 6.47 | 6.27 | 5.36 | 7.96 | 8.05 | 7.5 | 7.08 | 7.36 | 6.44 |
| | Ave. | 8.03 | 7.85 | 7.15 | 6.65 | 6.14 | 5.13 | 8.1 | 7.96 | 7.73 | 7.36 | 7.16 | 6.20 |
| $\sigma_t$/MPa | No. 1 | 6.16 | 5.89 | 5.15 | 4.19 | 3.08 | 1.10 | 6.14 | 6.08 | 5.89 | 4.85 | 4.02 | 1.90 |
| | No. 2 | 6.18 | 5.89 | 5.09 | 4.27 | 3.09 | 1.15 | 6.14 | 6.10 | 5.87 | 4.86 | 4.02 | 1.89 |
| | No. 3 | 6.14 | 5.88 | 5.18 | 4.15 | 3.08 | 1.08 | 6.11 | 6.07 | 5.93 | 4.85 | 4.00 | 1.89 |
| | Ave. | 6.16 | 5.89 | 5.14 | 4.20 | 3.09 | 1.11 | 6.13 | 6.08 | 5.9 | 4.85 | 4.01 | 1.90 |
| $V_P$/m/s | No. 1 | 4.69 | 4.35 | 3.45 | 2.44 | 2.38 | 1.35 | 5.27 | 4.35 | 3.77 | 3.23 | 3.14 | 1.61 |
| | No. 2 | 5.27 | 4.35 | 3.26 | 2.80 | 2.40 | 1.10 | 5.09 | 4.55 | 3.57 | 3.03 | 2.74 | 1.59 |
| | No. 3 | 5.26 | 4.35 | 3.13 | 2.56 | 2.33 | 1.04 | 4.84 | 4.75 | 3.37 | 2.83 | 2.86 | 1.49 |
| | Ave. | 5.08 | 4.35 | 3.28 | 2.60 | 2.37 | 1.16 | 5.07 | 4.55 | 3.57 | 3.03 | 2.91 | 1.56 |

### 3.3.1. Uniaxial Compressive Strength

The influence of different high-temperature treatments on the $UCS$ of specimens is shown in Figure 7. As the treatment temperature increases, the $UCS$ of specimens generally shows a nearly linear downward trend. This indicates that the mechanical behaviors of granite deteriorate significantly with the rise in temperature. When the heating temperature gradually rose from 200 to 600 °C, the value of $UCS$ of heated granite after water-cooling treatment was reduced to 88%, 75%, 60%, 51%, and 40% of that of granite specimens without heating treatment, respectively. The value of $UCS$ of granite specimens after air-cooling decreased to 86%, 77%, 73%, 64%, and 58%, respectively. As for the effect of the cooling route on the $UCS$ of granite specimens, it shows a different trend before and after 300 °C. In section A of Figure 7, the cooling rate has little effect on the $UCS$. However, in section B, the cooling rate had a great effect on the $UCS$ of granite specimens, and the influence becomes more obvious with the increase in treatment temperature. Compared with air cooling, the average value of the $UCS$ of granite specimens after water cooling from 400 to 600 °C decreased by 11.54, 11.07, and 17.46 MPa, respectively, and the reduction ranges were 15.97%, 17.58%, and 30.20%, respectively.

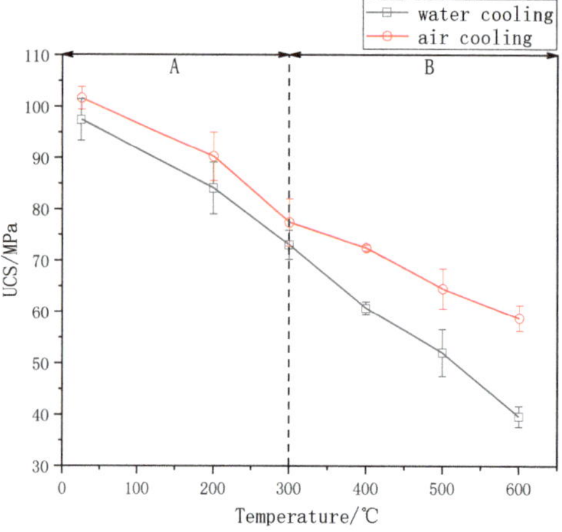


**Figure 7.** $UCS$ of granite specimens treated at different temperature methods.

### 3.3.2. Elasticity Modulus

Under the same cooling condition, the value of $E$ decreases with the increase in the treatment temperature of the granite specimen, and the decrease rate also gradually increases (Figure 8). In stage A in Figure 8, the value of $E$ of heated granite specimens after two different cooling methods changed little with temperature. However, in stages B and C in Figure 8, when the temperature rises from 200 to 600 °C, the value of $E$ of water-cooling granite specimens decreases to 97.76%, 89.04%, 82.81%, 76.46%, and 63.89% of that of original granite specimens, while the value of $E$ of air-cooling rock specimens decreases to 98.27%, 95.43%, 90.86%, 88.40%, and 76.54%, respectively. In addition, with the increase in treatment temperature, the value of $E$ of the granite specimens exposed to cooling water decreased more rapidly than that of granite exposed to air.

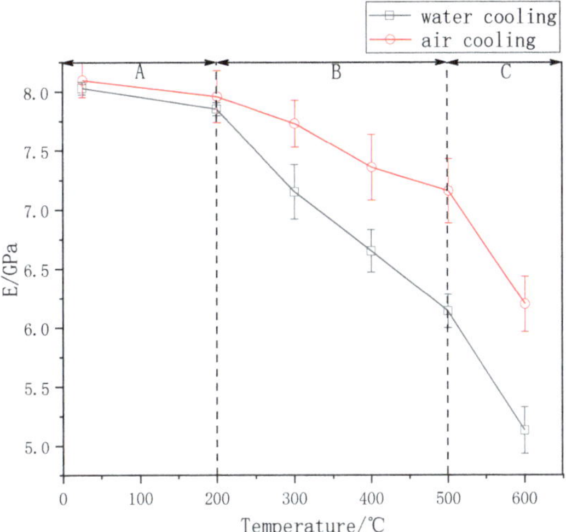


**Figure 8.** Relationships between temperature and elasticity modulus.

### 3.3.3. Tensile Strength

High temperature can significantly decrease the value of $\sigma_t$ of granite specimens with the increasing of treatment temperature. Compared to the air-cooling specimen, the $\sigma_t$ of the water-cooling granite specimen deteriorated more seriously (Figure 9). When the heating temperature rose from 200 to 600 °C, the value of $\sigma_t$ of the granite specimens after air-cooling decreased by 1%, 4%, 21%, 35%, and 70%, while the granite specimens in the water-cooling group decreased by 4%, 16%, 32%, 50%, and 82%, respectively.

### 3.4. Failure Patterns

### 3.4.1. Failure Patterns under Uniaxial Compression Tests

Failure patterns of heated granite after different cooling methods under uniaxial compression tests are shown in Table 2. The stress energy gradually accumulates in the heated granite during the loading process, and the granite specimens will suddenly break down accompanying a huge sound when it reaches a certain extent. The failure patterns of heated granite specimens are different after different temperature treatment processes. Above 300 °C, the failure pattern of specimens cooled by water is mainly shear failure. However, the failure pattern of specimens cooled by air is still mixed with tension failure with the heating temperature increasing to 400 °C, and the extent of the damage was lower than the water-cooling specimens.

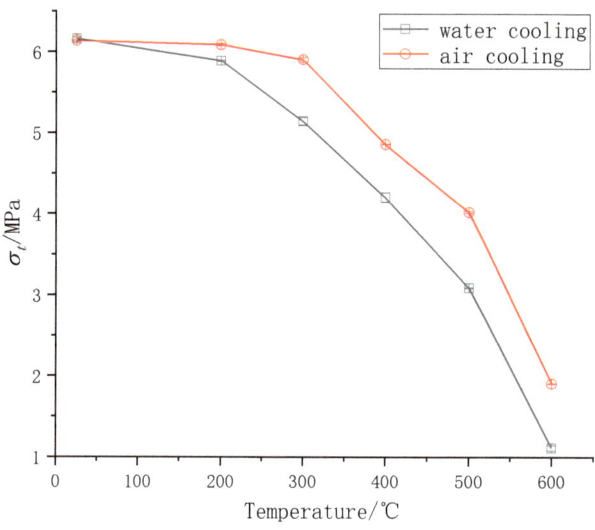

**Figure 9.** Relationships between temperature and tensile strength.

**Table 2.** Failure patterns of high-temperature granite under different cooling routes.

| Cooling Route | 25 °C | 200 °C | 300 °C | 400 °C | 500 °C | 600 °C |
|---|---|---|---|---|---|---|
| Water Cooling | | | | | | |
| Air Cooling | | | | | | |

Under the same cooling condition, the failure degree of the specimens increases with treatment temperature. The volume of the block formed from the granite specimen treated by higher temperature was smaller. The cooling rates also have a significant influence on the failure pattern of heated specimens. At the same treatment temperature, the failure degree of specimens cooled by water is larger than that of the air-cooling granite specimens.

### 3.4.2. Failure Patterns under Brazil Splitting Tests

Failure patterns under Brazil splitting tests of granite specimens treated with different temperature routes are shown in Table 3. Only a radial failure path through the loading point of the specimen is generated in the Brazilian splitting tests. The granite specimens generated fragments in the Brazilian splitting tests, and the higher the temperature, the greater the amount of fragments. Meanwhile, the fracture morphology and the roughness of the failure surface of the granite specimen treated with different temperature routes slightly changed. Under the same cooling condition, the fracture path becomes more tortuous, and the fracture surface becomes rougher because of the increase in treating temperature, which was manifested by the larger fracture aperture. The fracture path of water-cooling rock specimens is more tortuous, and the fracture surface is coarser than that of air-cooling granite specimens. Below 400 °C, the fracture path is relatively straight, and almost no fragment falls off from air-cooling granite specimens. Above 500 °C, the fracture path becomes obviously twisted. However, the fracture path of water-cooling granite specimens becomes obviously twisted at 300 °C.

**Table 3.** Failure patterns of heated granite underwent thermal shock under Brazil splitting tests.

| Cooling Method | 25 °C | 200 °C | 300 °C | 400 °C | 500 °C | 600 °C |
|---|---|---|---|---|---|---|
| Water Cooling |  | | | | | |
| Air Cooling |  | | | | | |

## 4. Discussion

### 4.1. Relationship between $V_P$ and Mechanical Parameters of Granite

It is costly and time-consuming to determine the mechanical properties of rock by mechanical tests, but the non-destructive rock testing technology can solve this problem [18]. Acoustic waves would refract when they encounter cracks in the rock, which would reduce the macroscopic P-wave velocity of rocks. Therefore, the $V_P$ can reflect the pore structure and damage of engineering materials and is used to assess the deterioration degree of engineering materials [54]. The P-wave velocity is also used to characterize the microstructure inside the rock as a non-destructive testing method [55,56]. The $V_P$ has been employed to predict the mechanical strength to reduce the testing cost of rock mechanical properties [21].

The relationship between the $V_P$ of heated granite specimens after two different cooling methods are discussed in this study. Figure 10 shows the relationship between $V_P$ and rock mechanical parameters ($UCS$, $E$, $\sigma_t$), and a great asymptote fitting relationship between them is obtained. The relationship between mechanical parameters and $V_P$ is as follows:

$$y = a - b * c^x \tag{1}$$

where $y$ is the mechanical parameter, $x$ is $V_P$, and $a, b, c$ are the fitting coefficients which may depend on rock properties, such as mineral composition, joints, etc.

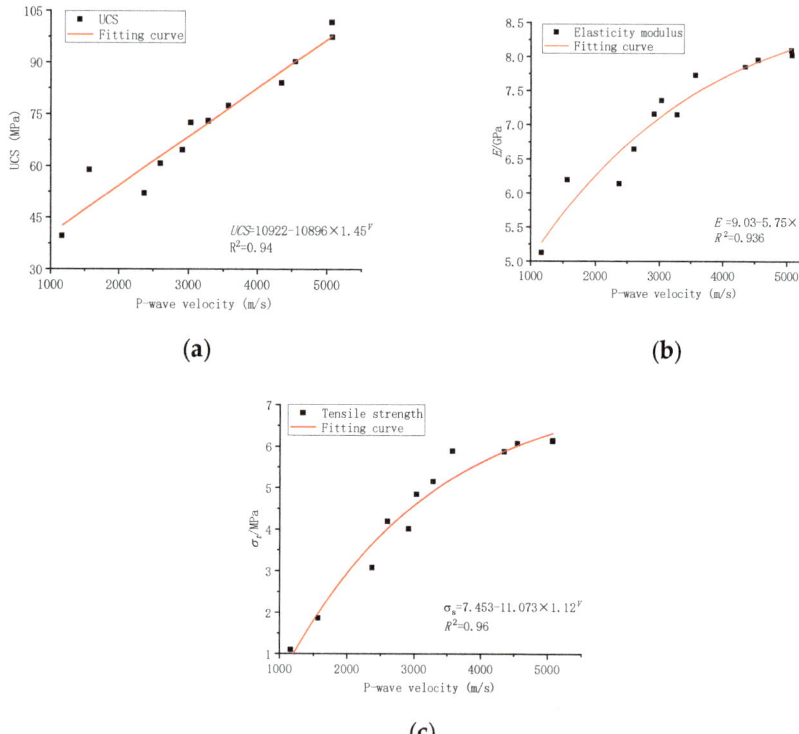

**Figure 10.** Relationship between $V_P$ and mechanical parameters. (**a**) Uniaxial compressive strength; (**b**) elasticity modulus; (**c**) tensile strength.

### 4.2. Effect of Temperature on Physical and Mechanics Behaviors

Thermal shock caused by heating and cooling treatment degrades the $V_P$ of granite specimens. In order to analyze the damage mechanism of temperature treatment on the $V_P$ of granite, the damage factor based on P-wave velocity was defined as follows [30]:

$$D_P = 1 - \frac{V_{PT}}{V_{P0}} \tag{2}$$

where $D_P$ is the damage factor of P-wave velocity, $V_{P0}$ represents the $V_P$ of the untreated specimen, and $V_{PT}$ is the $V_P$ of granite specimens measured at high temperatures of $T$.

The relationship between $D_P$ and temperature under different cooling methods is shown in Figure 11. With the increase in temperature, $D_P$ shows a continuous upward trend with an increasing rate, indicating that the density of microcracks constantly increases with heating temperature, and it is easier to form microcracks inside the granite specimens treated with a higher temperature.

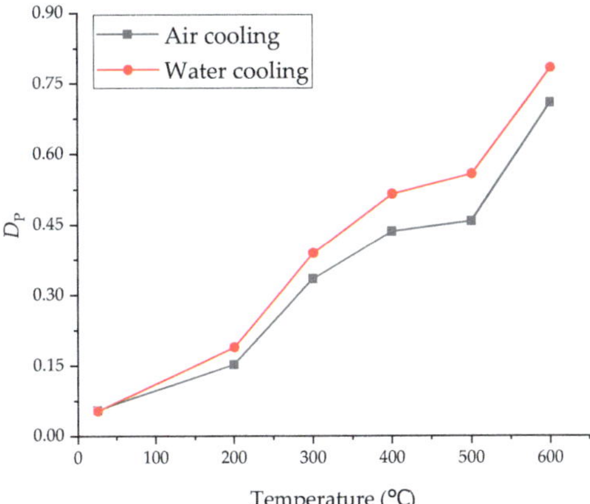


**Figure 11.** The relationship between $D_P$ and temperature under different cooling paths.

A high temperature will damage the mechanical properties of rock (Figures 7–9). In order to access the damage degree of the mechanical parameters of specimens under different thermal shock, the damage factor based on the mechanical parameters of specimens were defined as follows:

$$D_{UCS} = 1 - \frac{UCS_T}{UCS_0} \tag{3}$$

$$D_E = 1 - \frac{E_T}{E_0} \tag{4}$$

$$D_{\sigma_t} = 1 - \frac{\sigma_{sT}}{\sigma_{s0}} \tag{5}$$

where $D_{UCS}$, $D_E$, and $D\sigma_t$ represent the damage factor based on $UCS$, $E$, and $\sigma_t$, respectively. $UCS_0$, $E_0$, and $\sigma_{t0}$ represent the $UCS$, $E$, and $\sigma_t$ of original specimens, respectively, and $UCS_T$, $E_T$, and $\sigma_{tT}$ are the $UCS$, $E$, and $\sigma_t$ of granite specimens measured at a high temperature of $T$, respectively.

With the increase in temperature, the damage degree of mechanical behaviors of specimens increases with treatment temperature (Figure 12). Meanwhile, the cooling rate also has an important influence on the thermal damage degree based on the different mechanical properties of the granite specimen. In Figure 12a, when the temperature was below 300 °C, the $D_{UCS}$ of water cooling was slightly greater than that of air cooling. When the temperature was above 300 °C, the $D_{UCS}$ of water cooling was larger than that of air cooling, and the difference between them gradually increased with the treatment temperature. $D_E$ and $D\sigma_t$ of water-cooling specimens were always greater than that of air-cooling granite specimens, and the difference between $D_E$ and $D\sigma_t$ with two different cooling rates increased with the increase in treatment temperature (Figure 12b,c).

There are two main reasons for the damage of rock physical and mechanical parameters: (1) when the temperature of granite specimens changes dramatically, various mineral particles will expand unevenly. This inhomogeneous expansion can lead to thermal stress and further induce the generation and propagation of microcracks [34,57]. Meanwhile, the increase in treatment temperature degrades the mechanical strength of granite (Figures 7–9), and as a result, the specimens are more likely to rupture by a thermal stress-caused temperature gradient. Therefore, the increase in temperature further induces the internal microcracks in granite specimens. The presence of microfractures reduces the physical and mechanical behaviors of specimens. When the temperature is low, the thermal stress is not enough to degrade the mechanical behavior of the rock. So, when the temperature is lower than 200 °C, the $D_{UCS}$, $D_E$, and $D\sigma_t$. of the granite are relatively small, and the influence of the cooling method on the mechanical behavior of rock is not obvious. (2) Heat treatment makes the composition of granite change. On the one hand, the various types of water (attached water, bound water, and constitute ion water) within the specimens evaporate at different temperature ranges [15,16]. On the other hand, temperature changes the mineral composition of the rocks. For example, at 573 °C, quartz will undergo $\alpha/\beta$ transition [10,45,58]. As a result, the microcracks and micropore volume increase with temperature. Additionally, the water entering the fracture induced by thermal stress will further weaken the bonds between mineral particles, and the swelling force generated by the water boiling and vaporizing in the microcrack also provides the power for the expansion of the microcrack. Therefore, the damage degree of physical and mechanical parameters of the specimens subjected to water cooling is greater than that of air-cooling the specimens.

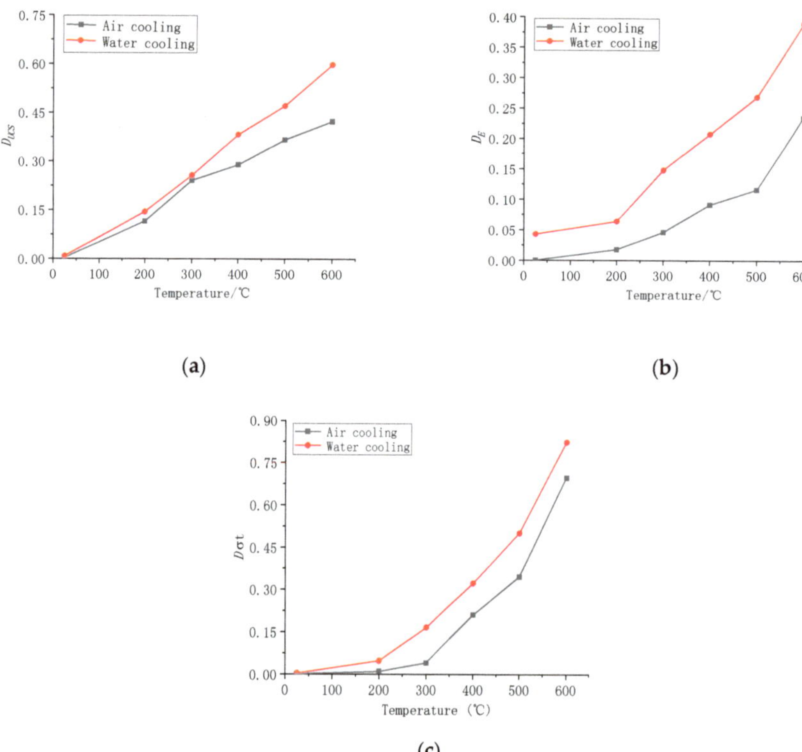


(a)

(b)

(c)

**Figure 12.** Relationship between temperature and thermal damage based on mechanical properties of granite. (**a**) $D_{UCS}$, (**b**) $D_E$, (**c**) $D\sigma_t$.

### 4.3. Effect of Thermal Shock on Failure Patterns of Rocks

The failure pattern of granite specimens treated with different temperature treatment routes changes greatly (Tables 2 and 3). The failure patterns of granite specimens conducted on uniaxial compression tests are obvious, so this part mainly discusses this. In order to quantify the failure patterns, the angle between the vertical direction of the granite specimens and the fracture surface was calculated, which is defined as the splitting angle, and the average splitting angle of rock specimens under different thermal shocks was obtained (Figure 13). It can be seen that the splitting angle gradually increases with temperature, and the rock failure pattern gradually transforms from longitudinal fracture to shear failure. This is because the number of microcracks in the rock and the degree of thermal damage increases, resulting in the decrease in the rock cohesion internal friction angle, and the continuous increase in splitting angle.

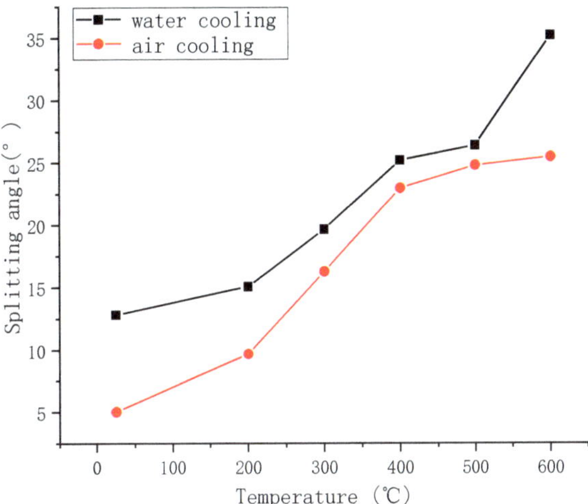

**Figure 13.** Relationship between rock splitting angle and temperature.

*4.4. Application of the Present Experimental Investigation*

During deep high-temperature drilling, cold drilling fluid continuously causes cooling shock to high-temperature sidewall rocks. In addition, the low temperature working fluid also continuously causes cooling shock to the high reservoir rocks during the long-term operation of the enhanced geothermal system [51]. Our studies show that cooling shock can deteriorate the mechanical behaviors of high-temperature granite. P-wave velocity, *UCS*, *E*, and $\sigma_t$ of heated granite after two cooling methods all showed a decreasing trend with temperature. Therefore, the cooling shock should be concerned in deep high-temperature rock engineering.

Our research confirms that thermal shock can generate a large number of microcracks in granite. Therefore, we can create an efficient heat recovery reservoir with lots of cracks by thermal stimulation. Meanwhile, considering that thermal shock can produce cracks in the reservoir, we can improve the permeability of the reservoir by intermittent cool working fluid injection.

Additionally, there is a good fitting relationship between P-wave velocity and rock mechanical parameters, so we can use the P-wave velocity test as a nondestructive rock testing technology to reduce cost and improve efficiency.

## 5. Conclusions

In the present article, the mechanical behaviors of granite subjected to two different thermal shocks have been studied based on mechanical experiments and an ultrasonic test technique. The main conclusions can be drawn as follows:

(1)  Thermal shock has a great influence on the physical and mechanical parameters of granite. $V_P$, *UCS*, *E*, and $\sigma_t$ of heated granite after two cooling methods all show a decreasing trend with temperature. There is a great asymptote fitting between mechanical strength and the P-wave of heated granite after two cooling methods.

(2)  High temperature causes uneven expansion of different minerals and further induces the development of microcracks. The degradation of physical and mechanical properties can be attributed to the generation and propagation of microcracks.

(3)  Water cooling further degrades the physical and mechanical behaviors of granite, compared with air cooling. Thermal shock caused by water cooling induces more microcracks, and water permeating into granite also further causes the propagation of microcracks.

(4)   Thermal shock can change the mechanical behavior of granite specimens. The splitting angle gradually increases with treatment temperature, and the rock failure pattern gradually transforms from longitudinal fracture to shear failure. A rapid cooling rate can exaggerate the damage degree of the granite specimens.

In the future, we will further study the corresponding quantitative relationship between temperature gradient and thermal stress, which can provide a basis to predict the damage degree of rock under the action of thermal stress.

**Author Contributions:** P.X. conceived and designed the experiment; H.T. and J.Z. provided a guide on experiments; P.X. wrote the manuscript; B.D. and G.C. provided advice on the writing of the abstract and conclusions; B.D., H.T., J.Z., G.C. and M.K. made a significant contribution to the revision of the manuscript. All authors have read and agreed to the published version of the manuscript.

**Funding:** This research was funded by The National Key Research and Development Programs of China, grant number 2019YFB1504204.

**Institutional Review Board Statement:** Not applicable.

**Informed Consent Statement:** Not applicable.

**Data Availability Statement:** Not applicable.

**Conflicts of Interest:** The authors declare that they have no known competing financial interests or personal relationships that could have appeared to influence the work reported in this paper.

## References


1.  Majorowicz, J.; Grasby, S.E. Deep Geothermal Heating Potential for the Communities of the Western Canadian Sedimentary Basin. *Energies* **2021**, *14*, 706. [CrossRef]
2.  Noorollahi, Y.; Shabbir, M.S.; Siddiqi, A.F.; Ilyashenko, L.K.; Ahmadi, E. Review of two decade geothermal energy development in Iran, benefits, challenges, and future policy. *Geothermics* **2019**, *77*, 257–266. [CrossRef]
3.  Mahlia, T.M.I.; Yanti, P.A.A. Cost efficiency analysis and emission reduction by implementation of energy efficiency standards for electric motors. *J. Clean. Prod.* **2010**, *18*, 365–374. [CrossRef]
4.  Isaka, B.L.A.; Gamage, R.P.; Rathnaweera, T.D.; Perera, M.S.A.; Chandrasekharam, D.; Kumari, W.G.P. An Influence of Thermally-Induced Micro-Cracking under Cooling Treatments: Mechanical Characteristics of Australian Granite. *Energies* **2018**, *11*, 13386. [CrossRef]
5.  Zhu, Z.; Tian, H.; Kempka, T.; Jiang, G.; Dou, B.; Mei, G. Mechanical Behaviors of Granite After Thermal Treatment Under Loading and Unloading Conditions. *Nat. Resour. Res.* **2021**, *30*, 2733–2752. [CrossRef]
6.  Li, K.; Bian, H.; Liu, C.; Zhang, D.; Yang, Y. Comparison of geothermal with solar and wind power generation systems. *Renew. Sust. Energy Rev.* **2015**, *42*, 1464–1474. [CrossRef]
7.  Bakhoda, H.; Almassi, M.; Moharamnejad, N.; Moghaddasi, R.; Azkia, M. Energy production trend in Iran and its effect on sustainable development. *Renew. Sust. Energy Rev.* **2012**, *16*, 1335–1339. [CrossRef]
8.  Zeng, Y.; Su, Z.; Wu, N. Numerical simulation of heat production potential from hot dry rock by water circulating through two horizontal wells at Desert Peak geothermal field. *Energy* **2013**, *56*, 92–107. [CrossRef]
9.  Zhao, Y.; Feng, Z.; Xi, B.; Wan, Z.; Yang, D.; Liang, W. Deformation and instability failure of borehole at high temperature and high pressure in Hot Dry Rock exploitation. *Renew. Energy* **2015**, *77*, 159–165. [CrossRef]
10. Siratovich, P.A.; Villeneuve, M.C.; Cole, J.W.; Kennedy, B.M.; Begue, F. Saturated heating and quenching of three crustal rocks and implications for thermal stimulation of permeability in geothermal reservoirs. *Int. J. Rock Mech. Min. Sci.* **2015**, *80*, 265–280. [CrossRef]
11. Zhu, M.; Yu, L.; Zhang, X.; Davarpanah, A. Application of Implicit Pressure-Explicit Saturation Method to Predict Filtrated Mud Saturation Impact on the Hydrocarbon Reservoirs Formation Damage. *Mathematics* **2020**, *8*, 1057. [CrossRef]
12. Gens, A.; Guimaraes, L.D.N.; Garcia-Molina, A.; Alonso, E.E. Factors controlling rock–clay buffer interaction in a radioactive waste repository. *Eng. Geol.* **2002**, *64*, 297–308. [CrossRef]
13. Davarpanah, A.; Shirmohammadi, R.; Mirshekari, B.; Aslani, A. Analysis of hydraulic fracturing techniques: Hybrid fuzzy approaches. *Arab. J. Geosci.* **2019**, *12*, 402. [CrossRef]
14. Zhu, Z.; Tian, H.; Chen, J.; Jiang, G.; Dou, B.; Xiao, P.; Mei, G. Experimental investigation of thermal cycling effect on physical and mechanical properties of heated granite after water cooling. *Bull. Eng. Geol. Environ.* **2020**, *79*, 2457–2465. [CrossRef]
15. Fox, D.B.; Sutter, D.; Beckers, K.F.; Lukawski, M.Z.; Koch, D.L.; Anderson, B.J.; Tester, J.W. Sustainable heat farming: Modeling extraction and recovery in discretely fractured geothermal reservoirs. *Geothermics* **2013**, *46*, 42–54. [CrossRef]
16. Fairhurst, C. Nuclear waste disposal and rock mechanics: Contributions of the Underground Research Laboratory (URL), Pinawa, Manitoba, Canada. *Int. J. Rock Mech. Min. Sci.* **2004**, *41*, 1221–1227. [CrossRef]

17. Géraud, Y.; Mazerolle, F.; Raynaud, S. Comparison between connected and overall porosity of thermally stressed granites. *J. Struct. Geol.* **1992**, *14*, 981–990. [CrossRef]
18. Zhu, Z.; Tian, H.; Mei, G.; Jiang, G.; Dou, B.; Xiao, P. Experimental investigation on mechanical behaviors of Nanan granite after thermal treatment under conventional triaxial compression. *Environ. Earth Sci.* **2021**, *80*, 46. [CrossRef]
19. Zhang, W.; Sun, Q.; Hao, S.; Geng, J.; Lv, C. Experimental study on the variation of physical and mechanical properties of rock after high temperature treatment. *Appl. Therm. Eng.* **2016**, *98*, 1297–1304. [CrossRef]
20. Vazquez, P.; Shushakova, V.; Gomez-Heras, M. Influence of mineralogy on granite decay induced by temperature increase: Experimental observations and stress simulation. *Eng. Geol.* **2015**, *189*, 58–67. [CrossRef]
21. Zhu, Z.; Ranjith, P.G.; Tian, H.; Jiang, G.; Dou, B.; Mei, G. Relationships between P-wave velocity and mechanical properties of granite after exposure to different cyclic heating and water cooling treatments. *Renew. Energy* **2021**, *168*, 375–392. [CrossRef]
22. Shen, Y.; Yuan, J.; Hou, X.; Hao, J.; Bai, Z.; Li, T. The strength changes and failure modes of high-temperature granite subjected to cooling shocks. *Geomech. Geophys. Geo-Energy Geo-Resour.* **2021**, *7*, 23. [CrossRef]
23. Kumari, W.G.P.; Ranjith, P.G.; Perera, M.S.A.; Chen, B.K.; Abdulagatov, I.M. Temperature-dependent mechanical behaviour of Australian Strathbogie granite with different cooling treatments. *Eng. Geol.* **2017**, *229*, 31–44. [CrossRef]
24. Peng, J.; Yang, S. Comparison of Mechanical Behavior and Acoustic Emission Characteristics of Three Thermally-Damaged Rocks. *Energies* **2018**, *11*, 2350. [CrossRef]
25. Martyushev, D.A.; Galkin, S.V.; Shelepov, V.V. The Influence of the Rock Stress State on Matrix and Fracture Permeability under Conditions of Various Lithofacial Zones of the Tournaisian–Fammenian Oil Fields in the Upper Kama Region. *Mosc. Univ. Geol. Bull.* **2019**, *74*, 573–581. [CrossRef]
26. Martyushev, D.A. Rock stress state influence on permeability of carbonate reservoirs. *Bull. Tomsk Polytech. Univ.* **2020**, *8*, 24–33.
27. David, C.; Menendez, B.; Darot, M. Influence of stress-induced and thermal cracking on physical properties and microstructure of La Peyratte granite. *Int. J. Rock Mech. Min. Sci.* **1999**, *36*, 433–448. [CrossRef]
28. Griffiths, L.; Heap, M.J.; Baud, P.; Schmittbuhl, J. Quantification of microcrack characteristics and implications for stiffness and strength of granite. *Int. J. Rock Mech. Min. Sci.* **2017**, *100*, 138–150. [CrossRef]
29. Fredrich, J.T.; Wong, T.F. Micromechanics of thermally induced cracking in three crustal rocks. *J. Geophys. Res. Solid Earth* **1986**, *91*, 12743–12764. [CrossRef]
30. Tang, Z.C.; Sun, M.; Peng, J. Influence of high temperature duration on physical, thermal and mechanical properties of a fine-grained marble. *Appl. Therm. Eng.* **2019**, *156*, 34–50. [CrossRef]
31. Chen, S.; Yang, C.; Wang, G. Evolution of thermal damage and permeability of Beishan granite. *Appl. Therm. Eng.* **2017**, *110*, 1533–1542. [CrossRef]
32. Khamrat, S.; Thongprapha, T.; Fuenkajorn, K. Thermal effects on shearing resistance of fractures in Tak granite. *J. Struct. Geol.* **2018**, *111*, 64–74. [CrossRef]
33. Peng, J.; Rong, G.; Cai, M.; Yao, M.; Zhou, C. Physical and mechanical behaviors of a thermal-damaged coarse marble under uniaxial compression. *Eng. Geol.* **2016**, *200*, 88–93. [CrossRef]
34. Wu, Q.; Weng, L.; Zhao, Y.; Guo, B.; Luo, T. On the tensile mechanical characteristics of fine-grained granite after heating/cooling treatments with different cooling rates. *Eng. Geol.* **2019**, *253*, 94–110. [CrossRef]
35. Zhao, F.; Shi, Z.; Sun, Q. Fracture Mechanics Behavior of Jointed Granite Exposed to High Temperatures. *Rock Mech. Rock Eng.* **2021**, *54*, 2183–2196. [CrossRef]
36. Wang, X.; Schubnel, A.; Fortin, J.; Gueguen, Y.; Ge, H. Physical properties and brittle strength of thermally cracked granite under confinement. *J. Geophys. Res. Solid Earth* **2013**, *118*, 6099–6112. [CrossRef]
37. Nasseri, M.H.B.; Schubnel, A.; Young, R.P. Coupled evolutions of fracture toughness and elastic wave velocities at high crack density in thermally treated Westerly granite. *Int. J. Rock Mech. Min. Sci.* **2007**, *44*, 601–616. [CrossRef]
38. Alm, O.; Jaktlund, L.L.; Shaoquan, K. The influence of microcrack density on the elastic and fracture mechanical-properties of stripa granite. *Phys. Earth Planet. Inter.* **1985**, *40*, 161–179. [CrossRef]
39. Zarei, V.; Mirzaasadi, M.; Davarpanah, A.; Nasiri, A.; Valizadeh, M.; Hosseini, M.J.S. Environmental Method for Synthesizing Amorphous Silica Oxide Nanoparticles from a Natural Material. *Processes* **2021**, *9*, 334. [CrossRef]
40. Zhu, Z.; Tian, H.; Mei, G.; Jiang, G.; Dou, B. Experimental investigation on physical and mechanical properties of thermal cycling granite by water cooling. *Acta Geotech.* **2020**, *15*, 1881–1893. [CrossRef]
41. Zhu, Z.; Tian, H.; Jiang, G.; Cheng, W. Effects of High Temperature on the Mechanical Properties of Chinese Marble. *Rock Mech. Rock Eng.* **2018**, *51*, 1937–1942. [CrossRef]
42. Sun, W.; Wu, S.; Xu, X. Mechanical behaviour of Lac du Bonnet granite after high-temperature treatment using bonded-particle model and moment tensor. *Comput. Geotech.* **2021**, *135*, 104132. [CrossRef]
43. Griffiths, L.; Lengline, O.; Heap, M.J.; Baud, P.; Schmittbuhl, J. Thermal Cracking in Westerly Granite Monitored Using Direct Wave Velocity, Coda Wave Interferometry, and Acoustic Emissions. *J. Geophys. Res. Solid Earth* **2018**, *123*, 2246–2261. [CrossRef]
44. Shao, Z.; Tang, X.; Wang, X. The influence of liquid nitrogen cooling on fracture toughness of granite rocks at elevated temperatures: An experimental study. *Eng. Fract. Mech.* **2021**, *246*, 107628. [CrossRef]
45. Yong, C.; Wang, C.Y. Thermally induced acoustic emission in westerly granite. *Geophys. Res. Lett.* **1980**, *7*, 1089–1092. [CrossRef]
46. Shao, S.; Wasantha, P.L.P.; Ranjith, P.G.; Chen, B.K. Effect of cooling rate on the mechanical behavior of heated Strathbogie granite with different grain sizes. *Int. J. Rock Mech. Min. Sci.* **2014**, *70*, 381–387. [CrossRef]

47. Zhang, F.; Zhang, Y.; Yu, Y.; Hu, D.; Shao, J. Influence of cooling rate on thermal degradation of physical and mechanical properties of granite. *Int. J. Rock Mech. Min. Sci.* **2020**, *129*, 104285. [CrossRef]
48. Heap, M.J.; Coats, R.; Chen, C.; Varley, N.; Lavallee, Y.; Kendrick, J.; Xu, T.; Reuschle, T. Thermal resilience of microcracked andesitic dome rocks. *J. Volcanol. Geoth. Res.* **2018**, *367*, 20–30. [CrossRef]
49. Lu, G.; Zhou, J.; Li, Y.; Zhang, X.; Gao, W. The influence of minerals on the mechanism of microwave-induced fracturing of rocks. *J. Appl. Geophys.* **2020**, *180*, 104123. [CrossRef]
50. Yang, Z.; Yang, S.; Tian, W. Peridynamic simulation of fracture mechanical behaviour of granite specimen under real-time temperature and post-temperature treatments. *Int. J. Rock Mech. Min. Sci.* **2021**, *138*, 104573. [CrossRef]
51. Zhang, B.; Tian, H.; Dou, B.; Zheng, J.; Chen, J.; Zhu, Z.; Liu, H. Macroscopic and microscopic experimental research on granite properties after high-temperature and water-cooling cycles. *Geothermics* **2021**, *93*, 102079. [CrossRef]
52. Walsh, J.B. The effect of cracks on the compressibility of rock. *J. Geophys. Res.* **1965**, *70*, 381–389. [CrossRef]
53. David, E.C.; Brantut, N.; Schubnel, A.; Zimmerman, R.W. Sliding crack model for nonlinearity and hysteresis in the uniaxial stress-strain curve of rock. *Int. J. Rock Mech. Min. Sci.* **2012**, *52*, 9–17. [CrossRef]
54. Heap, M.J.; Lavallee, Y.; Laumann, A.; Hess, K.U.; Meredith, P.G.; Dingwell, D.B.; Huismann, S.; Weise, F. The influence of thermal-stressing (up to 1000 degrees C) on the physical, mechanical, and chemical properties of siliceous-aggregate, high-strength concrete. *Constr. Build. Mater.* **2013**, *42*, 248–265. [CrossRef]
55. Jin, P.; Hu, Y.; Shao, J.; Zhao, G.; Zhu, X.; Li, C. Influence of different thermal cycling treatments on the physical, mechanical and transport properties of granite. *Geothermics* **2019**, *78*, 118–128. [CrossRef]
56. Freire-Lista, D.M.; Fort, R.; Varas-Muriel, M.J. Thermal stress-induced microcracking in building granite. *Eng. Geol.* **2016**, *206*, 83–93. [CrossRef]
57. Yang, F.; Wang, G.; Hu, D.; Liu, Y.; Zhou, H.; Tan, X. Calibrations of thermo-hydro-mechanical coupling parameters for heating and water-cooling treated granite. *Renew. Energy* **2021**, *168*, 544–558. [CrossRef]
58. Glover, P.; Baud, P.; Darot, M.; Meredith, P.G.; Boon, S.A.; Leravalec, M.; Zoussi, S.; Reuschlé, T. $\alpha/\beta$ phase transition in quartz monitored using acoustic emissions. *Geophys. J. Int.* **1995**, *120*, 775–782. [CrossRef]

*Article*

# The Effect of Perforation Spacing on the Variation of Stress Shadow

**Weige Han** [1,2], **Zhendong Cui** [3,4,5,*] **and Zhengguo Zhu** [1,2]

[1] State Key Laboratory of Mechanical Behavior and System Safety of Traffic Engineering Structures, Shijiazhuang Tiedao University, Shijiazhuang 050043, China; hanweige@stdu.edu.cn (W.H.); myztx@163.com (Z.Z.)

[2] Hebei Province Technical Innovation Center of Safe and Effective Mining of Metal Mines, Shijiazhuang 050043, China

[3] Key Laboratory of Shale Gas and Geoengineering, Institute of Geology and Geophysics, Chinese Academy of Sciences, Beijing 100029, China

[4] Innovation Academy for Earth Science, CAS, Beijing 100029, China

[5] College of Earth and Planetary Sciences, University of Chinese Academy of Sciences, Beijing 100049, China

[*] Correspondence: cuizhendong@mail.iggcas.ac.cn; Tel.: +86-010-8299-8295

**Citation:** Han, W.; Cui, Z.; Zhu, Z. The Effect of Perforation Spacing on the Variation of Stress Shadow. *Energies* **2021**, *14*, 4040. https://doi.org/10.3390/en14134040

Academic Editors: Junlong Shang, Chun Zhu and Manchao He

Received: 25 May 2021
Accepted: 29 June 2021
Published: 4 July 2021

**Publisher's Note:** MDPI stays neutral with regard to jurisdictional claims in published maps and institutional affiliations.

**Abstract:** When the shale gas reservoir is fractured, stress shadows can cause reorientation of hydraulic fractures and affect the complexity. To reveal the variation of stress shadow with perforation spacing, the numerical model between different perforation spacing was simulated by the extended finite element method (XFEM). The variation of stress shadows was analyzed from the stress of two perforation centers, the fracture path, and the ratio of fracture length to spacing. The simulations showed that the reservoir rock at the two perforation centers is always in a state of compressive stress, and the smaller the perforation spacing, the higher the maximum compressive stress. Moreover, the compressive stress value can directly reflect the size of the stress shadow effect, which changes with the fracture propagation. When the fracture length extends to 2.5 times the perforation spacing, the stress shadow effect is the strongest. In addition, small perforation spacing leads to backward-spreading of hydraulic fractures, and the smaller the perforation spacing, the greater the deflection degree of hydraulic fractures. Additionally, the deflection angle of the fracture decreases with the expansion of the fracture. Furthermore, the perforation spacing has an important influence on the initiation pressure, and the smaller the perforation spacing, the greater the initiation pressure. At the same time, there is also a perforation spacing which minimizes the initiation pressure. However, when the perforation spacing increases to a certain value (the result of this work is about 14 m), the initiation pressure will not change. This study will be useful in guiding the design of programs in simultaneous fracturing.

**Keywords:** hydraulic fracturing; perforation spacing; stress shadow; fracture path; extended finite element method (XFEM)

## 1. Introduction

Shale is the main target rock for global unconventional oil and gas production, and shale gas occupies an indispensable position in the world energy pattern [1]. However, shale reservoirs are dense. To improve the permeability of shale reservoirs and increase shale gas production, horizontal wells and hydraulic fracturing stimulation technology are widely used [2]. When hydraulic fracturing is applied to reservoir reconstruction, multiple perforations can be used for simultaneous fracturing, which can significantly increase fracturing capacity. However, simultaneous fracturing with adjacent perforations can result in a stress shadow phenomenon, which will affect the state of the geostress field, thereby affecting subsequent fractures propagation [3]. Fortunately, reasonable perforation spacing can effectively utilize the positive effect of stress shadow. Therefore, it is important to investigate the influence of perforation spacing on the variation of stress shadow.

In recent years, many scholars have considered the relationship between perforation spacing and stress shadow size through laboratory tests and numerical simulation. Zhou et al. [4] conducted atrue-triaxial hydraulic fracturing experiment to analyze the effect of stress shadow on hydraulic fractures by adjusting the notch distance and found a phenomenon of adjacent hydraulic fractures expanding in the opposite direction. To avoid the stress shadow effect, Morrill and Miskimins [5] obtained the optimum perforation spacing by combining multiple simulation parameters. Liu et al. [6] investigated the hydraulic fracture mutual interference using the finite element method (FEM), and put forward the optimization method of perforation spacing. In addition, the path of hydraulic fracturing is greatly affected by stress shadow [7]. Kresse et al. [8] analyzed the effect of perforation spacing on hydraulic fracture paths and stress shadow using the unconventional fracture model (UFM). Zeng and Yao [9] revealed the intersection pattern of natural and hydraulic fractures, and showed that perforation spacing affects fracture morphology. Weng et al. [10] adopted the UFM to study the interaction between stress shadow and fracture length under different perforation spacing. It was found that the stress shadow is negatively correlated with the perforation spacing. Vahab et al. [11] proposed a multi-state hydraulic fracturing treatment algorithm based on the extended finite element method (XFEM), and used the algorithm to conduct research on multi-perforation fracturing. Khoei et al. [12] studied the interaction between hydraulic fractures and natural fractures using the XFEM method. Furthermore, the change in stress intensity factor (SIF) can also reflect the effect of stress shadow on fracture propagation. Taghichian et al. [13] considered the influence of stress shadow on SIF, which dominates the change in SIF and leads to its substantial reduction.

In addition, perforation density is also a research hotspot. Qi et al. [14] researched the effect of perforation density on the evolution of fracture network by using a 2D coupled flow-stress-damage model. It was obtained that the interaction between micro-fractures and perforations becomes stronger with the increase in perforation density. Luo et al. [15] put forward the control method of variable density perforation and realized the balanced distribution of flow. Ghaderiet al. [16] considered the effect of fracture density on gas saturation distribution. When the fracture density is large, it will only reach a higher oil production rate in a short period. There is an optimum value for the effect of crack density on the oil recovery, and it also depends on reservoir properties. Three-dimensional numerical models were used to analyze the effect of perforation spacing on fracture paths. Additionally, it was found that non-uniform perforation spacing can promote crack propagation [17]. Meanwhile, the two adjacent fractures would deviate from each other under the influence of stress shadow [18]. At present, the same conclusion has been reached in the study of perforation spacing: the smaller the perforation spacing, the greater the size of the stress shadow. Furthermore, the existence of stress shadow will change the fracture paths, leading to the back propagation of adjacent fractures.

Recent studies have shown that perforation spacing, length, and density are the main parameters concerning the size of the stress shadow effect [8,10,16]. However, there are few studies on the change process of stress shadows. Revealing the variation rule of stress shadows is essential for optimizing fracturing design. Thus, two-dimensional numerical models with different perforation spacing were established. Subsequently, the XFEM was used to simulate the variation of stress shadow during the simultaneous fracturing of two adjacent perforations. The change process of stress shadows under different perforation spacing was analyzed.

## 2. Numerical Analysis Method

### 2.1. Principle of the XFEM

By increasing the degrees of freedom and enhancing functions, the XFE Mimplements discontinuity. In addition, the fracture is described by the jump function and the level set function [19].

The displacement vector function can be represented as follows:

$$u = \sum_{I=1}^{N} N_I(x) \left[ H(x)a_I + \sum_{\alpha=1}^{4} F_\alpha(x)b_I^\alpha + u_I \right],$$

(1)

where the nodal function is denoted by $N_I(x)$. $a_I$ and $b_I^\alpha$ represent the joint improvement in the degrees of freedom of the elements penetrated by cracks and the crack tip element, respectively. The Heaviside step function $H(x)$ can be applied to describe fractures [20]. The asymptotic displacement function $F_\alpha(x)$ can be represented as follows:

$$\{F_\alpha(r,\theta)\}_{\alpha \in \{1,4\}} = \sqrt{r}\left\{ \sin\frac{\theta}{2}, \cos\frac{\theta}{2}, \sin\frac{\theta}{2}\sin\theta, \sin\frac{\theta}{2}\cos\theta \right\}$$

(2)

The XFEM has been described in detail previously [21].

### 2.2. Fluid-Mechanical Coupling Principle

The mechanical equilibrium equation of rocks can be obtained from the principle of virtual work.

$$\int_V \sigma\,\delta_\varepsilon dV = \int_S T\,\delta_v dS + \int_V f\,\delta_v dV + \int_V e\rho_w g\delta_v dV,$$

(3)

where $V$ and $S$ are the integral space and integral space surface, respectively; $\delta_\varepsilon$ and $\delta_v$ represent virtual strain field and virtual velocity field, respectively; $e$ is the rockporosity; and $T$ is the external surface force [21].

### Fluid Flow Model

Fracture propagation is primarily driven by the fluid pressure generated by fracturing fluid acting on the fracture surface. Supposing the fluid is continuous and incompressible, tangential flow and normal flow can represent fluid flow in fractures.

The tangential flow on the surface of the fracture element can be simulated by the Newtonian model, which can be expressed as follows [21]:

$$Q = -\frac{d^3}{12\mu}\nabla p,$$

(4)

The normal flow corresponds to the engineering phenomenon of leak-off, which can be considered as the volume rate at which fluid flows into the simulated area unit. It is represented as follows:

$$\begin{cases} q_t = c_t(p_i - p_t) \\ q_b = c_b(p_i - p_b) \end{cases},$$

(5)

where $q$ represents the flow rates; $p$ is the pore pressure of different surfaces; $c$ is theleak-off coefficients; subscripts t and d represent the top and bottom surfaces of a cracked element, respectively [22].

### 2.3. Failure Criterion

This study used the maximum principal stress criterion to simulate fracture.

$$f = \left\{ \frac{\langle \sigma_{max} \rangle}{\sigma_{max}^0} \right\}.$$

(6)

where $\sigma_{max}^0$ is the maximum allowable principal stress. The symbol $\langle \rangle$ represents the Macaulay bracket with the usual interpretation. When $f$ is greater than 1 within the tolerance range, the damage is initiated.

The damage evolution criterion was represented by damage variable $D$ based on the effective displacement. It can be expressed as follows:

$$D = \frac{\delta_m^f \left( \delta_m^{max} - \delta_m^0 \right)}{\delta_m^{max} \left( \delta_m^f - \delta_m^0 \right)},$$

(7)

where $\delta_m^0$ and $\delta_m^f$ are the effective displacement at damage initiation and the effective displacement at complete failure, respectively. $\delta_m^{max}$ refers to the maximum value of the effective displacement. The author has given a detailed description previously [21].

## 3. Model Validation

The author has previously used the Kristonovich-Geertsma-de Klerk (KGD) model to verify the method, so this paper verifies the accuracy of the model by comparing with the results of previous studies. Wu et al. [23] and Zhang et al. [24] established two perforations for fracturing in the horizontal wellbore. Based on the input data of Wu et al.'s [25] model, the simulation results were obtained in this paper (Figure 1). As shown in Figure 1, our model is relatively accurate. However, when the fracture is about to propagate to the boundary, there is a slight difference between the two results, which may be caused by the boundary effect.

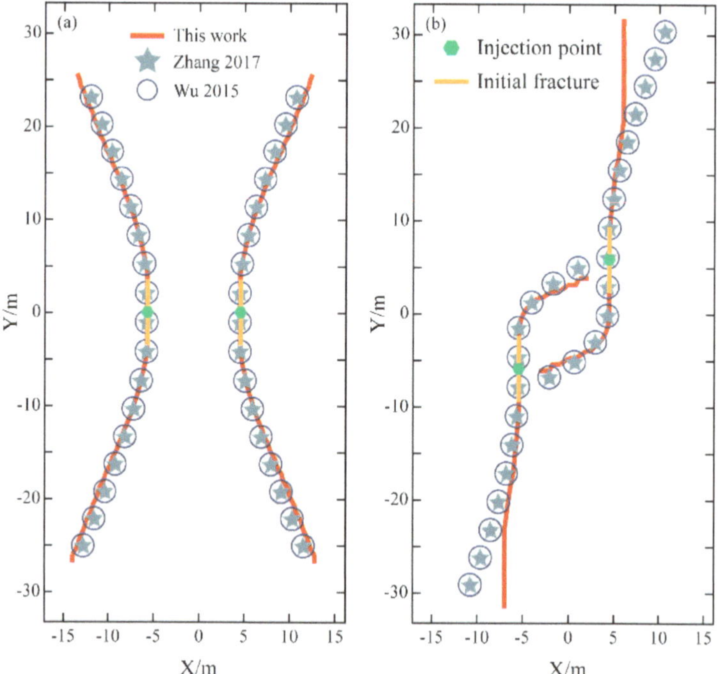


**Figure 1.** Compared with the fracture paths of Zhang et al. [24] and Wu et al. [25]. (**a**) Two parallel fractures. (**b**) Two offset fractures.

## 4. Numerical Model

Considering the variation of stress shadow at different perforation spacing, a 50 m square model containing two perforations of 1 m in length was established, as shown in Figure 2. Perforation spacing was defined as variable $L$. $L$ was chosen as 2, 4, 6, 8, 10, 12, 14, 16, 18, 20, 22, and 24 m, respectively. Assuming that the model is isotropic and the perforation spacing is the only variable, the influence of the perforation spacing on the stress shadow is focused on.

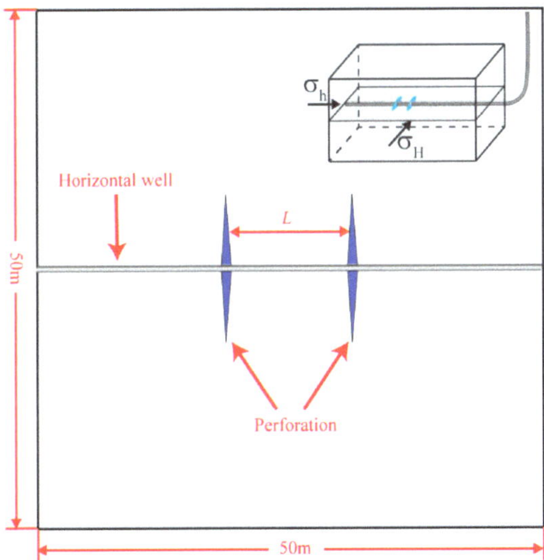


**Figure 2.** Geometric model: contains two perforations (perforation spacing: *L*, length: 1 m, direction: perpendicular to minimum principal stress).

The in situ stress balance and fracturing fluid injection were achieved by setting the geostatic analysis step and soils analysis step. The fracturing time was 60 s. Additionally, the pore fluid response was set to transient consolidation. In addition, the model used aCPE4P element with a total of 10,000 in ABAQUS simulation software. The fracturing fluid was injected through the C-flow function. The simulation parameters (Table 1) are based on the data of laboratory experiments and articles [21].

**Table 1.** Simulation parameters.

| Parameter | Units | Value |
| --- | --- | --- |
| Elastic modulus $E$ | GPa | 27 |
| Poisson's ratio | - | 0.23 |
| Tensile strength | MPa | 10 |
| Maximum horizontal stress $\sigma_H$ | MPa | 18 |
| Minimum horizontal stress $\sigma_h$ | MPa | 12 |
| Fluid leak off | m/Pa·s | $1 \times 10^{-14}$ |
| Void ratio | % | 5 |
| Permeability | m/s | $1 \times 10^{-7}$ |
| Injection rate | m$^3$/s | 0.01 |
| Fluid viscosity | Pa·s | $1 \times 10^{-3}$ |

## 5. Results

### 5.1. Stress Distribution

The stress shadow has a great influence on the stress distribution. Therefore, the horizontal $S_{11}$ stress contour was extracted under different perforation spacing, and the stress distribution characteristics of different perforation spacing under the same reservoir parameters and injection rate were analyzed. Figure 3 illustrates the horizontal stress distribution under different perforation spacing when the injection time is 10 s. According to the contour, although the injection rate and reservoir parameters are identical, there are obvious differences in the fracture path and stress distribution with different perforation spacing at the same fracturing time. Under different perforation spacing, this contour shows that tension stress concentration occurs at the hydraulic fracture tips, and fracture initiate by tension. However, there is compressive stress on the sides of hydraulic fractures, owing to the stress shadow. The stress shadow zone is formed when the additional compressive stresses intersect in the middle of two hydraulic fractures. At the same injection time, with the increase in perforation spacing, the compressive stress concentration area (blue area) keeps separating and gradually gathers towards the perforation, which demonstrates that the size of the stress shadow reduces with the increase in perforation spacing.

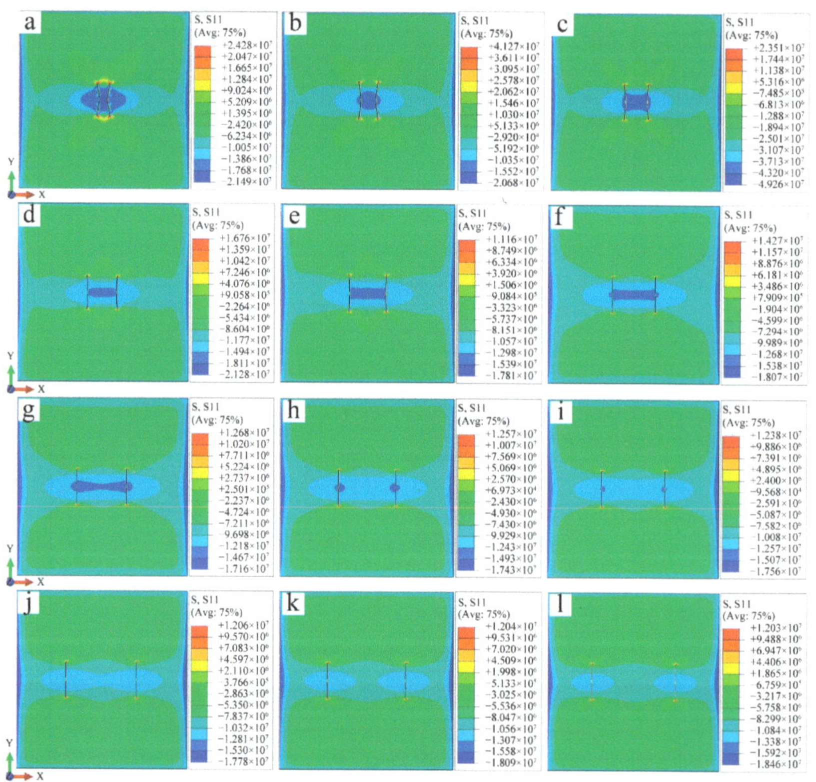

**Figure 3.** Horizontal stress contours with different perforation spacing at injection time of 10 s. There is a concentrated zone of compressive stress between the two perforations, which decreases with the increase in perforation spacing. (**a–l**) represents perforation spacing from 2 m–24 m, respectively.

## 5.2. Stress at the Center of Two Perforations

According to Figure 3, it can be found that the middle of the two perforations is a stress shadow zone. Therefore, detecting the stress variation in the middle of the two perforations can intuitively indicate the change process of stress shadows. Hence, the horizontal $S_{11}$ stress of the two perforation center points was extracted, and the process of stress variation at the point throughout the whole fracturing process was observed, as shown in Figure 4. According to Figure 4, it can be found that the two perforation center points are in the state of compressive stress during the whole fracturing process. Before the beginning of fracturing, the compressive stress at the center point under different perforation spacing is 7 MPa, which is the initial ground stress. As the fracturing progresses, the compressive stress at the center point increases first and then decreases. That is to say, with the propagation of the fracture, the compressive stress between two perforations increases continuously until the hydraulic fracture extends to a certain length, and then the compressive stress begins to decrease. That is, the stress shadow effect begins to weaken, which indicates that the size of the stress shadow is affected by the fracture length. The size of the stress shadow changes with the propagation of hydraulic fractures.

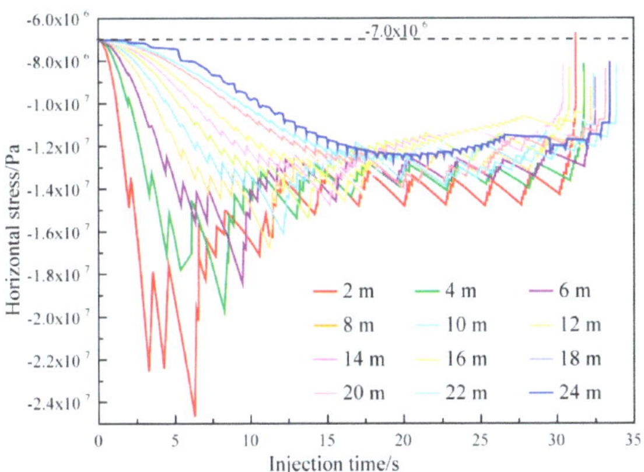


**Figure 4.** Horizontal stress at two perforation centers. The stress at this position is the in situ stress (7 MPa) before the beginning of fracturing. With the progress of fracturing, the stress first increases and then decreases.

Since the stress at the center point of the two perforations constantly changes, the size of the stress shadow effect can be characterized by the magnitude of the compressive stress at this point. Therefore, the maximum value of the compressive stress at the center point was extracted during the fracturing process under different perforation spacing. Figure 5 shows the maximum value of the compressive stress and the time to reach the maximum value. With the increase in perforation spacing, the maximum compressive stress value decreases, while the time to reach the maximum compressive stress increases. This shows that the size of the stress shadow effect becomes larger with decreasing perforation spacing. When the perforation spacing is large, it takes a longer injection time to produce mutual interference between the two perforations, which is directly related to the fracture length.

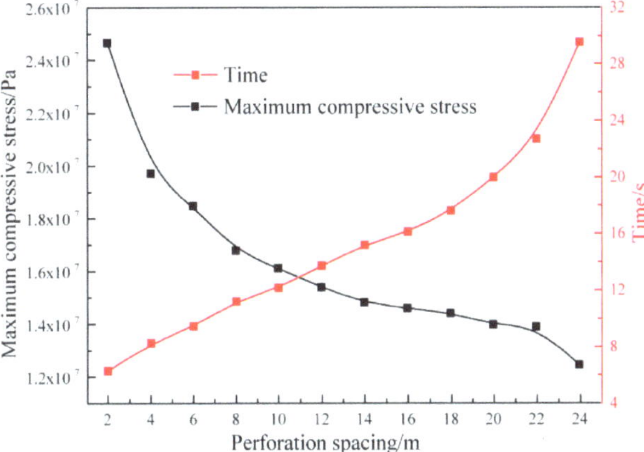


**Figure 5.** Variation curve of maximum compressive stress at the center of two perforations with perforation spacing. The black and red lines indicate the value of maximum compressive stress and the time to reach the maximum compressive stress, respectively. With the increase in perforation spacing, the maximum compressive stress decreases and the time increases.

When the compressive stress at the center reaches the maximum value, the stress shadow effect can be considered to be the strongest, and the fracture length at this time can be considered as the key parameter affecting the stress shadow. The ratio of fracture length to perforation spacing has an important influence on the stress shadow.

### 5.3. Fracture Propagation Path

Figure 6 shows the final fracture propagation path with different perforation spacing. When two perforations are fractured at the same time, the stress shadow alters the in situ stress field distribution, which leads to the two fractures diverging from the maximum horizontal stress direction extension, showing a reverse expansion trend [26]. With the increase in perforation spacing, the deflection degree of the fracture reduces and the stress shadow weakens. According to the time at which the compressive stress reaches the maximum value obtained in Figure 5, fracture paths with different perforation spacing before that time were obtained, which are shown as the grey shadows in Figure 6. When the fracture propagates to the boundary of the gray shaded area in Figure 6, the compressive stress reaches the maximum value, and the stress shadow is the strongest. Thus, it can be found that the fracture length, corresponding to the maximum compressive stress at the two perforation centers, increases as the perforation spacing increases. This also explains the fact that the maximum compressive stress value reduces with the enlargement of perforation spacing in Figure 5, and the time to reach the maximum compressive stress increases. After the fracture propagates to the boundary of the gray shaded area, the fracture spacing increases while the size of the stress shadow is gradually decreasing. Afterward, the in situ stress field again dominates thefracture propagation.

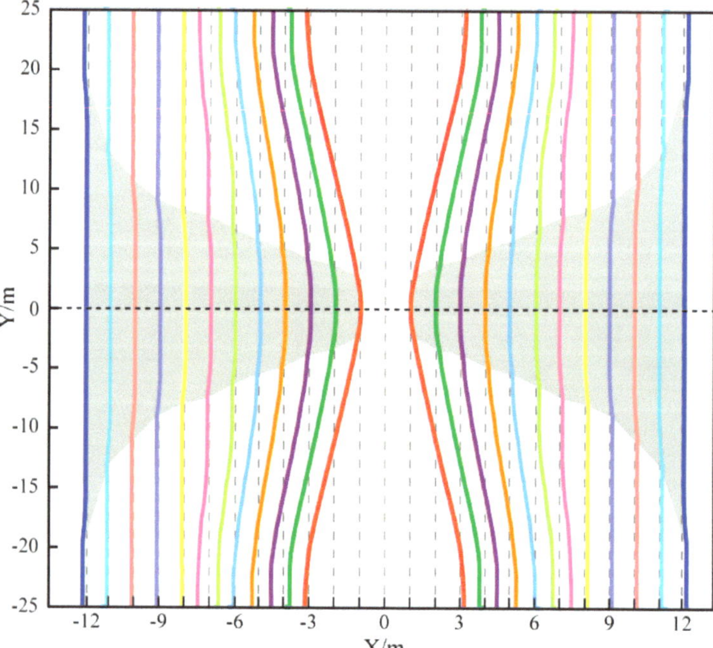

**Figure 6.** Fracture propagation paths with different perforation spacing. Different colors represent different perforation spacing. Due to the isotropic model, fractures corresponding to the two perforations are symmetrically distributed. When hydraulic fractures propagate to the boundary of the grey shadow zone, the stress shadow effect is the strongest.

### 5.4. Fracture Length

The fracture length was extracted when the fracture propagates to the boundary of the gray shaded area in Figure 6. Since the stress shadow effect is the strongest at this time, it is crucial to study the relationship between the fracture length and the perforation spacing to reveal the stress shadow effect.

Figure 7 shows the length of fracture with different perforation spacing at this time. The fracture length is positively related to the perforation spacing when the stress shadow effect is strongest. The ratio of fracture length to perforation spacing was defined as R. The R-value under different perforation spacing was obtained. When the stress shadow effect is the strongest, the R-value under different perforation spacing is distributed between 2 and 4, with an average value of 2.5. Therefore, a conclusion can be drawn that the maximum size of the stress shadow effect is obtained when the fracture length is 2.5 times the length of the perforation spacing.

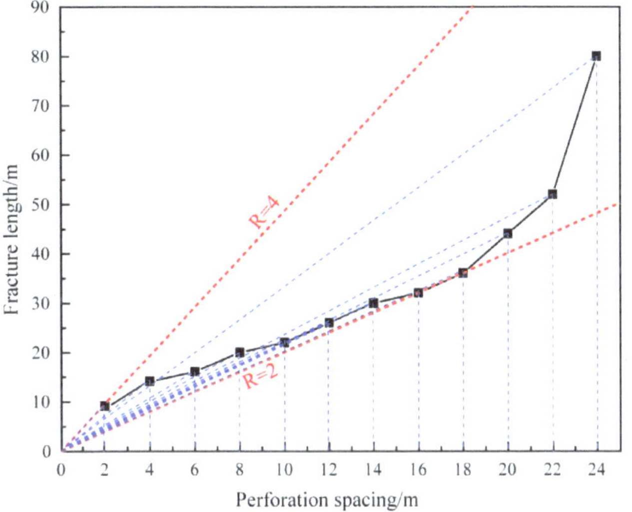


**Figure 7.** Fracture length with the strongest stress shadow effect under different perforation spacing. R represents the ratio of fracture length to perforation spacing. R has a strong concentration, mainly between 2 and 4.

## 6. Discussion

In this work, the variation process of stress shadow under different perforation spacing is deeply analyzed and the compressive stress value of the two perforation center points directly represents the change in the stress shadow effect. It can be more intuitive to study the change rule of the stress shadows effect with the fracture propagation in the whole fracturing process. At the same time, the ratio of perforation spacing to fracture length is proposed when the stress shadow effect is strongest, which has great significance for guiding fracturing design.

According to Figure 6, the two hydraulic fractures are separated from each other, so the reservoir area between the two perforations is never fractured. According to this feature, the fracturing design can be optimized for step fracturing or re-fracturing the reservoir to increase fracture connectivity [3]. According to the perforation distribution in Figure 8, the fracturing sequence can be designed, first fracturing ①③⑤, and then fracturing ②④. After the completion of fracturing work, it can be re-perforated in the middle of the two perforations to achieve re-fracturing, to increase the distribution area of the fracture network and optimize fracturing.

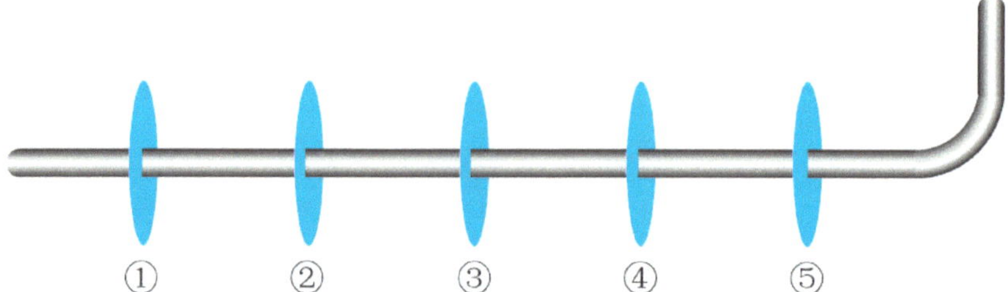

**Figure 8.** Perforation distribution.

According to the fracturing design method proposed above, XFEM was used to verify it. Five initial fractures were arranged as perforations and numbered (Figure 9a). First fracturing ①③⑤, the first fracture path was obtained (Figure 9b). Under the effect of stress shadow, fractures ① and ⑤ expand backward, while fracture ③ expands along the direction of maximum principal stress. The middle of the fracture is always the unfractured zone. Therefore, after the completion of the first fracturing, the perforations ② and ④ can be re-fractured to obtain the fracture path after the second fracturing (Figure 9c). During the second fracturing, the hydraulic fracture path deflects under the influence of the pre-existing hydraulic fractures. Both fractures ② and ④ will deflect towards fracture ③, and eventually converge to form a fracture network. Meanwhile, the secondary fracturing can activate the primary fracturing silent fracture, so that the fracture of the first fracturing can be reactivated and expanded, further increasing the fracture distribution area. In addition, after the second fracturing, there is still an unfractured zone in the middle of the two perforations. Therefore, a new perforation can be inserted between the two perforations to perform multi-batch fracturing, although the number of fracturing processes in multi-batch fracturing needs to be studied further. This fracturing design scheme can effectively obtain a complex fracture network and improve the production rate.

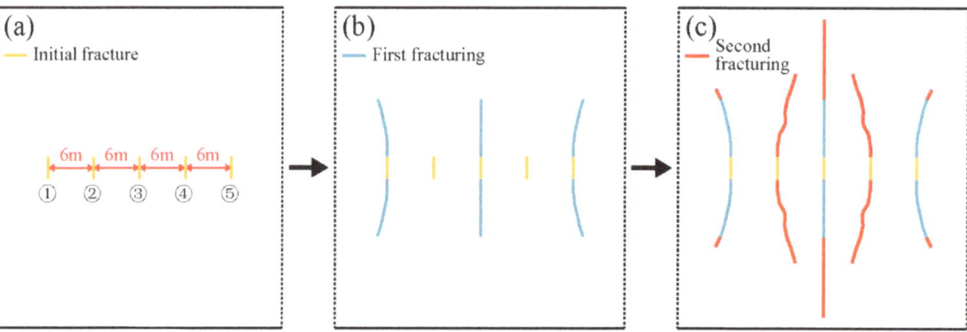

**Figure 9.** Secondary fracturing results. (**a**) Perforation diagram. (**b**) Fracture path after first fracturing. (**c**) Fracture path after the second fracturing.

However, a higher pump pressure is required for secondary fracturing. The initiation pressure of the above model is ② ④ > ③ > ① ⑤. Therefore, the subsequent crack initiation is more difficult due to the effect of stress shadow, especially the superposition of stress shadow of multiple cracks. According to the research results, the optimal perforation spacing can be determined in fracturing design. Firstly, the two perforations are fractured at the same time with a small pumping rate. After the fracture stops expanding, a large pumping rate is carried out in the middle of the two perforations to generate new fractures and reactivate the existing fractures to increase the fracturing area. Since the proppant is not considered in this model, the competition between perforations can be characterized by the fracture width (Figure 10).

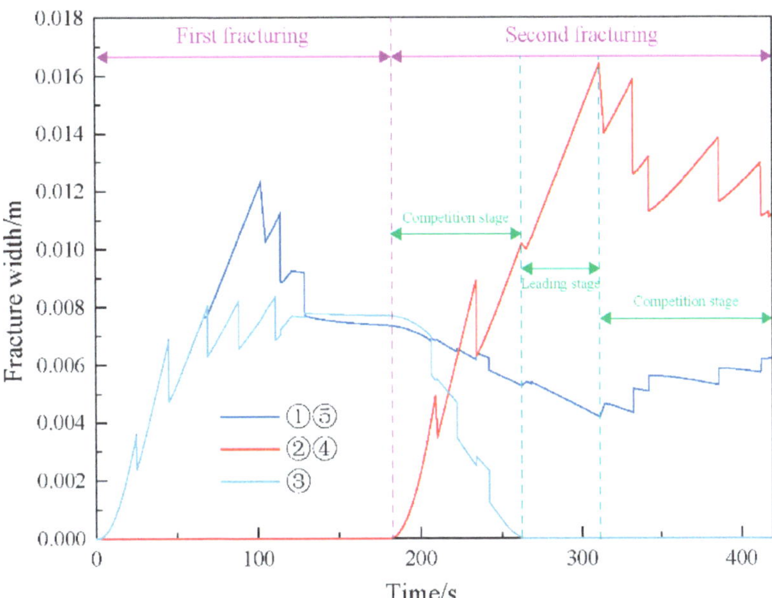



**Figure 10.** Fracture width. There is a competitive tendency between perforation first and subsequent fracturing.

During the first fracture, perforations ① and ⑤ inhibited the opening of perforation ③, resulting in a smaller fracture width. With the development of secondary fracturing, the growth of the second hydraulic fracture has a great impact on the width of the pre-existing fracture, and the pre-existing fracture tends to close, so there is a higher requirement for proppant. After the second fracturing, it can be divided into three stages according to the width of the fracture. The first stage is a competitive stage, where secondary fracture initiation and increasing fracture width lead to the closure of pre-existing fractures. Perforation ③ has a significant tendency to close and the width of the fracture decreases rapidly, thus requiring higher proppant strength. In the second stage, perforation ③ was completely closed. In the third stage, the fracture width of perforations ① and ⑤ is competitive with that of secondary fractures, where the proppant is subjected to fatigue loading. As a result, higher proppant strength is required for the initial fracture to resist the impact of subsequent fractures on the previous fracture.

Many scholars have proved that the propagation of hydraulic fractures will change the distribution of in situ stress fields. Roussel et al. [3] discussed the range of the stress reversal zone and proposed that it was not suitable for re-perforation in this region, whereas the stress reversal caused by the stress shadow also has a positive effect. As shown in Figure 11a, the in situ stress field between the two fractures changed from vertical to horizontal after fracturing. Therefore, when refracturing between two perforations, the new perforation can be set to be angled with the existing perforation. Especially for layered rock masses, the maximum principal stress is perpendicular to the bedding plane (in the Barnett shale). It is generally believed that an angle of 30 degrees between perforation and maximum principal stress can obtain a better fracturing effect. The secondary in situ stress field between the perforations after fracturing is roughly orthogonal to the perforations (Figure 11b), so the parallel of the perforation angle and in situ stress field will have a better fracturing effect during the secondary fracturing. A multi-batch variable angle perforation fracturing method can be proposed according to the research results.

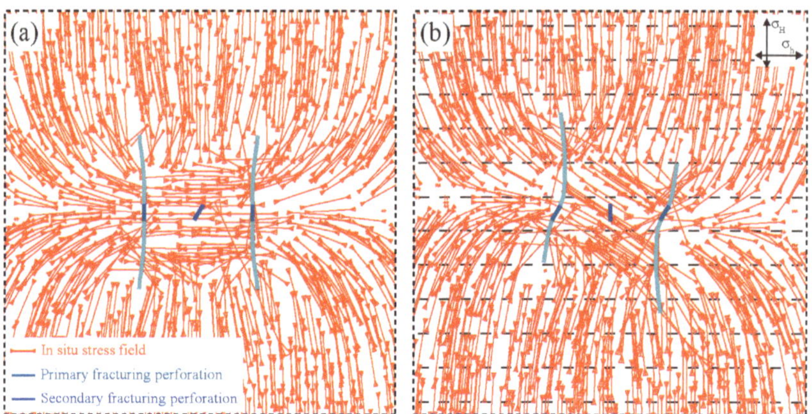


**Figure 11.** Distribution of secondary stress field. (**a**) Parallel perforating in the direction of maximum principal stress. (**b**) Perforated at 30 degrees with maximum principal stress.

Fracture initiation pressure is an essential key parameter in the field fracturing construction [27], which has an important influence on the selection of construction equipment and the control of injection rate in hydraulic fracturing design [28]. Therefore, the study of fracture initiation pressure is crucial for oil and gas production. Existing studies show that fracture trace angle [29], port number [30], the pressurization rate [31] and viscosity of fracturing fluid [32] all affect fracture initiation pressure. However, there are few studies on the influence of perforation spacing on initiation pressure. Therefore, in this work, the initiation pressure was extracted, and the effect of different perforation spacing on reservoir initiation pressure was obtained (Figure 12). With the decrease in perforation spacing, the initiation pressure increases. Thus is because, when the perforation spacing is small, the fracture initiation must overcome both the in situ stress field and the secondary stress field induced by the stress shadow, which results in larger initiation pressure. Furthermore, when the perforation spacing increases to a particular value, the stress shadow decreases and the initiation mainly overcomes the in situ stress and tensile strength. Additionally, the in situ stress and strength parameters are fixed, so the initiation pressure no longer changes. However, it is interesting that the initiation pressure does not decrease monotonically with the perforation spacing. Before the perforation spacing reaches a certain fixed value, there is already a perforation spacing which can minimize the initiation pressure (the simulation results of this paper show that the minimum initiation pressure is found when the perforation spacing is 10 m). Therefore, if we can find the perforation spacing with the minimum fracturing pressure before the initiation pressure is constant. Then, under this perforation spacing, the expansion of two hydraulic fractures will be affected by the stress shadow, resulting in a backward expansion phenomenon, increasing the expansion area of the hydraulic fracture. Meanwhile, there is a small initiation pressure. This takes advantage of the positive effects of stress shadows.

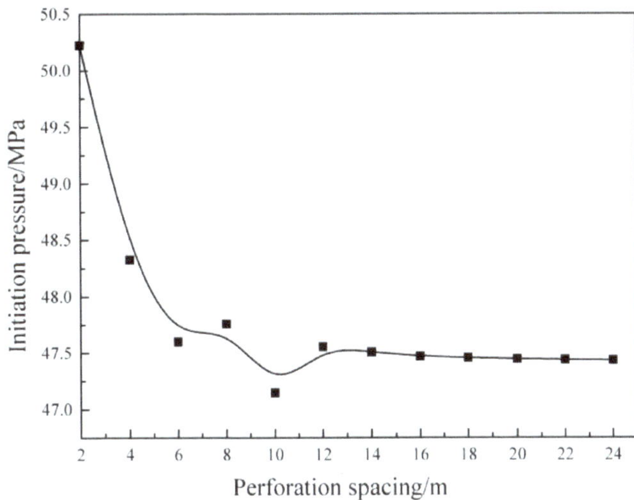

**Figure 12.** Relation curve between initiation pressure and perforation spacing. With the increase in perforation spacing, the initiation pressure decreases and it will not change when the spacing reaches a certain value.

The heterogeneity of shale is not considered in this study. However, shale, as a sedimentary rock, is not only inhomogeneous in terms of mineral particles but also heterogeneous in terms of laminar structure [33]. Both affect the fracture propagation path [34–36]. Han et al. [21] proposed a method to create a mineral heterogeneity model. Using this model, we compared the effects of isotropy and mineral heterogeneity on fracture propagation paths (Figure 13). Mineral heterogeneity determines the initiation sequence of fractures and increases the roughness of fractures. Of course, more impact results need to be studied. Additionally, modeling methods for heterogeneous materials with complex microstructure can also be used [37].

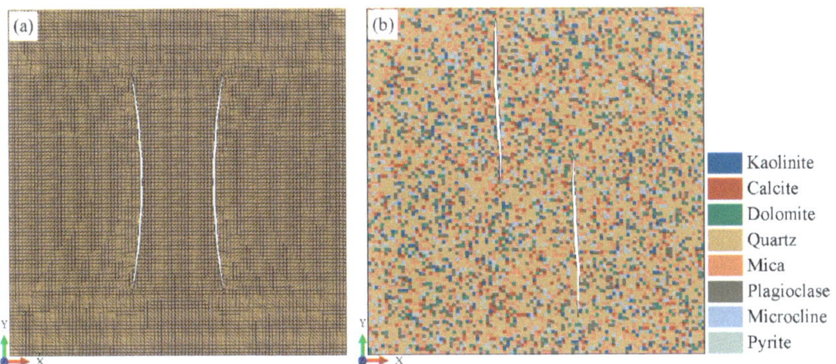


**Figure 13.** Comparison of fracture paths in two media. (**a**) Isotropic model. (**b**) Mineral heterogeneity model.

Meanwhile, natural fractures control the permeability of rocks [38,39]. Therefore, the natural fracture has become an indispensable part of hydraulic fracturing research. Based on this, a discrete fracture network (DFN) model was proposed to characterize the distribution of natural fractures [40]. A large number of scholars have researched the effect of natural fractures on hydraulic fracture propagation based on the DFN model and put forward three interaction modes [41,42]. The current research only considers the effects of natural fracturing or mineral heterogeneity on hydraulic fracturing, lacking a combination of both. Therefore, the next study will consider the hydraulic fracture propagation characteristics under a combination of mineral heterogeneity and natural fracture.

## 7. Conclusions

In this work, the variation process of stress shadow under different perforation spacing was analyzed by the XFEM. According to the stress distribution between two perforations, the compressive stress values at the center of two perforations were extracted. The change process of the stress shadow effect was quantitatively analyzed according to the stress value. In addition, the effect of the relationship between fracture length and perforation spacing on stress shadows was examined. The study reached the following conclusions:

1. The rock reservoir between the two perforations is always in a state of compressive stress. Additionally, the perforation spacing is small while the compressive stress value is large. The stress shadow effect can be quantitatively analyzed by using the compressive stress values at the two perforation centers.
2. The size of the stress shadow varies with the fracturing process. Moreover, the size is more likely to be maximized at smaller perforation spacing.
3. As the two fractures deviate from each other, fracture spacing increases continuously. When it enlarges to a particular value, the stress shadow effect begins to weaken until the in situ stress becomes the dominant factor again.
4. The stress shadow effect will be strongest if the fracture length is 2.5 times the length of the perforation spacing.
5. There is a perforation spacing that minimizes the initiation pressure, so this spacing can be used as a fracture design value. Furthermore, the positive effects of the stress shadow can be exploited by changing the fracturing sequence and controlling the perforation angle, thus increasing the fracture area and productivity.

**Author Contributions:** Conceptualization, Z.C.; Methodology, Z.Z.; Visualization, W.H.; Writing—original draft, W.H.; Writing—review and editing, W.H., Z.C. and Z.Z. All authors have read and agreed to the published version of the manuscript.

**Funding:** This work was funded by the Second Tibetan Plateau Scientific Expedition and Research Program (STEP) (Grant No. 2019QZKK0904), National Key Research and Development Project (Grant No. 2019YFC1509705), National Natural Science Foundation of China (Grant No. 41972296), Science and Technology Project of State Grid Cooperation of China (Grant No. SGSDJYOOSJJS1900042) and National Natural Science Foundation of China (Grant No. 51978424).



## References

1. Yuan, J.; Luo, D.; Feng, L. A review of the technical and economic evaluation techniques for shale gas development. *Appl. Energy* **2015**, *148*, 49–65. [CrossRef]
2. Middleton, R.S.; Gupta, R.; Hyman, J.D.; Viswanathan, H.S. The shale gas revolution: Barriers, sustainability, and emerging opportunities. *Appl. Energy* **2017**, *199*, 88–95. [CrossRef]
3. Roussel, N.P.; Sharma, M.M.J.S.P.; Operations. Optimizing fracture spacing and sequencing in horizontal-well fracturing. *SPE Prod. Oper.* **2011**, *26*, 173–184. [CrossRef]
4. Zhou, J.; Zeng, Y.; Jiang, T.; Zhang, B. Laboratory scale research on the impact of stress shadow and natural fractures on fracture geometry during horizontal multi-staged fracturing in shale. *Int. J. Rock Mech. Min. Sci.* **2018**, *107*, 282–287. [CrossRef]
5. Morrill, J.; Miskimins, J.L. Optimization of Hydraulic Fracture Spacing in Unconventional Shales. In Proceedings of the SPE Hydraulic Fracturing Technology Conference, The Woodlands, TX, USA, 1 January 2012; p. 12.
6. Liu, C.; Liu, H.; Zhang, Y.; Deng, D.; Wu, H. Optimal spacing of staged fracturing in horizontal shale-gas well. *J. Pet. Sci. Eng.* **2015**, *132*, 86–93. [CrossRef]
7. Yu, Y.; Zhu, W.; Li, L.; Wei, C.; Yan, B.; Li, S. Multi-fracture interactions during two-phase flow of oil and water in deformable tight sandstone oil reservoirs. *J. Rock Mech. Geotech. Eng.* **2020**, *12*, 821–849. [CrossRef]
8. Kresse, O.; Weng, X.; Gu, H.; Wu, R. Numerical Modeling of Hydraulic Fractures Interaction in Complex Naturally Fractured Formations. *Rock Mech. Rock Eng.* **2013**, *46*, 555–568. [CrossRef]

9.    Zeng, Q.; Yao, J. Numerical simulation of fracture network generation in naturally fractured reservoirs. *J. Nat. Gas Sci. Eng.* **2016**, *30*, 430–443. [CrossRef]

10.   Weng, X.; Kresse, O.; Chuprakov, D.; Cohen, C.-E.; Prioul, R.; Ganguly, U. Applying complex fracture model and integrated workflow in unconventional reservoirs. *J. Pet. Sci. Eng.* **2014**, *124*, 468–483. [CrossRef]

11.   Vahab, M.; Khalili, N. X-FEM Modeling of Multizone Hydraulic Fracturing Treatments Within Saturated Porous Media. *Rock Mech. Rock Eng.* **2018**, *51*, 3219–3239. [CrossRef]

12.   Khoei, A.R.; Vahab, M.; Hirmand, M. An enriched–FEM technique for numerical simulation of interacting discontinuities in naturally fractured porous media. *Comput. Methods Appl. Mech. Eng.* **2017**, *331*, 197–231. [CrossRef]

13.   Taghichian, A.; Hashemalhoseini, H.; Zaman, M.; Beheshti Zavareh, S. Propagation and aperture of staged hydraulic fractures in unconventional resources in toughness-dominated regimes. *J. Rock Mech. Geotech. Eng.* **2018**, *10*, 249–258. [CrossRef]

14.   Qi, M.; Li, M.; Li, Y.; Guo, T.; Gao, S. Numerical simulation of horizontal well network fracturing in glutenite reservoir. *Oil Gas Sci. Technol. Rev. d'IFP Energ. Nouv.* **2018**, *73*. [CrossRef]

15.   Luo, W.; Li, H.-T.; Wang, Y.-Q.; Wang, J.-C.; Zhu, S.-Y. A New Completion Methodology to Improve Oil Recovery for Horizontal Wells Completed in Highly Heterogeneous Reservoirs. *Arab. J. Sci. Eng.* **2014**, *39*, 9227–9237. [CrossRef]

16.   Ghaderi, S.M.; Clarkson, C.R.; Kaviani, D. Investigation of Primary Recovery in Low-Permeability Oil Formations: A Look at the Cardium Formation, Alberta (Canada). *Oil Gas Sci. Technol. Rev. d'IFP Energ. Nouv.* **2013**, *69*, 1155–1170. [CrossRef]

17.   Peirce, A.; Bunger, A. Interference Fracturing: Nonuniform Distributions of Perforation Clusters That Promote Simultaneous Growth of Multiple Hydraulic Fractures. *SPE-172500-PA* **2015**, *20*, 384–395. [CrossRef]

18.   Salimzadeh, S.; Usui, T.; Paluszny, A.; Zimmerman, R.W. Finite element simulations of interactions between multiple hydraulic fractures in a poroelastic rock. *Int. J. Rock Mech. Min. Sci.* **2017**, *99*, 9–20. [CrossRef]

19.   Belytschko, T.; Black, T. Elastic crack growth in finite elements with minimal remeshing. *Int. J. Numer. Methods Eng.* **1999**, *45*, 601–620. [CrossRef]

20.   Stolarska, M.; Chopp, D.L.; Moës, N.; Belytschko, T. Modelling crack growth by level sets in the extended finite element method. *Int. J. Numer. Methods Eng.* **2001**, *51*, 943–960. [CrossRef]

21.   Han, W.; Cui, Z.; Zhang, J. Fracture path interaction of two adjacent perforations subjected to different injection rate increments. *Comput. Geotech.* **2020**, *122*, 103500. [CrossRef]

22.   ABAQUS. Analysis Techniques, Modeling Abstractions. In *Abaqus Documentation*; Version 2017; ABAQUS: Paris, France, 2017.

23.   Wu, R.; Kresse, O.; Weng, X.; Cohen, C.-E.; Gu, H. Modeling of Interaction of Hydraulic Fractures in Complex Fracture Networks. In Proceedings of the SPE Hydraulic Fracturing Technology Conference, The Woodlands, TX, USA, 1 January 2012; p. 14.

24.   Zhang, Z.; Li, X.; He, J.; Wu, Y.; Li, G. Numerical study on the propagation of tensile and shear fracture network in naturally fractured shale reservoirs. *J. Nat. Gas Sci. Eng.* **2017**, *37*, 1–14. [CrossRef]

25.   Wu, K.; Olson, J.E. Simultaneous Multifracture Treatments: Fully Coupled Fluid Flow and Fracture Mechanics for Horizontal Wells. *SPE-172500-PA* **2015**, *20*, 337–346. [CrossRef]

26.   Nagel, N.B.; Sanchez-Nagel, M.A.; Zhang, F.; Garcia, X.; Lee, B. Coupled Numerical Evaluations of the Geomechanical Interactions Between a Hydraulic Fracture Stimulation and a Natural Fracture System in Shale Formations. *Rock Mech. Rock Eng.* **2013**, *46*, 581–609. [CrossRef]

27.   Crosby, D.G.; Rahman, M.M.; Rahman, M.K.; Rahman, S.S. Single and multiple transverse fracture initiation from horizontal wells. *J. Pet. Sci. Eng.* **2002**, *35*, 191–204. [CrossRef]

28.   Haimson, B.; Fairhurst, C. Initiation and extension of hydraulic fractures in rocks. *Soc. Pet. Eng. J.* **1967**, *7*, 310–318. [CrossRef]

29.   Huang, J.; Griffiths, D.V.; Wong, S.-W. Initiation pressure, location and orientation of hydraulic fracture. *Int. J. Rock Mech. Min. Sci.* **2012**, *49*, 59–67. [CrossRef]

30.   Yang, H.; Wang, R.; Zhou, W.; Li, L.; Chen, F. A study of influencing factors on fracture initiation pressure of cemented sliding sleeve fracturing. *J. Nat. Gas Sci. Eng.* **2014**, *18*, 219–226. [CrossRef]

31.   Cheng, Y.; Zhang, Y. Experimental study of fracture propagation: The application in energy mining. *Energies* **2020**, *13*, 1411. [CrossRef]

32.   Guo, Y.; Deng, P.; Yang, C.; Chang, X.; Wang, L.; Zhou, J. Experimental investigation on hydraulic fracture propagation of carbonate rocks under different fracturing fluids. *Energies* **2018**, *11*, 3502. [CrossRef]

33.   Yang, H.L.; Shen, R.C.; Fu, L. Composition and mechanical properties of gas shale. *Pet. Drill. Tech.* **2013**, *41*, 31–35.

34.   Han, W.; Cui, Z.; Tang, T.; Zhang, J.; Wang, Y. Effects of different bedding plane strength on crack propagation process under three points bending. *J. China Coal Soc.* **2019**, *44*, 3022–3030. (In Chinese)

35.   Janiszewski, M.; Shen, B.; Rinne, M. Simulation of the interactions between hydraulic and natural fractures using a fracture mechanics approach. *J. Rock Mech. Geotech. Eng.* **2019**, *11*, 1138–1150. [CrossRef]

36.   Cui, Z.; Han, W. In situ scanning electron microscope (SEM) observations of damage and crack growth of shale. *Microsc. Microanal.* **2018**, *24*, 107–115. [CrossRef] [PubMed]

37.   Meng, Q.-X.; Xu, W.-Y.; Wang, H.-L.; Zhuang, X.-Y.; Xie, W.-C.; Rabczuk, T. DigiSim—An Open Source Software Package for Heterogeneous Material Modeling Based on Digital Image Processing. *Adv. Eng. Softw.* **2020**, *148*, 102836. [CrossRef]

38.   Wang, L.; Liu, J.-F.; Pei, J.-L.; Xu, H.-N.; Bian, Y. Mechanical and permeability characteristics of rock under hydro-mechanical coupling conditions. *Environ. Earth Sci.* **2015**, *73*, 5987–5996. [CrossRef]

39. Li, B.; Bao, R.; Wang, Y.; Liu, R.; Zhao, C. Permeability Evolution of Two-Dimensional Fracture Networks During Shear Under Constant Normal Stiffness Boundary Conditions. *Rock Mech. Rock Eng.* **2021**, *54*, 409–428. [CrossRef]
40. Zhang, F.; Nagel, N.; Lee, B.; Sanchez-Nagel, M. The Influence of Fracture Network Connectivity on Hydraulic Fracture Effectiveness and Microseismicity Generation. In Proceedings of the 47th U.S. Rock Mechanics/Geomechanics Symposium, San Francisco, CA, USA, 1 January 2013; p. 13.
41. Zhang, F.; Dontsov, E.; Mack, M. Fully coupled simulation of a hydraulic fracture interacting with natural fractures with a hybrid discrete-continuum method. *Int. J. Numer. Anal. Methods Geomech.* **2017**, *41*, 1430–1452. [CrossRef]
42. Hou, B.; Chen, M.; Cheng, W.; Diao, C. Investigation of Hydraulic Fracture Networks in Shale Gas Reservoirs with Random Fractures. *Arab. J. Sci. Eng.* **2016**, *41*, 2681–2691. [CrossRef]

*Article*

# Experimental Investigation of the Dynamic Tensile Properties of Naturally Saturated Rocks Using the Coupled Static–Dynamic Flattened Brazilian Disc Method

**Xinying Liu, Feng Dai, Yi Liu *, Pengda Pei and Zelin Yan**

State Key Laboratory of Hydraulics and Mountain River Engineering, College of Water Resource and Hydropower, Sichuan University, Chengdu 610065, China; liuxinying1@stu.scu.edu.cn (X.L.); fengdai@scu.edu.cn (F.D.); 2019323060008@stu.scu.edu.cn (P.P.); yanzelin2@stu.scu.edu.cn (Z.Y.)
* Correspondence: liuyi616@scu.edu.cn; Tel.: +86-138-8203-2278

**Abstract:** In a naturally saturated state, rocks are likely to be in a stress field simultaneously containing static and dynamic loads. Since rocks are more vulnerable to tensile loads, it is significant to characterize the tensile properties of naturally saturated rocks under coupled static–dynamic loads. In this study, dynamic flattened Brazilian disc (FBD) tensile tests were conducted on naturally saturated sandstone under static pre-tension using a modified split-Hopkinson pressure bar (SHPB) device. Combining high-speed photographs with digital image correlation (DIC) technology, we can observe the variation of strain applied to specimens' surfaces, including the central crack initiation. The experimental results indicate that the dynamic tensile strength of naturally saturated specimens increases with an increase in loading rate, but with the pre-tension increases, the dynamic strength at a certain loading rate decreases accordingly. Moreover, the dynamic strength of naturally saturated sandstone is found to be lower than that of natural sandstone. The fracture behavior of naturally saturated and natural specimens is similar, and both exhibit obvious tensile cracks. The comprehensive micromechanism of water effects concerning the dynamic tensile behavior of rocks with static preload can be explained by the weakening effects of water on mechanical properties, the water wedging effect, and the Stefan effect.

**Keywords:** coupled static–dynamic loads; naturally saturated rocks; flattened Brazilian disc; fracturing behavior; dynamic tensile strength

**Citation:** Liu, X.; Dai, F.; Liu, Y.; Pei, P.; Yan, Z. Experimental Investigation of the Dynamic Tensile Properties of Naturally Saturated Rocks Using the Coupled Static–Dynamic Flattened Brazilian Disc Method. *Energies* **2021**, *14*, 4784. https://doi.org/10.3390/en14164784

Academic Editor: Joel Sarout

Received: 12 July 2021
Accepted: 4 August 2021
Published: 6 August 2021

**Publisher's Note:** MDPI stays neutral with regard to jurisdictional claims in published maps and institutional affiliations.

## 1. Introduction

Deep rocks are usually faced with complicated geoenvironments with multifield or multiscale coupling, such as thermomechanical coupling, groundwater disturbance, and static–dynamic load coupling [1–5]. Rocks in deep engineering are not only subjected to static loads caused by gravity stress and tectonic stress, but also affected by dynamic loads, generally induced by earthquakes, excavation, blasting, and drilling [6–9]. There is a general consensus that the water evidently weakens the mechanical properties of rocks, which may pose potential hazards to rock engineering [10–13]. Moreover, rocks are quite vulnerable to tensile loads, and their tensile strength is exceedingly low compared to their compressive strength [14,15]. Therefore, under simultaneous dynamic and static loads, characterizing the tensile behavior of naturally saturated rock is crucial for the stability assessment and construction safety of underground rock structures.

For brittle rocks, the object of existing studies mainly concerns the tensile properties under pure static or dynamic loadings, by conducting direct or indirect tensile tests [16–18]. For the direct tensile tests, the stress concentration always occurs around the gripping areas, causing partial damage to rock specimens. Moreover, it is difficult to keep the clamping device and rock specimen axis coincident, which induces bending moments in the specimen and causes measurement deviation. To avoid such fatal defects in direct tensile tests, some indirect methods have been proposed, and the Brazilian disc (BD) test is recommended by the International Society for Rock Mechanics (ISRM) to gauge the tensile strength of rocks [19–23]. For the BD testing method, the appearance of a central crack is a very eventful factor to validly calculate the tensile strength of rocks [24]. Since high stress concentration always appears around the contact area between the specimen's periphery and the loading plates in BD experiments, observation of the central crack's initiation cannot always be guaranteed [25]. To meet the prerequisite of central crack initiation in BD tests, researchers made some modifications to the loading devices, including using a curved loading jaw and adding cushions to the loading plates [17]. However, these modifications greatly complicate the BD tests. Alternatively, another convenient method is to change the shape of the BD specimen. To reasonably change the specimen's shape, Wang et al. [26] cut the two ends of the disc into a platform shape—namely, a flattened Brazilian disc (FBD) specimen—so as to reliably ensure the central crack's initiation. By selecting a proper loading angle ($2\alpha$) between $20°$ and $30°$, it can be well guaranteed to observe the central crack's initiation in the FBD testing method [27]. Subsequently, the FBD method was proposed to conduct dynamic tensile tests, and it has proven suitable to gauge the dynamic tensile strength of rocks via this indirect method [28,29].

Previous studies have revealed that for brittle materials, the capacity for resisting external force differs significantly between pure static or dynamic loads; that is, the dynamic strength is significantly greater than the static strength [7,8,25]. Additionally, a great deal of literature on rocks related to the tensile response devotes attention to static or dynamic tensile loading tests, with little attention paid to coupled dynamic–static tensile loading tests. Zhou et al. [30] explored the dynamic behavior of BD specimens under simultaneous dynamic and static loads, and pointed out that the dynamic tensile strength decreases with pre-static load. Wu et al. [31] applied coupled static–dynamic loads to BD specimens and revealed that the dynamic tensile strength goes up with the loading rate, and that the influence of pre-tension on total tensile strength can be negligible. Recently, Pei et al. [32] proposed a novel flattened Brazilian disc (FBD) testing method for rocks under simultaneous dynamic and static loads, measuring the tensile behavior of sandstone at a certain pre-static load and loading rate through experiments and numerical simulations.

The investigations reviewed above were focused on natural rocks, yet the tensile behavior of naturally saturated rocks under coupled dynamic–static loading has never been reported. Since underground water significantly affects the physical and mechanical properties of rocks, sufficient attention should be applied to the dynamic tensile response of saturated sandstone at various static pre-tensions [5,33,34]. In this study, the dynamic FBD test method was adopted for naturally saturated sandstone under static pre-compression via a modified SHPB system. Under conditions of simultaneous dynamic and static loading, the tensile properties and fracturing behavior of naturally saturated sandstone were investigated, and the comprehensive effects of water on the dynamic tensile properties of naturally saturated FBD sandstone are discussed. The structure of this paper is as follows: there is a brief introduction of the SHPB system, the test method for rocks under simultaneous dynamic and static loads, and the preparation of the naturally saturated FBD specimen in Section 2; Section 3 reports the experimental results, including dynamic force equilibrium, central crack initiation, and dynamic tensile strength; Section 4 discusses the micromechanism of water effects on the dynamic tensile properties and fracturing behavior of naturally saturated sandstone; the summary is presented in Section 5.

## 2. Methodology

### 2.1. Testing Apparatus

As exhibited in Figure 1, the coupled static–dynamic FBD tests were conducted using a modified SHPB device. The modified SHPB device was made of a system of bars—namely, the striker bar, the incident bar, and the transmitted bar—an axial pre-tension loading unit, and a data acquisition system. The density of the above low-alloy steel bars was $7800 \text{ kg/m}^3$, and all of the bars had a consistent diameter of 50 mm. Since the elastic modulus $E$ of the bars was 211 GPa, the longitudinal wave velocity was calculated by the formula ($c_0 = \sqrt{E/\rho}$)—namely, 5201 m/s. By utilizing small copper discs as shapers, the waveform of stress wave generated by the impact of striker bar can be shaped into a ramped wave, with the dynamic force equilibrium on both platforms of the FBD specimen being facilitated accordingly. As shown in Figures 1 and 2a, the FBD specimens were first clamped among the incident bar and the transmitted bar. Driven by a hydraulic chamber, the transmitted bar moved towards the incident end and was fixed by the flange; thus, the specific pre-load was successfully applied to the FBD specimens. The four tied rods were used to tighten the reaction plates to avoid bending moments in the bars. After the pre-static load was applied, the incident wave could be generated to propagate along the bars by the impact test. Due to the rapid reflection and transmission of the stress wave in the interface between the bars and the specimen, the stress waves in the transmitted bar were generated by the conversion of incident waves. With two sets of strain gauges properly mounted on the bars, in the form of a strain signal, the stress waves can be recorded without overlap. In line with the one-dimensional wave theory, as long as the strain history on bars is measured, the dynamic force history on both sides of specimens can be determined as follows [35]:

$$P_1(t) = AE(\varepsilon_0 + \varepsilon_i(t) + \varepsilon_r(t)) \tag{1}$$

$$P_2(t) = AE(\varepsilon_0 + \varepsilon_t(t)) \tag{2}$$

where $P_1(t)$ and $P_2(t)$ are the dynamic force at the incident and transmitted platforms of the specimen, respectively; $\varepsilon_0$, $\varepsilon_i(t)$, $\varepsilon_r(t)$, and $\varepsilon_t(t)$ are the strains caused by the pre-load, incident wave, reflected wave, and transmitted wave, respectively; and $A$ and $E$ denote the cross-sectional area and the elastic modulus of the bars, respectively.

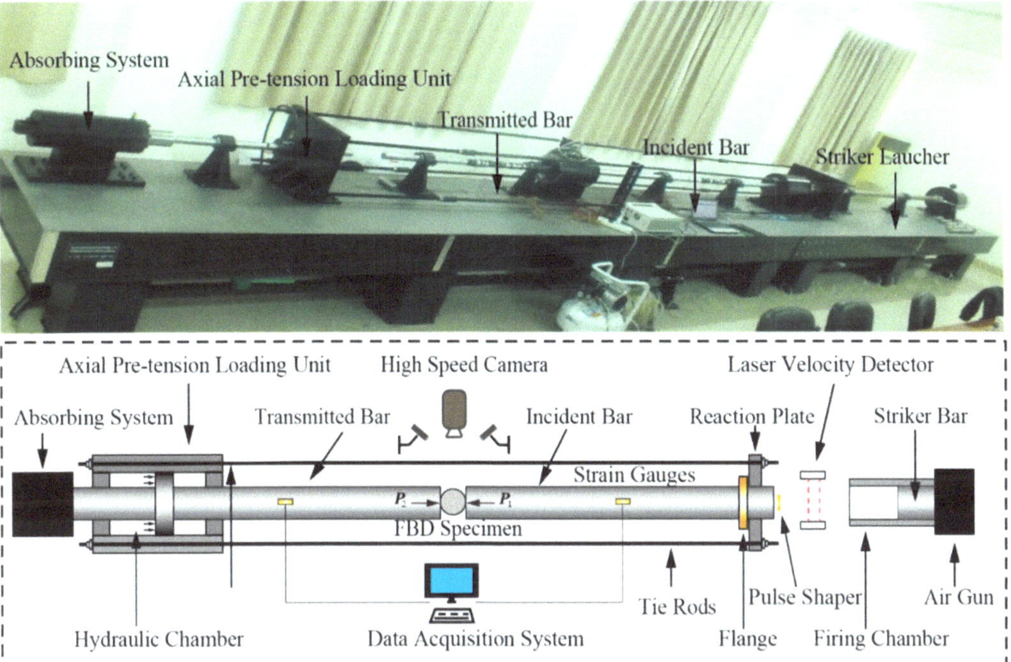


**Figure 1.** Modified split-Hopkinson pressure bar device.

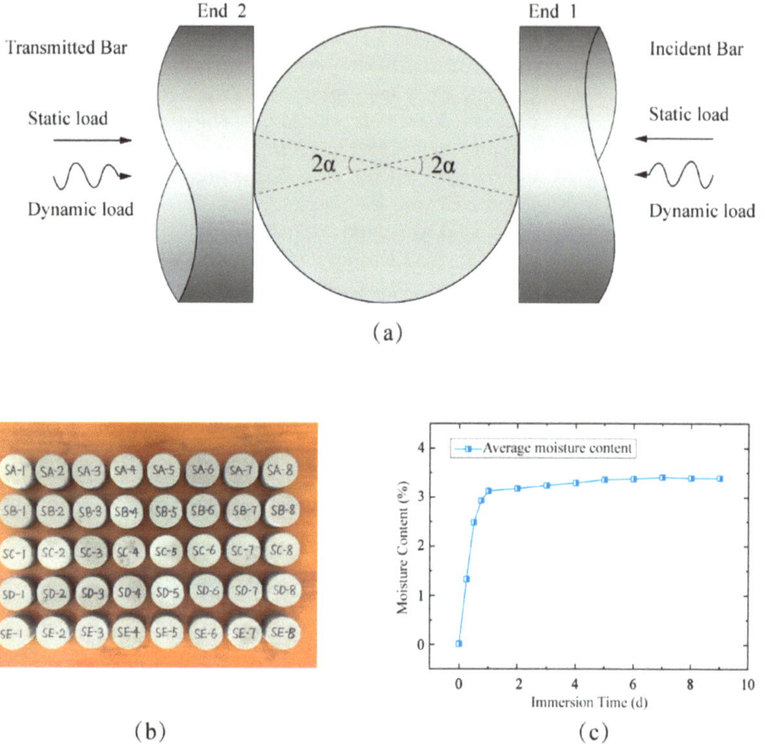



**Figure 2.** (**a**) Schematic of the FBD specimens under simultaneous dynamic and static loading. (**b**) Naturally saturated FBD specimens employed in our tests. (**c**) Curve of average moisture content in FBD specimens versus immersion time.

To capture the progressive fracture process during the SHPB experiment, we used digital image correlation (DIC) technology, which is based on the calculation of deformation from the reference state to subsequently deformed states on the surface of the specimen using digital images from a high-speed camera [36,37]. The high-speed camera was set at 180,000 frames per second, and provided successive digital images of specimens with a 256- $\times$ 256-pixel resolution. For the sake of improving the calculation accuracy of the DIC algorithm, random speckle patterns were painted on the facade of the specimens, which were used to calculate the strain field by tracking their movement.

### 2.2. The Coupled Static–Dynamic FBD Test Method

Considering the pivotal matter in question for the BD tests mentioned above, the coupled static–dynamic FBD testing method newly improved by Pei et al. [32] was applied. Under the condition of simultaneous dynamic and static loading, the advantages of using FBD specimens to measure tensile response are as follows: (1) due to the two-platform design of the disc, the stress concentration can be avoided by increasing the contact area between the FBD specimens and the force device; (2) the two parallel platforms in FBD specimens can ensure a consistent contact area throughout the entire loading process; (3) under dynamic impact loads, it can be ensured that the central crack's initiation is observed with the proper loading angle of $20° < 2\alpha < 30°$ [26,27]. In this study, to ensure that the stress in the disc center was larger than that of any other places in the FBD specimens, so as to ensure the feasibility of the FBD method, we chose a loading angle of $2\alpha = 25°$ when designing the FBD specimens [26].

Due to the existence of two platforms on the FBD specimens, it was no longer suitable to calculate tensile strength by using traditional BD calculation formulae. According to the Griffith criterion, the tensile strength of FBD specimens with certain loading angles can be determined by adding a coefficient to the traditional BD formula [32]:

$$\sigma_t = k\frac{2P}{\pi DB} \tag{3}$$

where $\sigma_t$ is the measured tensile strength; $P$ is the maximum load; $B$ and $D$ are the thickness and diameter of the FBD specimens, respectively; and $k$ is the coefficient with a certain loading angle, in which the coefficient $k$ equals 0.9445 for a loading angle $2\alpha = 25°$ [32]. Therefore, the dynamic tensile strength in our tests can be determined as follows:

$$\sigma_t = 0.9445\frac{2P}{\pi DB} \tag{4}$$

### 2.3. Specimen Preparation and Test Scheme

In FBD tests, in order to reduce the dispersion of the test results, all of the FBD specimens were taken from a same sandstone block, which was fine-grained homogeneous sandstone taken from Neijiang, Sichuan Province, China. The FBD specimens had a consistent diameter and thickness of 87 mm and 43.5 mm, respectively. The surface roughness of the specimens was maintained within 0.02 mm, and the non-parallelism of both platforms was less than 0.05 mm. The naturally saturated specimen was prepared via the immersion tests [38–40], and all of the prepared FBD specimens are shown in Figure 2b. These FBD specimens were first immersed in pure water for 6 h, and then taken out and weighed after the surfaces of the specimens were wiped off with a towel. The above operations were repeated until the quality difference of the saturated specimens between the last two cycles did not exceed 0.1 g. Rock specimens after final immersion treatment were considered to have reached the naturally saturated state, and the last weighing was considered to be the quality of the naturally saturated rock FBD specimen. The curve of average moisture content in sandstone versus immersion time is shown in Figure 2c, in which the water content of sandstone can be determined as follows:

$$w_s = \frac{m_s - m_d}{m_d} \tag{5}$$

where $w_s$ is the naturally saturated water content; and $m_s$ and md are the mass of the FBD specimen in its naturally saturated state and natural state, respectively.

As revealed in Figure 2c, it takes almost 9 days for the natural sandstone specimen to reach a naturally saturated state. The average density and longitudinal wave speed of the naturally saturated FBD specimens were 2387 kg/m$^3$ and 2570 m/s, respectively.

In this study, the quasistatic loading tests were first carried out for naturally saturated FBD specimens to determine the static tensile strength, using the MTS-815 rock-testing system to provide a reference for selecting appropriate pre-static loads. The typical load–displacement curve is shown in Figure 3a. The tensile strength is determined by the stress $P_{\max}$ of the curve, and just at this time the first crack appears in the central disc, as shown in Figure 3b. Table 1 lists the quasistatic test results. The average static tensile strength of the naturally saturated sandstone was 5.5 MPa. In order to explore the influence of static preloads on the dynamic tensile properties of naturally saturated sandstone, five groups of tests are considered (i.e., groups SA, SB, SC, SD, and SE). For the five test groups, the static pre-tension loads were 0 MPa, 1.1 MPa, 2.2 MPa, 3.3 MPa and 4.4 MPa, corresponding to the static pre-tension ratios of 0, 0.2, 0.4, 0.6 and 0.8, respectively. For the dynamic loading under each static preload, eight impacting tests were conducted to cover a wide range of loading rates (50~350 GPa/s).

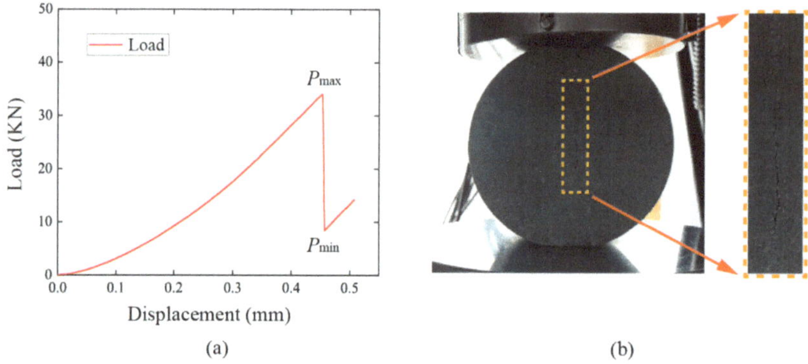

**Figure 3.** Splitting tests on naturally saturated FBD specimens under quasistatic loads: (**a**) load–displacement curve; (**b**) fracture behavior at disc center.

**Table 1.** Quasistatic test results of naturally saturated sandstone specimens.

| Specimen No. | Peak Force (KN) | Tensile Strength (MPa) |
|:---:|:---:|:---:|
| S-1 | 35.41 | 5.62 |
| S-2 | 34.01 | 5.40 |
| S-3 | 36.49 | 5.79 |
| S-4 | 33.54 | 5.32 |
| S-5 | 35.05 | 5.56 |
| Average | 34.9 | 5.5 |

## 3. Experimental Results

### 3.1. Dynamic Force Equilibrium

Given the fact that ensuring balance of the dynamic forces at both ends of the FBD specimen is the crucial prerequisite for ignoring the inertial effect and conducting the valid quasistatic data analysis method in the experiments [35,41], we firstly checked the force equilibrium in the dynamic impact tests by adapting a pulse-shaping technique, which can convert incident waves into ramped waveforms.

Figure 4a depicts the force history for a typical coupled static–dynamic FBD test (specimen SC-1). By shifting the time zeros of stress waves to the bar–specimen interface, and using the abovementioned Equations (1) and (2), the dynamic force histories on both side of the specimen can be determined, which are denoted as black curve In and red curve Re for incident force history and reflected force history, respectively, in Figure 4a. The dynamic force history on the incident platform of the specimen is the superposition of the incident and reflected force (represented as blue curve In + Re), while the force history on the transmitted side of the specimen is represented as green curve Tr. For rock materials, after 3–4 stress wave reverberations inside the specimen, the force on both sides of the specimen becomes consistent [42]. Since the P-wave velocity of naturally saturated sandstone is ~2570 m/s and the diameter of FBD is 87 mm, the time required for dynamic force equilibrium is ~102–135 μs. As shown in Figure 4a, during the time from the initial equilibrium moment (represented as $t_{equil}$) to the peak stress moment (represented as $t_f$)—namely, 105–211 μs—the dynamic force at both platforms of the FBD specimen is approximately equal, indicating that the dynamic force equilibrium of the FBD specimen is achieved in our tests. From Equation (4), the maximum stress in the tensile stress history can be adopted to measure the tensile strength. Moreover, as shown in Figure 4b, the slope of the pre-peak linear region of the stress history curve can be used to determine the loading rate.

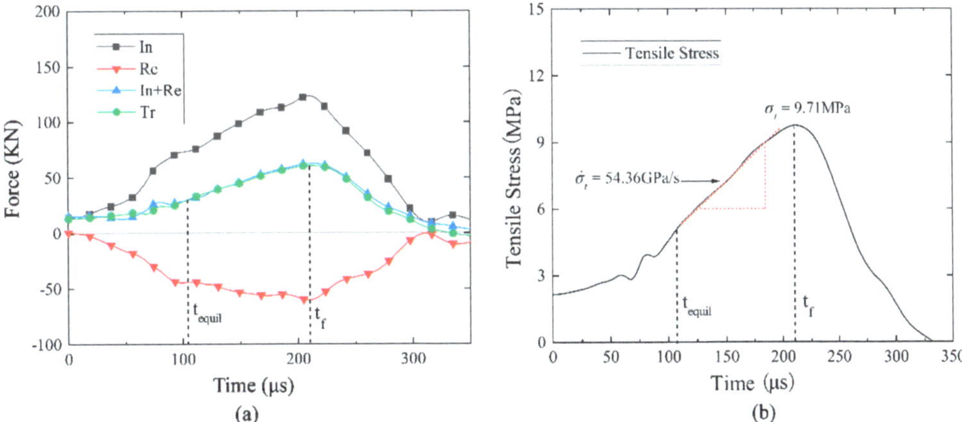

**Figure 4.** (**a**) Force histories of the bars for specimen SC-1. (**b**) Tensile stress history of specimen SC-1 under coupled static–dynamic loading.

### 3.2. Central Crack Initiation

As mentioned above, the central crack's initiation should be carefully checked before the interpretation of the results. Via virtual high-speed photography, the progressive failure processes of FBD specimens under the condition of coupled static–dynamic loading were completely recorded. To measure the progressive evolution of strain applied to the surface of the specimen, high-speed photographs were analyzed using DIC technology. Figure 5 depicts the typical progressive fracturing processes of specimen SC-1, and the interval of the legend; that is, the strain range is $(-10 \sim 30) \times 10^{-3}$. The first frame (at 0 μs) depicts the initial state of the FBD specimen when the incident ramped wave first appears on the interface of the incident bar and the specimen. At this moment, no obvious strain and macroscopic cracks can be observed on the specimen surface. As loading continues, a tensile crack first appears at the disc center, and a corresponding strain concentration band appears. Note that the tensile stress history also reaches the maximum stress at the same time (214.5 μs). Subsequently, as the strain concentration bands form and expand, the central tensile crack horizontally propagates along the loading direction, and secondary cracks also initiate from the edges of the platform during the post-peak stage. Eventually, two strain concentration banks in the shape of an "X" are formed (275 μs), leading to the final failure of the FBD specimen. For the whole failure process, the initiation of the central crack was watched in the FBD specimen, and thus the important factor for a valid method was guaranteed in our tests.

### 3.3. Influence of Loading Rate on the Dynamic Tensile Strength

By subtracting the pre-static tensile stress from the total tensile strength, the experimental results—namely, dynamic tensile strength—are shown in Table 2. Figure 6a–e exhibits the curves of dynamic tensile strength versus loading rate for the naturally saturated FBD specimens with distinct pre-tension ratios of 0, 0.2, 0.4, 0.6, and 0.8, respectively. Without considering the pre-static load, the dynamic tensile strength increases with the loading rate. Taking the pre-tension ratio of 0.4 as a typical case, with the loading rate increases from 54.36 GPa/s to 330.01 GPa/s, the dynamic tensile strength increases from 7.51 MPa to 17.47 MPa, showing a significant loading rate dependency.

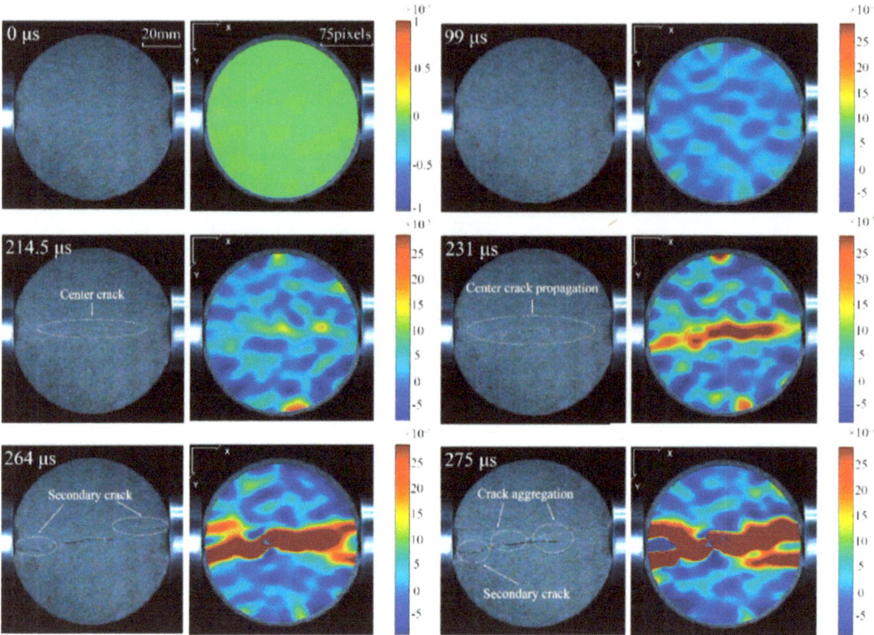

**Figure 5.** Progressive fracture process of specimen SC-1 under simultaneous dynamic and static loading.

**Table 2.** Coupled static–dynamic FBD test results of naturally saturated sandstone specimens.

| Pre-Tension Ratio | Specimen No. | Loading Rate (GPa/s) | Dynamic Tensile Strength (MPa) |
|---|---|---|---|
| | SA-1 | 56.57 | 8.45 |
| | SA-2 | 73.93 | 10.83 |
| | SA-3 | 95.23 | 11.59 |
| 0.0 | SA-4 | 241.27 | 16.29 |
| | SA-5 | 138.96 | 13.67 |
| | SA-6 | 187.53 | 14.24 |
| | SA-7 | 210.43 | 15.73 |
| | SA-8 | 274.03 | 17.22 |
| | SB-1 | 54.97 | 7.90 |
| | SB-2 | 81.80 | 11.14 |
| | SB-3 | 141.80 | 13.63 |
| 0.2 | SB-4 | 202.47 | 15.51 |
| | SB-5 | 207.32 | 15.00 |
| | SB-6 | 249.47 | 16.52 |
| | SB-7 | 342.58 | 18.73 |
| | SB-8 | 318.16 | 17.95 |
| | SC-1 | 54.36 | 7.51 |
| | SC-2 | 102.41 | 11.06 |
| | SC-3 | 154.50 | 13.53 |
| 0.4 | SC-4 | 176.01 | 14.26 |
| | SC-5 | 128.61 | 12.72 |
| | SC-6 | 283.03 | 16.56 |
| | SC-7 | 330.01 | 17.47 |
| | SC-8 | 251.10 | 15.81 |

**Table 2.** *Cont.*

| Pre-Tension Ratio | Specimen No. | Loading Rate (GPa/s) | Dynamic Tensile Strength (MPa) |
|---|---|---|---|
| 0.6 | SD-1 | 54.02 | 7.05 |
| | SD-2 | 67.31 | 9.44 |
| | SD-3 | 124.20 | 11.79 |
| | SD-4 | 208.41 | 14.37 |
| | SD-5 | 210.74 | 14.83 |
| | SD-6 | 177.22 | 14.13 |
| | SD-7 | 300.60 | 16.54 |
| | SD-8 | 258.68 | 15.61 |
| 0.8 | SE-1 | 58.37 | 6.73 |
| | SE-2 | 109.05 | 10.40 |
| | SE-3 | 157.40 | 12.56 |
| | SE-4 | 211.20 | 14.25 |
| | SE-5 | 226.02 | 14.46 |
| | SE-6 | 198.81 | 13.34 |
| | SE-7 | 333.09 | 16.31 |
| | SE-8 | 283.24 | 15.39 |

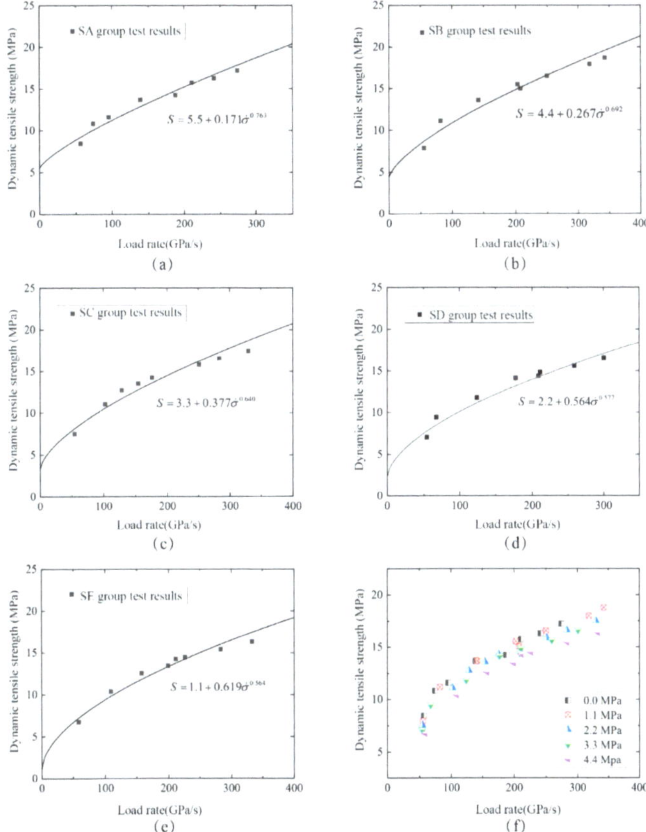


**Figure 6.** The variation of dynamic tensile strength versus loading rate at a specified static pre-tension: (**a**) 0 MPa for group SA; (**b**) 1.1 MPa for group SB; (**c**) 2.2 MPa for group SC; (**d**) 3.3 MPa for group SD; and (**e**) 4.4 MPa for group SE. Panel (**f**) shows the relationship between dynamic tensile strength and loading rate at certain pre-tension ratios.

To fit the empirical function of the dynamic tensile strength and the loading rate at a certain static pre-tension ratio, the exponential function ($S = \sigma_r + A\dot{\sigma}^B$) was used to fit the experimental results [32]:

$$\begin{cases} S = 5.5 + 0.171\dot{\sigma}^{0.763}, & R^2 = 0.975 \ \left(\sigma_{pre}/\sigma_{st} = 0\right) \\ S = 4.4 + 0.267\dot{\sigma}^{0.693}, & R^2 = 0.975 \ \left(\sigma_{pre}/\sigma_{st} = 0.2\right) \\ S = 3.3 + 0.377\dot{\sigma}^{0.640}, & R^2 = 0.971 \ \left(\sigma_{pre}/\sigma_{st} = 0.4\right) \\ S = 2.2 + 0.564\dot{\sigma}^{0.577}, & R^2 = 0.983 \ \left(\sigma_{pre}/\sigma_{st} = 0.6\right) \\ S = 1.1 + 0.619\dot{\sigma}^{0.564}, & R^2 = 0.984 \ \left(\sigma_{pre}/\sigma_{st} = 0.8\right) \end{cases} \quad (6)$$

where $S$, $\sigma_r$, and $\sigma_{st}$ are the dynamic tensile strength, residual tensile strength, and static tensile strength of naturally saturated FBD specimens, respectively; $\dot{\sigma}$ is the dynamic loading rate; $A$ and $B$ are the correlation coefficients varied with pre-tension ratios; and $\sigma_{pre}$ is the pre-tension stress. Using Equation (6), the dynamic tensile strength of naturally saturated sandstone under simultaneous dynamic and static loading can be predicted.

### 3.4. Influence of Pre-Tension on the Dynamic Tensile Strength

As revealed in Figure 6f, dynamic tensile strength not only displays rate dependency, but is also affected by the pre-static load. To demonstrate, the dynamic tensile strength and pre-tension ratio curves of specimen SC-1 at typical loading rates are exhibited in Figure 7. It can be observed that as the static pre-tension ratio increases, the dynamic strength at a specific loading rate consistently decreases accordingly. For instance, under a typical loading rate of 200 GPa/s (green curve in Figure 7), as the pre-tension ratio increases from 0 to 0.8 at an interval of 0.2, the strength reductions of the naturally saturated specimen are 0.34 MPa, 0.40 MPa, 0.30 MPa, and 0.81 MPa, respectively. Note that there are evident turning points at the pre-tension ratio of 0.6. The test results can perhaps be interpreted by the opening and generation of microcracks caused by static pre-tension [32]. In the wake of the increase of the pre-static load, more closed cracks open and more microcracks accumulate in the naturally saturated specimen, which decreases the dynamic tensile strength. When the pre-tension ratio exceeds 0.6, the microcracking begins to accelerate, inducing further reduction of the dynamic tensile properties.

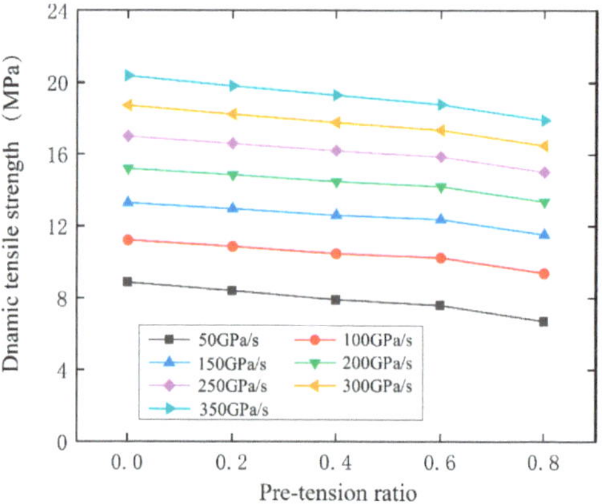


**Figure 7.** Dynamic tensile strength versus pre-tension ratio curve of naturally saturated FBD specimens under different loading rates.

## 4. Discussion

### 4.1. Comparison of the Dynamic Tensile Strength between Natural and Naturally Saturated Sandstone

To compare the dynamic tensile strength at natural and naturally saturated states, the experimental data of coupled static–dynamic FBD tests from Pei et al. [32] are cited, and the interpolation is applied to obtain the dynamic tensile strength in the loading rate range of 50–350 GPa/s with various pre-tensions (i.e., 1.1 MPa, 2.2 MPa, 3.3 MPa, and 4.4 MPa). As shown in Figure 8a,b, rocks in both natural and naturally saturated states, at various loading rates and pre-tension ratios, show a similar tendency for the dynamic tensile strength—namely, it increases with increasing loading rate, while decreasing with an increase in the pre-tension ratio. Figure 9a–e illustrates the variation of the dynamic tensile strength in natural and naturally saturated sandstone versus loading rate. Moreover, we can find that the dynamic tensile strength of naturally saturated sandstone is lower than that of natural sandstone in the event of a similar loading rate, indicating the overall degradation in the mechanical properties of the naturally saturated sandstone. To quantitatively analyze the rate effect of tensile strength, the rate effect difference is presented as the derived difference in the tensile strength fitting formula concerning the loading rate for natural and naturally saturated rocks, as exhibited in Table 3, in which the tensile strength-fitting formulae of natural and naturally saturated rocks are taken from Pei et al. [32] and Equation (6), respectively. The upper and lower limits of rate effect difference are calculated as follows:

$$
\begin{cases}
\left| \Delta \dfrac{\partial S}{\partial \dot{\sigma}_t} \right|_{\text{upper}} = \left| \dfrac{\partial S_{\text{sat}}}{\partial \dot{\sigma}_t} - \dfrac{\partial S_{\text{nat}}}{\partial \dot{\sigma}_t} \right|_{\max} \\[2ex]
\left| \Delta \dfrac{\partial S}{\partial \dot{\sigma}_t} \right|_{\text{lower}} = \left| \dfrac{\partial S_{\text{sat}}}{\partial \dot{\sigma}_t} - \dfrac{\partial S_{\text{nat}}}{\partial \dot{\sigma}_t} \right|_{\min}
\end{cases}
\tag{7}
$$

where $\left| \Delta \dfrac{\partial S}{\partial \dot{\sigma}_t} \right|_{\text{upper}}$ and $\left| \Delta \dfrac{\partial S}{\partial \dot{\sigma}_t} \right|_{\text{lower}}$ are the upper and lower limits of rate effect difference, respectively; and $\dfrac{\partial S_{\text{sat}}}{\partial \dot{\sigma}_t}$ and $\dfrac{\partial S_{\text{nat}}}{\partial \dot{\sigma}_t}$ are the derivatives of the tensile strength fitting formula concerning the loading rate for natural and naturally saturated rocks, respectively—namely, the rate effect for natural and naturally saturated rocks.

**Table 3.** Rate effect difference of natural and naturally saturated sandstone specimens.

| Pre-Tension Ratio | Upper Limit of Rate Effect Difference (%) | Lower Limit of Rate Effect Difference (%) |
|---|---|---|
| 0.0 | 7.40 | 0.78 |
| 0.2 | 9.09 | 1.51 |
| 0.4 | 8.36 | 0.39 |
| 0.6 | 7.99 | 0.75 |
| 0.8 | 17.14 | 2.25 |

According to Figure 8c, it can be found that when the pre-tension ratio is less than 0.6, the upper and lower limits of rate effect difference are relatively stable. However, when the pre-tension ratio reaches 0.8, the upper and lower limits of rate effect difference increase sharply, from 7.99% to 17.14% and 0.75% to 2.25%, respectively. The above phenomenon reveals that the presence of pore water can influence the rate effect of rock to some extent, and that the opening and generation of microcracks caused by a pre-tension ratio of 0.8 increases the influence of pore water on the rate effect of the tensile strength of rocks.

Combining high-speed photographs with DIC technology, the progressive fracture process of specimen NC-1 (from Pei et al. [32]) is shown in Figure 10, which shares a similar loading rate and pre-tension rate with specimen SC-1. The range of the legend in Figure 10—that is, the strain range—is $(-10\text{~}30) \times 10^{-3}$. At 258.5 μs, the specimen cracks first at the center of the disc, and an evident strain concentration bank occurs. As loading continues, secondary cracks appear on the edges of the two flat ends. Subsequently, with the propagation and coalescence of cracks, two strain concentration banks in the shape of an "X" are gradually formed, leading to the final failure. Compared with Figure 5, it can be observed that the progressive fracture processes of naturally saturated and natural specimens are similar; they both display central crack initiation and the occurrence of secondary cracks at the ends of the specimens. Finally, both specimens fail accompanied by two strain concentration banks in the shape of an "X".

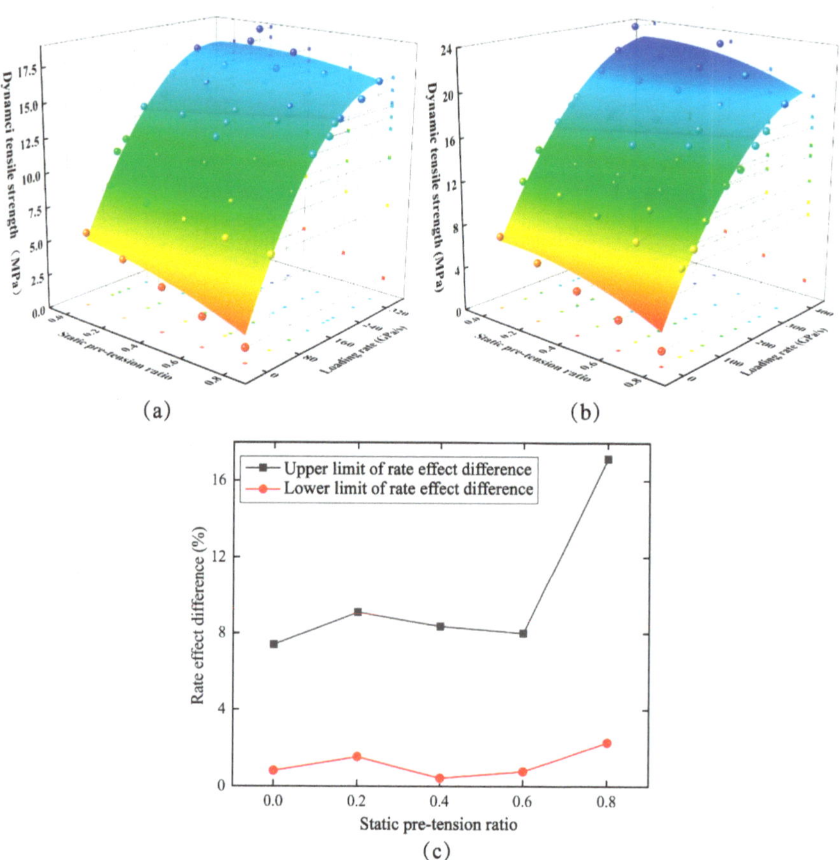

**Figure 8.** (**a**) The diversification of dynamic tensile strength under various external load conditions for natural rocks. (**b**) The diversification of dynamic tensile strength under various external load conditions for naturally saturated rocks. (**c**) The curve of the upper and lower limits of the rate effect difference of natural and naturally saturated rocks versus static pre-tension ratio.

### 4.2. Micromechanism of Dynamic Tensile Strength Reduction Induced by Water Effects

The weakening effects of water have been widely researched and recognized with respect to the physical and mechanical properties of rock materials [9–13]. The softening coefficient ($K$) was introduced to evaluate the weakening effect of water on the dynamic tensile strength of sandstone under various external forces, while $k$ is delimited as the specific value of the tensile strength of the naturally saturated sandstone relative to that of the natural sandstone:

$$K = \frac{f}{F} \tag{8}$$

where $f$ and $F$ represent the tensile strength of sandstone in naturally saturated and natural states, respectively. Thus, the softening coefficient $K_0$ of static tensile strength can be calculated—namely, $K_0 = 0.803$. This can be explained by the fact that the hydrophilic clay minerals in rocks are partially dissolved in the pore water, resulting in reduced cementation strength between the mineral particles, while the friction coefficient between mineral particles will be reduced after saturation, inducing a reduction in the load-bearing capacity.

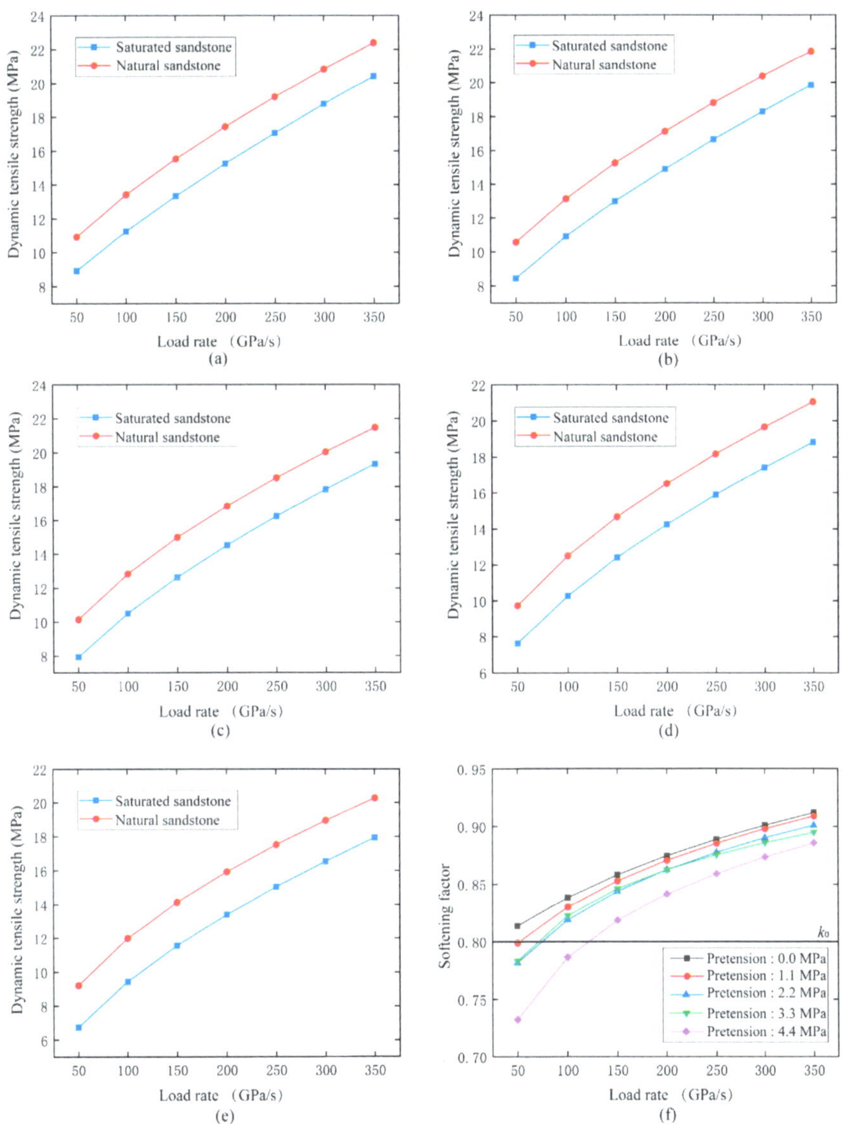


**Figure 9.** The variation in dynamic tensile strength versus loading rate for naturally saturated and natural sandstone under a specified static pre-tension: (**a**) 0 MPa, (**b**) 1.1 MPa, (**c**) 2.2 MPa, (**d**) 3.3 MPa for, and (**e**) 4.4 MPa. (**f**) The relationship between softening factor and loading rate under different pre-tensions.

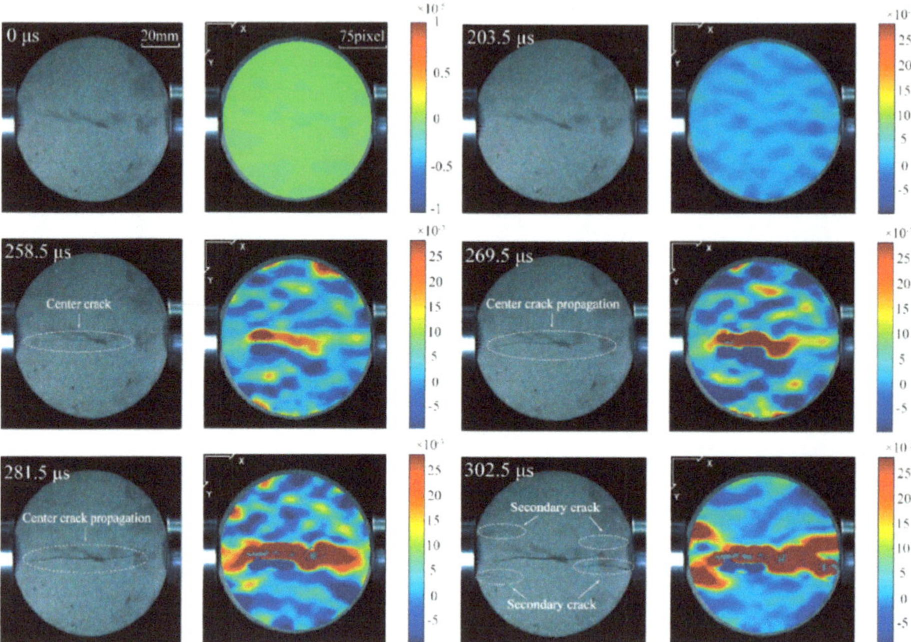

**Figure 10.** Progressive fracture process of a dry FBD specimen under coupled static–dynamic loading.

Figure 9f shows the variation of the softening coefficient versus the loading rate under different pre-tensions. It can be found that, under lower loading rates, the softening coefficient of naturally saturated rocks under pre-static loads is less than the static softening coefficient $K_0$. Moreover, the larger the pre-static load, the smaller the softening coefficient, which indicates that the pre-static load can weaken the dynamic tensile strength of naturally saturated rocks. This can be interpreted, as regards the weakening effect of water on pre-static load, as follows: Firstly, the pre-static load causes compressive deformation of FBD specimens, resulting in the generation of pore water pressure and the migration of pore water to the crack tip in the saturated pores. Moreover, the wedging effect in the saturated pores [34,42], which is induced by pore water under external loads, further promotes the propagation of microcracks, leading to a reduction in tensile strength, as shown in Figure 11a. In addition, when the pre-tension ratio is 0.6 and the loading rate is 54 GPa/s, the time of central crack initiation in naturally saturated rocks and natural rocks is 214.5 µs (Figure 5) and 258.5 µs (Figure 10), respectively, which proves that the pre-static load can weaken the dynamic tensile properties of naturally saturated rocks.

It can be clearly seen from Figure 9f that the softening coefficient gradually increases with the increase in loading rate from—that is, the weakening effects induced by water decrease with loading rate—and the Stefan effect can be used to explained this phenomenon. As shown in Figure 11c, when two plates with a radius of $r$ and a distance of $h$ are separated at a speed $\dot{h}$, the viscous fluid between the plates will generate a reaction force Ps to prevent the two plates from separating. According to the Stefan effect, for saturated rocks, the pore water is considered to be the viscous fluid, while the mineral particles on both sides of the pores are equivalent to the two plates, as shown in Figure 11b,c [43–45]. When naturally saturated rocks are subjected to dynamic loads, much like saturated rocks, the mineral particles bear tensile stress and tend to separate from one another. The higher the loading rate, the faster the separation speed of mineral particles, resulting in a larger reaction force to restrain the separation of the mineral particles. Therefore, the softening coefficient decreases with loading rate.

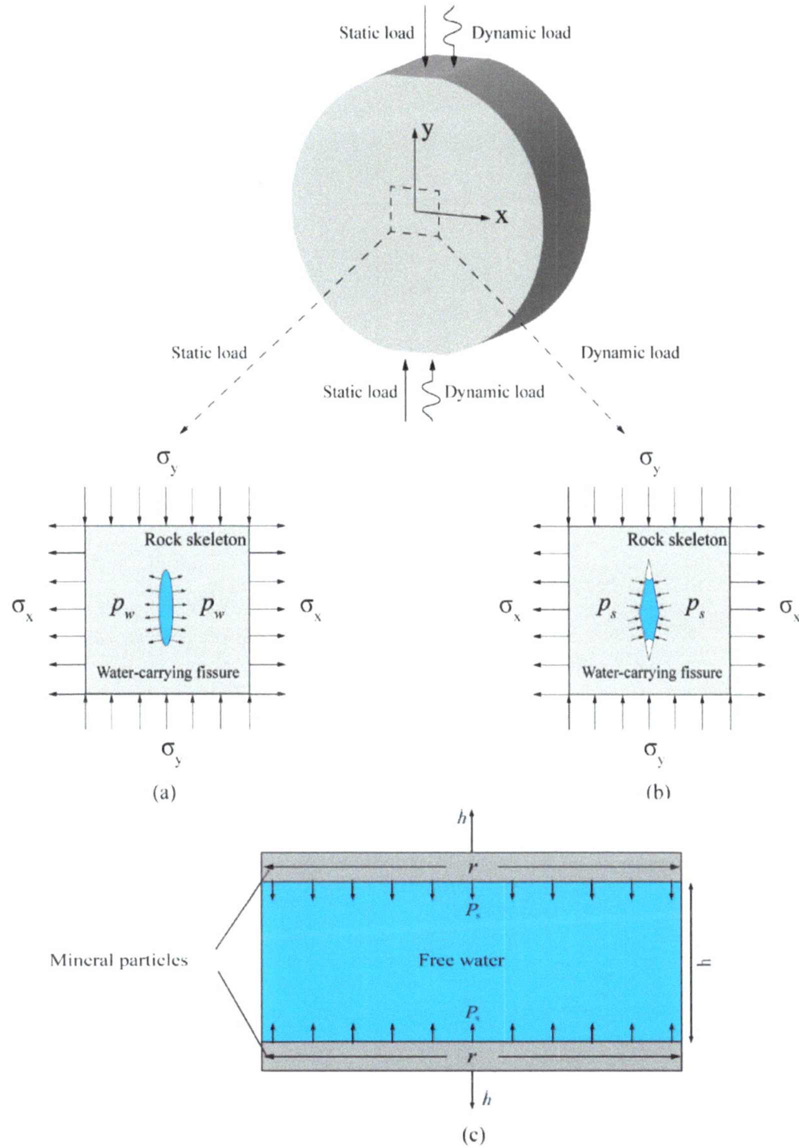



**Figure 11.** Pore water effect on rock under coupled static–dynamic loading: (**a**) pore water pressure under pre-static load promotes crack growth; (**b**) Stefan effect hinders crack growth under dynamic load; (**c**) a schematic diagram of the Stefan effect.

## 5. Conclusions

In this study, coupled static–dynamic FBD tests were conducted to explore the tensile behavior of naturally saturated sandstone using a modified SHPB device. Five groups of specimens with pre-tension ratios of 0, 0.2, 0.4. 0.6, and 0.8 were tested under a wide loading rate range of 50–300 GPa/s. By careful use of a pulse-shaping technique, the dynamic force equilibrium of the FBD specimens was achieved, and the initiation of the central crack was also observed using high-speed photography and DIC technology. Our experimental results comprehensively revealed the influences of loading rate and pre-static load on the tensile failure mechanism of naturally saturated sandstone. The dynamic tensile strength of naturally saturated sandstone increases with an increase in loading rate at a certain pre-static load, showing an obvious loading rate dependency. Under similar loading rate conditions, the dynamic tensile strength of naturally saturated sandstone decreases with increasing pre-static load, indicating that the pre-tension weakens the dynamic tensile strength of naturally saturated sandstone. Compared with the natural sandstone specimens, the tensile strength of naturally saturated sandstone specimens was lower at a certain pre-tension ratios and loading rates, while both specimens displayed a similar tensile failure process. The comprehensive micromechanism of water's effects on the dynamic tensile failure of naturally saturated sandstone was discussed, and explained by the weakening effects of water on mechanical properties, the water-wedging effect, and the Stefan effect.

**Author Contributions:** Conceptualization, X.L. and P.P.; methodology, X.L.; software, Y.L.; validation, F.D. and Y.L.; formal analysis, X.L.; investigation, P.P.; resources, F.D.; data curation, P.P.; writing—original draft preparation, X.L.; writing—review and editing, Z.Y.; visualization, Y.L.; supervision, F.D.; project administration, F.D.; funding acquisition, F.D. All authors have read and agreed to the published version of the manuscript.

**Funding:** This research was funded by the National Natural Science Foundation of China, grant number 52039007 and 51779164.

**Data Availability Statement:** All data included in this study are available upon request from the corresponding author.

**Acknowledgments:** The authors are grateful for the financial support from the National Natural Science Foundation of China (No. 52039007 and No. 51779164).

**Conflicts of Interest:** The authors declare no conflict of interest.

# References

1. Xiao, P.; Zheng, J.; Dou, B.; Tian, H.; Cui, G.; Kashif, M. Mechanical behaviors of granite after thermal shock with different cooling rates. *Energies* **2021**, *14*, 3721. [CrossRef]
2. Du, H.; Dai, F.; Xu, Y.; Yan, Z.; Wei, M. Mechanical Responses and failure mechanism of hydrostatically pressurized rocks under combined compression-shear impacting. *Int. J. Mech. Sci.* **2020**, *165*, 105219. [CrossRef]
3. Xiong, F.; Zhu, C.; Jiang, Q. A novel procedure for coupled simulation of thermal and fluid flow models for rough-walled rock fractures. *Energies* **2021**, *14*, 951. [CrossRef]
4. Liu, Y.; Dai, F.; Fan, P.; Xu, N.; Dong, L. Experimental investigation of the influence of joint geometric configurations on the mechanical properties of intermittent jointed rock models under cyclic uniaxial compression. *Rock Mech. Rock Eng.* **2017**, *50*, 1453–1471. [CrossRef]
5. Wong, L.N.Y.; Maruvanchery, V.; Liu, G. Water effects on rock strength and stiffness degradation. *Acta Geotech.* **2016**, *11*, 713–737. [CrossRef]
6. Jiang, R.; Dai, F.; Liu, Y.; Li, A. Fast marching method for microseismic source location in cavern-containing rockmass: Performance analysis and engineering application. *Engineering* **2021**, *7*, 1023–1034. [CrossRef]
7. Feng, P.; Dai, F.; Liu, Y.; Xu, N.; Du, H. Coupled effects of static-dynamic strain rates on the mechanical and fracturing behaviors of rock-like specimens containing two unparallel fissures. *Eng. Fract. Mech.* **2019**, *207*, 237–253. [CrossRef]
8. Liu, Y.; Dai, F. A damage constitutive model for intermittent jointed rocks under cyclic uniaxial compression. *Int. J. Rock Mech. Min.* **2018**, *103*, 289–301. [CrossRef]
9. Liu, Y.; Dai, F.; Feng, P.; Xu, N. Mechanical behavior of intermittent jointed rocks under random cyclic compression with different loading parameters. *Soil Dyn. Earthq. Eng.* **2018**, *113*, 12–24. [CrossRef]
10. Dang, W.; Wu, W.; Konietzky, H.; Qian, J. Effect of shear-induced aperture evolution on fluid flow in rock fractures. *Comput. Geotech.* **2019**, *114*, 103152. [CrossRef]
11. Zhao, Z.; Yang, J.; Zhang, D.; Peng, H. Effects of wetting and cyclic wetting–drying on tensile strength of sandstone with a low clay mineral content. *Rock Mech. Rock Eng.* **2017**, *50*, 485–491. [CrossRef]
12. Zhao, Z.; Yang, J.; Zhou, D.; Chen, Y. Experimental investigation on the wetting-induced weakening of sandstone joints. *Eng. Geol.* **2017**, *225*, 61–67. [CrossRef]
13. Koudelka, P.; Fila, T.; Rada, V.; Zlamal, P.; Sleichrt, J.; Vopalensky, M.; Kumpova, I.; Benes, P.; Vavrik, D.; Vavro, L.; et al. In-situ X-ray differential micro-tomography for investigation of water-weakening in quasi-brittle materials subjected to four-point bending. *Materials* **2020**, *13*, 1405. [CrossRef] [PubMed]

14. Liu, Y.; Dai, F.; Dong, L.; Xu, N.; Feng, P. Experimental investigation on the fatigue mechanical properties of intermittently jointed rock models under cyclic uniaxial compression with different loading parameters. *Rock Mech. Rock Eng.* **2018**, *51*, 47–68. [CrossRef]
15. Wei, M.; Dai, F.; Xu, N.; Zhao, T. Stress intensity factors and fracture process zones of ISRM-suggested chevron notched specimens for mode I fracture toughness testing of rocks. *Eng. Fract. Mech.* **2016**, *168*, 174–189. [CrossRef]
16. Guo, R.; Ren, H.; Zhang, L.; Long, Z.; Jiang, X.; Wu, X.; Wang, H. Direct dynamic tensile study of concrete materials based on mesoscale model. *Int. J. Impact Eng.* **2020**, *143*, 103598. [CrossRef]
17. Li, D.; Wong, L.N.Y. The Brazilian disc test for rock mechanics applications: Review and new insights. *Rock Mech. Rock Eng.* **2013**, *46*, 269–287. [CrossRef]
18. Imani, M.; Nejati, H.R.; Goshtasbi, K. Dynamic response and failure mechanism of Brazilian disk specimens at high strain rate. *Soil Dyn. Earthq. Eng.* **2017**, *100*, 261–269. [CrossRef]
19. Zhao, S.; Zhang, Q.; Lesiuk, G. Dynamic crack propagation and fracture behavior of pre-cracked specimens under impact loading by split Hopkinson pressure bar. *Adv. Mater. Sci. Eng.* **2019**, *2019*, 1–11. [CrossRef]
20. Yan, Z.; Dai, F.; Liu, Y.; Wei, M.; You, W. New insights into the fracture mechanism of flattened Brazilian disc specimen using digital image correlation. *Eng. Fract. Mech.* **2021**, *252*, 107810. [CrossRef]
21. Liu, Y.; Dai, F.; Xu, N.; Zhao, T. Cyclic flattened Brazilian disc tests for measuring the tensile fatigue properties of brittle rocks. *Rev. Sci. Instrum.* **2017**, *88*, 083902. [CrossRef] [PubMed]
22. Shi, X.; Liu, D.A.; Yao, W.; Shi, Y.; Tang, T.; Wang, B.; Han, W. Investigation of the anisotropy of black shale in dynamic tensile strength. *Arab. J. Geosci.* **2018**, *11*, 1–11. [CrossRef]
23. Suggested ISRM. Method for determining tensile strength of rock material. *Int. J. Rock Mech. Min. Sci. Geomech. Abstr.* **1978**, *15*, 99–103. [CrossRef]
24. Swab, J.J.; Yu, J.; Gamble, R.; Kilczewski, S. Analysis of the diametral compression method for determining the tensile strength of transparent magnesium aluminate spinel. *Int. J. Fract.* **2011**, *172*, 187–192. [CrossRef]
25. Wong, L.N.Y.; Zou, C.; Cheng, Y. Fracturing and failure behavior of Carrara marble in quasi-static and dynamic Brazilian disc tests. *Rock Mech. Rock Eng.* **2014**, *47*, 1117–1133. [CrossRef]
26. Wang, Q.; Jia, X.; Kou, S.; Zhang, Z.; Lindqvist, P.A. The flattened Brazilian disc specimen used for testing elastic modulus, tensile strength and fracture toughness of brittle rocks: Analytical and numerical results. *Int. J. Rock Mech. Min.* **2004**, *41*, 245–253. [CrossRef]
27. Wang, Q.; Li, W.; Song, X. A method for testing dynamic tensile strength and elastic modulus of rock materials using SHPB. *Pure Appl. Geophys.* **2006**, *163*, 1091–1100. [CrossRef]
28. Chen, R.; Dai, F.; Qin, J.; Lu, F. Flattened Brazilian disc method for determining the dynamic tensile stress-strain curve of low strength brittle solids. *Exp. Mech.* **2016**, *53*, 1153–1159. [CrossRef]
29. Liu, Y.; Dai, F.; Xu, N.; Zhao, T.; Feng, P. Experimental and numerical investigation on the tensile fatigue properties of rocks using the cyclic flattened Brazilian disc method. *Soil Dyn. Earthq. Eng.* **2018**, *105*, 68–82. [CrossRef]
30. Zhou, Z.; Li, X.; Zou, Y.; Jiang, Y.; Li, G. Dynamic Brazilian tests of granite under coupled static and dynamic loads. *Rock Mech. Rock Eng.* **2014**, *47*, 495–505. [CrossRef]
31. Wu, B.; Chen, R.; Xia, K. Dynamic tensile failure of rocks under static pre-tension. *Int. J. Rock Mech. Min.* **2015**, *80*, 12–18. [CrossRef]
32. Pei, P.; Dai, F.; Liu, Y.; Wei, M. Dynamic tensile behavior of rocks under static pre-tension using the flattened Brazilian disc method. *Int. J. Rock Mech. Min.* **2020**, *126*, 104208. [CrossRef]
33. Liu, Y.; Dai, F. A review of experimental and theoretical research on the deformation and failure behavior of rocks under cyclic loads. *J. Rock Mech. Geotech. Eng.* **2021**. [CrossRef]
34. Zhao, Z.; Guo, T.; Ning, Z.; Dou, Z.; Dai, F.; Yang, Q. Numerical modeling of stability of fractured reservoir bank slopes subjected to water-rock interactions. *Rock Mech. Rock Eng.* **2018**, *51*, 2517–2531. [CrossRef]
35. Du, H.; Dai, F.; Liu, Y.; Xu, Y.; Wei, M. Dynamic response and failure mechanism of hydrostatically pressurized rocks subjected to high loading rate impacting. *Soil Dyn. Earthq. Eng.* **2020**, *129*, 105927. [CrossRef]
36. Yan, Z.; Dai, F.; Liu, Y.; Du, H.; Luo, J. Dynamic strength and cracking behaviors of single-flawed rock subjected to coupled static–dynamic compression. *Rock Mech. Rock Eng.* **2020**, *53*, 4289–4298. [CrossRef]
37. Zhou, X.; Wang, Y.; Zhang, J.; Liu, F. Fracturing behavior study of three-flawed specimens by uniaxial compression and 3D digital image correlation: Sensitivity to brittleness. *Rock Mech. Rock Eng.* **2019**, *52*, 691–718. [CrossRef]
38. Guo, J.; Feng, G.; Qi, T.; Wang, P.; Yang, J.; Li, Z.; Bai, J.; Du, X.; Wang, Z. Dynamic mechanical behavior of dry and water saturated igneous rock with acoustic emission monitoring. *Shock. Vib.* **2018**, *2018*, 2348394. [CrossRef]
39. Kahraman, S. The correlations between the saturated and dry P-wave velocity of rocks. *Ultrasonics* **2007**, *46*, 341–348. [CrossRef]
40. Wang, S.; Li, H.; Wang, W.; Li, D.; Yang, W. Experimental study on strength, acoustic emission, and energy dissipation of coal under naturally and forcedly saturated conditions. *Adv. Civ. Eng.* **2018**, *2018*, 1049802. [CrossRef]
41. Dai, F.; Xia, K. Loading Rate Dependence of tensile strength anisotropy of barre granite. *Pure Appl. Geophys.* **2010**, *167*, 1419–1432. [CrossRef]
42. Zhou, Z.; Cai, X.; Ma, D.; Chen, L.; Wang, S.; Tan, L. Dynamic tensile properties of sandstone subjected to wetting and drying cycles. *Constr. Build. Mater.* **2018**, *182*, 215–232. [CrossRef]

43. Wang, Y.; Zhou, X.; Kou, M. Three-dimensional numerical study on the failure characteristics of intermittent fissures under compressive-shear loads. *Acta Geotech.* **2019**, *14*, 1161–1193. [CrossRef]
44. Rossi, P. A physical phenomenon which can explain the mechanical behavior of concrete under high strain rates. *Mater. Struct.* **1991**, *24*, 422–424. [CrossRef]
45. Zheng, D.; Li, Q. An explanation for rate effect of concrete strength based on fracture toughness including free water viscosity. *Eng. Fract. Mech.* **2004**, *71*, 2319–2327. [CrossRef]

*Article*

# Research on Strong Ground Pressure of Multiple-Seam Caused by Remnant Room Pillars Undermining in Shallow Seams

Dan Yu [1,2], Xiaoyong Yi [2,*], Zhimeng Liang [2], Jinfu Lou [3] and Weibing Zhu [1,2,*]

[1]  State Key Laboratory of Coal Resources and Safe Mining, China University of Mining and Technology, Xuzhou 221116, China; ts20020179p21@cumt.edu.cn

[2]  School of Mines, China University of Mining and Technology, Xuzhou 221116, China; liangmeng1019@163.com

[3]  China Coal Technology and Engineering Group Co., Ltd., Coal Mining Research Institute, Beijing 100013, China; loujinfu@tdkcsj.com

*  Correspondence: TS18020059A3TM1@cumt.edu.cn (X.Y.); cumtzwb@cumt.edu.cn (W.Z.)

**Abstract:** Numerous room-and-pillar mining goaf are apparent in western China due to increasing small coal mining activities, which causes the collapse of the overlying coal pillars and the occurrence of strong ground pressure on the longwall face and surface subsidence. In this study, Yuanbao Bay Coal Mine, Shuozhou, Shanxi, was selected to study the collapse of the overlying coal pillars on the longwall face and reveal the mechanism of the pillar collapse and the disaster-causing mechanism caused by strong ground pressure. Results show that the dynamic collapse process of coal pillars is relatively complicated. First, the coal pillars on both sides of the goaf are destroyed and destabilized, followed by the adjacent coal pillars, which eventually cause a large-scale collapse of the coal pillars. This results in a large-scale cut-off movement of the overlying strata, and the large impact load that acts on the longwall face causes an unmovable longwall face support. Moreover, the roof weighting is severe when strong ground pressure occurs on the longwall face, causing local support jammed accidents. Furthermore, the data of each measurement point of the strata movement inside the ground borehole significantly increases, and the position of the borescope peeping error holes in the ground drill hole rise steeply. The range of movement of the overlying strata increases instantaneously, and the entire strata begin to move. Research on the mechanism of strong ground pressure can effectively prevent mine safety accidents and avoid huge economic losses.

**Keywords:** undermining; room pillar; multiple-seam; dynamic load pressure; pillar collapse; strata movement

**Citation:** Yu, D.; Yi, X.; Liang, Z.; Lou, J.; Zhu, W. Research on Strong Ground Pressure of Multiple-Seam Caused by Remnant Room Pillars Undermining in Shallow Seams. *Energies* **2021**, *14*, 5221. https://doi.org/10.3390/en14175221

Academic Editor: Adam Smoliński

Received: 4 August 2021
Accepted: 21 August 2021
Published: 24 August 2021

**Publisher's Note:** MDPI stays neutral with regard to jurisdictional claims in published maps and institutional affiliations.

## 1. Introduction

Coal, which is one of the main energy sources of China, plays a significant role in China's energy consumption structure. Particularly, western China became the concentrated place of numerous and increasing mining activities in the past decades. Most mining operations in these regions occur in shallow-buried and close-distance coal seams [1]. Existing and upper coal seams with simple geological conditions have been mined due to the intensive mining activities in recent years, and the mining process of the close-distance lower coal seam has been inevitably affected in secondary mining. This multiple-seam mining [2–5] causes an increase in the overburden movement range and surface damage. Moreover, the abnormal stress field and movement of the overlying strata in the stope causes the appearance of abnormal ground pressure, resulting in the increasing attention that close-distance coal seam mining has received in recent years [6,7].

The main problem in the Yuanbao Bay Mine in Shuozhou, Shanxi, is ensuring the stability of the overlying coal pillars affected by undermining [2,3,8–12]. Several theoretical and numerical studies have been conducted to address this problem and evaluate the stability of coal pillars, including the modeling and calculation of coal pillar bearings [13–15], determination of the mechanical characteristics of coal pillars under long-term bearing action [16–18], investigation of the function evaluation method of the coal pillar stability [19], and the identification of the main factors affecting the stability of coal pillars [20,21].

The overlying coal seam goaf is affected by lower seam undermining if it is a room-and-pillar mined-out area. Moreover, the coal pillar collapse mechanism, the influence of this collapse on the working face of the lower coal seam, and the surface subsidence [22] when the overlying coal pillar loses stability have been extensively studied by several experts and scholars.

Therefore, the large-area support crushing is crucial and necessary due to the movement and instability of the overburden structure. Some scholars systematically analyzed the large-area support crushing event of stope in China and found that the articulated rock block structure formed by the key strata of the overburden lacks stability [23] and causes the large-area support crushing event in a stope. Therefore, the working resistance [24] of the working face support should be at a high level to prevent support crushing. Meanwhile, Wang et al. [25] investigated the principles behind the destruction of loose confined aquifers in mining and revealed the important role of the unconsolidated confined aquifer (UCA) in the process of load transfer due to its liquidity and timely replenishment [26]. On the other hand, Huang et al. [27] focused on the number of key strata in shallow-buried and close-distance coal seams and determined that different overlying strata structures were formed after one or two layers were broken. Additionally, the structure formed by one broken layer was carried together in the form of hinged arches and arch shells, while the structure of two broken layers was formed in a variety of combinations of steps and masonry, and the composite overburden structure jointly bears the load of the overburden. Zhu et al. [28] studied the roof damage in room-and-pillar mined-out areas caused by longwall mining in the lower coal seam and observed that different width-to-height ratios of coal pillars lead to different modes of collapse failure [29] and provided reference values for the development trend of the cracks after collapse of different width-to-height ratios of coal pillars during mining underneath the goaf. Wang et al. [30,31] have done great research on automatically formed roadway (AFR) and found that the new mining method satisfies the surrounding rock control requirement, which improves the safety and economic impact of coal mining. Feng et al. [32], using the key stratum theory, put forward the concepts of key stratum breaking distance and advancing mining coal pillars to address the problem of coal seam mining above the knife-pillar goaf and formed the knife-pillar goaf overmining [33–35] feasibility judgment theory and method. Wang et al. [36] comprehensively considered the factors influencing the stability of the interburden and adopted the safety factor method to evaluate the stability of the interburden in order to ensure the safe mining of a coal seam that overlies the knife-pillar goaf area. Dychkovskyi R. et al. [37] conducted research into stress-strain state of the rock mass condition in the process of the operation of double-unit longwalls in order to substantiate changes in stress-strain state of rock mass in the process of long-pillar mining with the help of double-unit longwalls while evaluating stress of a mine field in terms of Lvivvuhillia SE mine.

However, most of the existing research is only aimed at the collapse of coal pillar groups under static load. There is limited research on the stability of coal pillars and the dynamic collapse process of coal pillar groups under the influence of mining. Additionally, there is only one layer of the room-and-pillar mined-out region in overburden, which cannot show the dynamic collapse process. This limits the investigation into the combination collapse of the coal pillar and the roof during the dynamic collapse process. Meanwhile, new measurement method is applied to the strata movement, optical fiber is employed during strata movement monitoring to detect the displacement through strain [38–40]. However, the cumulative strain calculation displacement causes a certain error. Based on the aforementioned limitations of the previous studies, this study considers the Yuanbao Bay 6107 longwall face [2,3,41] as the study location in order to investigate the stability of the coal pillar in undermining conditions using the adopted experimental method. By adopting the three key strata shift measurement point layout, this experiment captured the violent settlement caused by strong ground pressure, then studied the mechanism of the dynamic collapse of the coal pillar group through physical simulation, and finally, deeply analyzed the mechanism of the disaster through a simplified model. This research achieved the goal of monitoring the movement characteristics [42] of the overlying strata and studying the disaster-causing mechanism of strong ground pressure as well as monitoring the movement characteristics of the overlying strata and studying the disaster mechanism caused by strong ground pressure. Moreover, this study represents an important contribution to the prevention and control of support crushing disasters by putting forward effective measures and countermeasures to guarantee the safety of mine personnel, equipment, and facilities by predicting the initial location of an occurrence of strong ground pressure and carrying out effective prevention and control through methods like hydraulic fracturing to solve the vulnerabilities in advance.

## 2. Basic Conditions of Longwall Face

There are four key strata in the overburden of the 6107 longwall face, all of which are medium sandstone, based on the borehole columnar section 2101. From bottom to top, the lowest region has a burial depth of 148.9 m and a thickness of 7.75 m, which exist in the inferior key stratum between the No. 6 and No. 4 coal seams. The next layer has a burial depth of 126.3 m and an inferior key stratum thickness of 8.8 m. Then, the succeeding layer has a burial depth of 76.7 m, and the thickness of the inferior key stratum is 5.5 m. Finally, the main key stratum has a burial depth of 49.5 m and a thickness of 10.5 m.

The mining method adopted in this study is one-time full-height longwall retreating mining with the 6107 longwall face having an advancing length of 500 m, a width of 240 m, an average coal seam thickness of 3.5 m, an inclination angle of 4° to 8° (with an average of 6°), and a coal seam burial depth of approximately 150 m. The No. 6 coal seam is overlaid by the room-and-pillar mining goaf formed by small coal kilns before integrating No. 3 and No. 4 coal seams. There is still a certain bearing capacity based on the preliminary exploration. Combining the actual survey results on the site, the coal pillar widths of No. 3 and No. 4 coal seams are approximately 5–7 m, and the goaf of them are about 4-6 m.

The main problem with mining in the 6107 longwall face of Yuanbao Bay Mine is the threat of collapse of the coal pillars in the room-and-pillar mined-out area in the overlying 3 and 4 coal goaf. If the overlying goaf is affected by the undermining of the No. 6 coal seam longwall faces, it will cause a large-scale collapse that will result in the overall breaking movement of the overburden, inducing the occurrence of dynamic loading ground pressure disasters on the longwall face.

## 3. Characteristics of Strong Ground Pressure

### 3.1. Location and Situation of Strong Ground Pressure

A strong ground pressure event occurred on August 12. The head and tail positions of the 6107 longwall face were advanced to 318 and 325 m (with an average distance of 322 m), respectively. The pressure on support Nos. 16 to 135 exceeded 40 MPa, and all safety valves were opened. Among them, support Nos. 110 to 135 and Nos. 80 to 110 had a stroke of support movable column of 60 and 40 cm, respectively. In addition, support Nos. 16 to 80 had no stroke. The supports were crushed, and the advance of the longwall face was stopped for several days. Figure 1 shows the range of the longwall face support crushing and the position of the newly developed fractures [43].

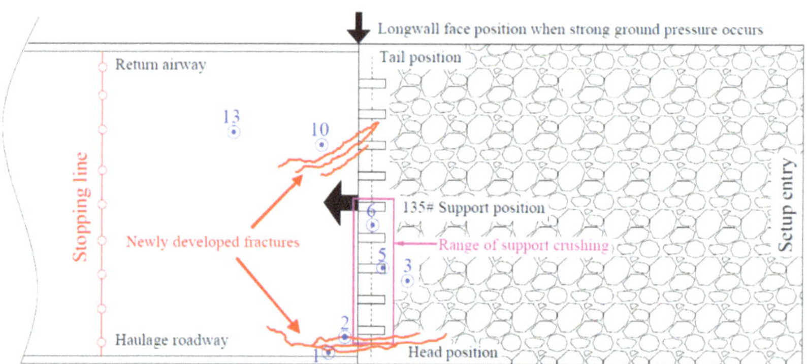


**Figure 1.** Plan view of the position of the newly developed fractures and the range of longwall face support crushing.

On the day of the occurrence of strong ground pressure, a detailed survey of the development of the surface fractures was carried out. According to the survey results, it was found that two groups of newly developed surface fractures appeared at the position, 50 m in front of the longwall face, and each group of fractures consisted of 3–4 stepped fractures, as shown in Figure 1. The first group of fractures was approximately located near holes No. 1 and No. 2, and the second group was approximately located near hole No. 10. The direction of the fractures was roughly along the longwall face advancing direction. The surface fractures near hole No. 2 developed as shown in Figure 2. The maximum width of these surface fractures was 1.0 m, and the maximum displacement of the steps was 0.8 m.

**Figure 2.** Surface fracture development near hole No. 2.

### 3.2. Characteristics of Unstable Overburden Movement under Strong Ground Pressure

To further understand the characteristics of the overburden movement, it was decided to arrange multiple strata movement measuring points at the key stratum position as required. A strong ground pressure phenomenon was observed during the monitoring, and the strata movement characteristics when strong ground pressure occurred were captured. Holes No. 6 and 10 were chosen as the internal strata displacement installation holes based on the results of the borehole columnar section, key stratum identification, and the investigation results of the longwall face drill hole. Specifically, the position of hole No. 6 was in the center of the range of the newly developed fractures and used to illustrate the overburden movement process, the No. 6 drill hole installation plan, as is shown in Figure 3a, and the remaining holes (i.e., 1, 2, 3, 5, and 13) of grouting used for the borescope peeping.

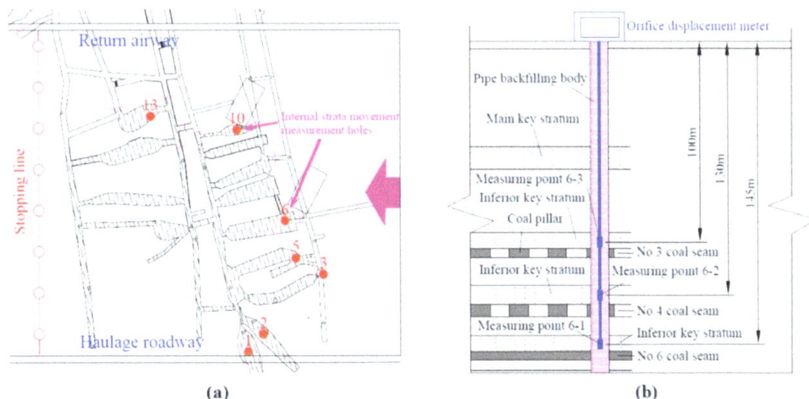



**Figure 3.** Internal strata movement installation plan of the longwall face. (**a**) Installation plan of the internal strata movement measurement and observation holes; (**b**) Diagram of the internal strata movement measurement hole No. 6 installation.

A schematic diagram of the internal strata movement measurement hole No. 6 installation is shown in Figure 3b, where each measuring point represents the movement state of each key stratum and its controlled strata. The alternative relationship between each measuring point sinking amount and the distance between the 6107 longwall face and hole No. 6 is shown in Figure 4.

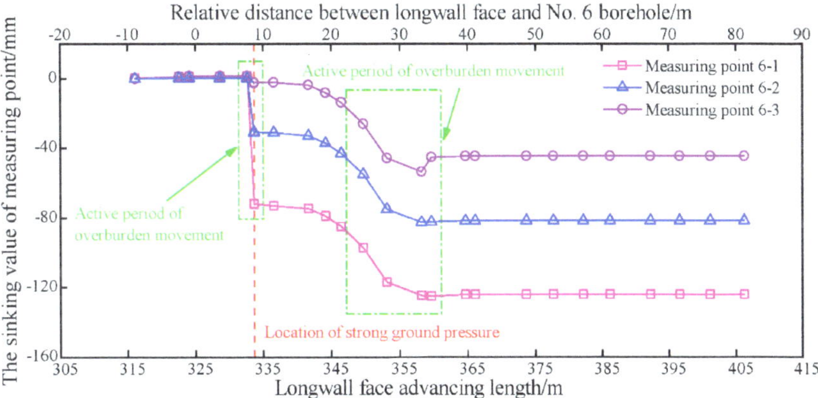



**Figure 4.** Relationship between each measuring point sinking amount and the distance between the 6107 longwall face and hole No. 6.

Figure 4 shows a large sinking amount of measuring points 6-1 and 6-2 with sinking increments of 70 and 30 mm, fast change process, and short and rapid time when the longwall face is pushed through the hole for about 8 m, indicating that the first step subsidence occurred in the strata corresponding to the two measuring points that are in the active period, which is short, of the overburden movement. All measuring points change slowly for a long period when the longwall face is pushed through the hole at approximately 28 m. The total increments of its subsidence were 120, 80, and 40 mm, with the relative increments of 50, 50, and 40 mm, respectively. This indicates that the strata corresponding to the three measurement points had secondary step subsidence with a similar amount of subsidence, an active overburden movement, and a relatively longer active period. After these two active periods, the sinking value did not change significantly.

The ground pressure on the longwall face was extremely intense during the large-scale movement of the overburden. In addition, the time and location of the sinking experienced by hole No. 6 had a one-to-one correspondence with the time and location of the strong ground pressure on the longwall face, indicating that the active period of the overburden movement and the occurrence of the strong ground pressure on the longwall face are inseparable.

### 3.3. Results of the Drill Hole Borescope Peeping

Borehole No. 1 is within the range of newly developed fractures, reflecting the detailed process of the overlying strata movement, as shown in Figure 5.

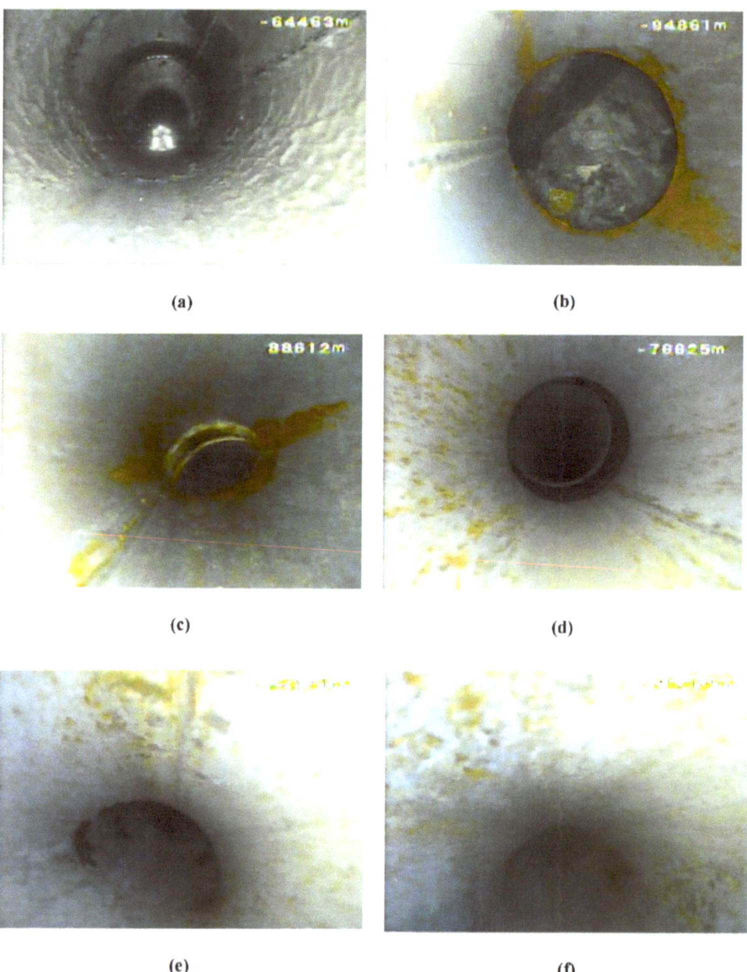

**Figure 5.** Observation results of borehole No. 1. (**a**) Water accumulation location at −64.46 m; Error hole position at (**b**) −94.86 m, (**c**) −88.61 m, (**d**) −78.82 m, (**e**) −73.17 m, and (**f**) −35.20 m.

The observations started when borehole No. 1 was 53 m away from the longwall face. Results from the observations show that the probe was lowered to −64.46 m, wherein water accumulated in the hole, and cracks under the hole did not develop. Therefore, the probe was stopped, as shown in Figure 5a. Additionally, the probe was lowered to approximately −94.86 m when the longwall face was 20 m from the hole. Error holes in the hole were found, and the distance between them was relatively large, as shown in

Figure 5b. Moreover, the position of the staggered hole in the hole remained unchanged at −94.86 m when the longwall face was 17 to 0 m. We found error holes when the probe was lowered to a position of approximately −88.61 m, and the longwall face was pushed through the hole at 5 m. The distance of the error holes was not large, but the probe was not able to pass, and the descending was forced to stop, as shown in Figure 5c. During the 6 to 27 m period when the longwall face pushed through the hole, the position of the error hole did not change. When the longwall face was pushed through 33 m, it was found that when the probe was lowered to about −78.82 m, the drill hole was slightly wrong, the casing fell off, the probe could not pass, and the detection ended, as shown in Figure 5d. The position of the error hole remained unchanged from 39 to 43 m when the longwall face was pushed through the borehole. The probe went down to approximately −73.17 m, and the error hole occurred when the longwall face was pushed through the borehole for more than 48 m. The error hole was not completely closed, but the probe could not pass through, so it stopped descending, as shown in Figure 5e. In the 53–58 m period, the position of the error hole remained unchanged when the longwall face was pushed through the borehole. On the other hand, the probe was lowered to −35.20 m, and the error hole occurred when the longwall face was pushed through the hole at 64 m, as shown in Figure 5f. It was of little use to continue the detection because the position of the error hole was already above the main key stratum. Thus, the borescope observation was complete.

According to the above observation results, the relationship of the depth and the relative distance between the longwall face and hole No. 1 and the position of the staggered hole can be obtained, as shown in Figure 6. The first misalignment of the hole coincides with the location of the strong ground pressure, indicating that the large movement of the overlying strata below the inferior key stratum is the main cause of the strong ground pressure. In the subsequent mining engineering of the longwall face, the position of the error hole has a small step rise, and finally, the position of the error hole near the loose layer no longer changes. The observation results indicate that the entire overburden movement has a stepwise upward trend.

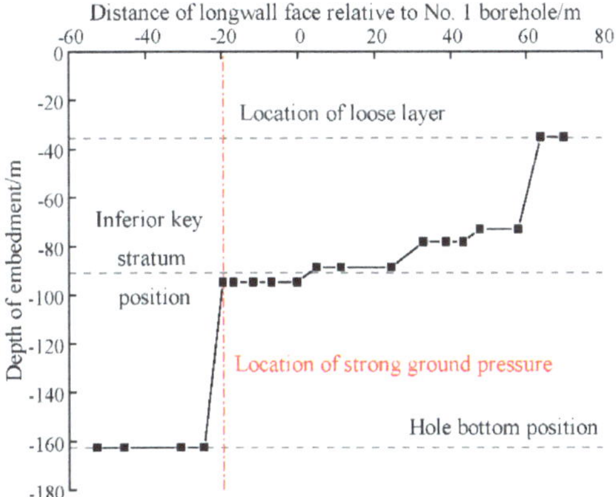


**Figure 6.** Relationship between the depth of misaligned holes in No. 1 borehole and the distance of longwall face relative to No. 1 borehole.

## 4. Dynamic Collapse Process of Room Pillars and the Disaster Mechanism caused by Strong Ground Pressure

*4.1. Scheme Design of the Similar Simulation Model*

The occurrence of the strong ground pressure disaster on the 6107 longwall face was caused by the overall movement of the overburden. Therefore, we employed a similar simulation [44] to study the disaster mechanism of the strong ground pressure on the 6107 longwall face of Yuanbao Bay Mine caused by the overlying room pillar collapse.

The purpose of this simulation is: (1) to simulate the overlying room pillar dynamic collapse process on the 6107 longwall face; (2) to simulate the movement characteristics of the overlying strata after the coal pillar collapse; and (3) to analyze the disaster mechanism based on the collapse process and the characteristics of the overlying strata movement.

We selected a two-dimensional model frame with a dimension of 1.3 m × 0.12 m × 1 m for the simulation based on the mining conditions and occurrence characteristics of the longwall face. Based on the same conditions, the geometric similarity ratio of the model was 1:100, Poisson's ratio similarity ratio was 1:1, density similarity ratio was 1:1.6, and stress similarity ratio was 1:160. The height of the model was 65 cm. A corresponding load of approximately 80 m thickness overlying rock was applied to simulate the actual burial depth, equivalently replaced by a uniform load of 12.5 kPa according to the stress similarity ratio. A schematic of the model is shown in Figure 7.

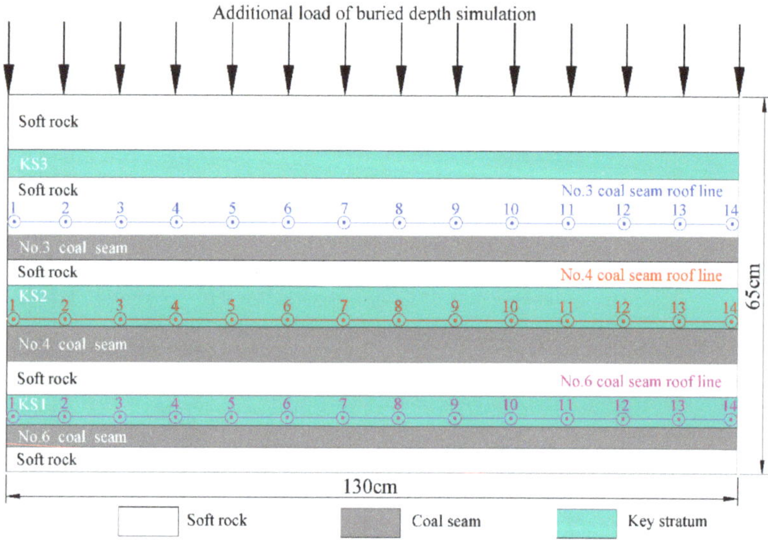


**Figure 7.** Schematic diagram of similar simulation model.

There are three strata for the overburden according to the borehole columnar section 2101 and the key stratum identification results. The number of key strata was also set to three layers, and the others were replaced by soft rocks whose structure is the interbedded mudstone and sandy mudstone in the bedrock. The mechanical properties of these two rock formations are very similar, so they can be approximately reckoned as the same material to simplify the structure of the overlying rock and reflect the main control effect of the key strata [6]. The thickness of each layer is 1.0–3.5 m; after homogenization, each layer thickness is 2 cm, as shown in Table 1.

**Table 1.** Proportion table of similar materials.

| No. | Lithology | M | Material/kg | | | | t | S |
| --- | --- | --- | --- | --- | --- | --- | --- | --- |
| | | | Sand | Calcium Carbonate | Plaster | Water | | |
| 11 | Soft rock | 773 | 21.84 | 2.18 | 0.94 | 2.78 | 10 | 5 |
| 10 | KS3 | 337 | 10.92 | 1.09 | 2.55 | 1.62 | 5 | 1 |
| 9 | Soft rock | 773 | 21.84 | 2.18 | 0.94 | 2.78 | 10 | 5 |
| 8 | No. 3 coal seam | 773 | 8.74 | 0.87 | 0.37 | 1.08 | 4 | 1 |
| 7 | Soft rock | 773 | 8.74 | 0.87 | 0.37 | 1.11 | 4 | 2 |
| 6 | KS2 | 437 | 15.29 | 1.15 | 2.68 | 2.12 | 7 | 1 |
| 5 | No. 4 coal seam | 773 | 13.10 | 1.31 | 0.56 | 1.61 | 6 | 1 |
| 4 | Soft rock | 773 | 13.10 | 1.31 | 0.56 | 1.66 | 6 | 3 |
| 3 | KS1 | 437 | 10.92 | 0.82 | 1.91 | 1.52 | 5 | 1 |
| 2 | No. 6 coal seam | 773 | 8.74 | 0.87 | 0.37 | 1.08 | 4 | 1 |
| 1 | Floor | 673 | 8.74 | 1.02 | 0.44 | 1.13 | 4 | 2 |

Note: t is the thickness of strata, m; S is the stratification of strata; M is the matching number of material.

Observation marks were placed on the roof of each coal seam to understand the subsidence of the coal roof in detail. There were 14 measuring points in each roof line with a 10 cm interval between them.

The goaf areas of the No. 3 and No. 4 coal seams were designed according to the actual engineering background ratios of mined to unmined width, 1.8 and 2.8, respectively. Generally considered, the widths of the No. 3 coal seam goaf and the coal pillar were 7 and 3 cm, respectively. Meanwhile, the widths of the No. 4 coal seam goaf and the coal pillar were 8 and 2 cm, respectively. Both 6.5 and 6 m width of coal pillars were left on the model boundaries of the No. 3 and No. 4 coal seams, respectively, to avoid the influence of the model boundary on the goaf during the mining period.

The No. 6 coal seam mining length was 90 cm, with 10 cm width boundary coal pillars left on both sides. It had a mining step of 5 cm, a total of 18 excavations, and the time interval between each excavation of half an hour was used for the strata to stabilize. Then, we employed high-speed and real-time photography.

*4.2. Dynamic Collapse Process of Room Pillars*

The state of the No. 6 coal seam roof changes significantly, eventually breaks, and the amount of subsidence increases significantly up to 45 mm when the longwall face advances 70 cm. Then, the advancing distance continues to increase, while the maximum subsidence value of the roof of the No. 6 coal seam no longer increased. However, the lateral range of its movement subsidence increases from 90 to 120 cm, as shown in Figure 8a. Results show that the No. 4 coal seam roof sinks later than the No. 6 coal seam. The No. 4 coal seam roof drastically sinks and breaks after the coal pillar becomes unstable at 70 cm, and the sinking amount suddenly increases to approximately 70 mm. The amount of coal roof subsidence remains unchanged as the longwall face continuously advances, while the subsidence range increased slightly, as shown in Figure 8b. Moreover, the amount of roof subsidence of the No. 3 coal seam increases when the advancing distance is within 60 cm. On the other hand, the advancing distance of the longwall face is 70 cm when the coal roof breaks and sinks, which is the same position as the roof-breaking movement of the No. 6 and No. 4 coal seams. The amount of subsidence increases up to approximately 90 cm, indicating that the roof of the No. 3 coal seam moves violently after the overburden movement. Moreover, subsequent mining has little effect on the sinking movement of the No. 3 coal seam roof, as shown in Figure 8c.

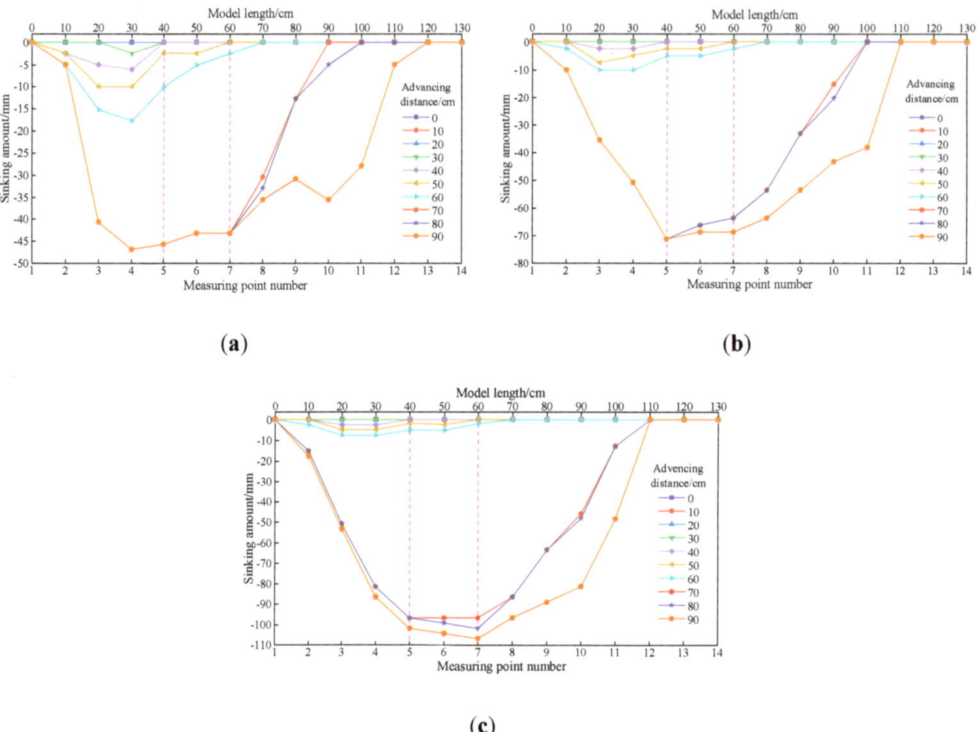

**Figure 8.** Coal seam roof subsidence of (**a**) No. 6, (**b**) No. 4, and (**c**) No. 3 during model excavation.

Figure 9 shows the failure sequence and the detailed process of the dynamic collapse of coal pillars of overlying No. 3 and No. 4 coal seams on the longwall face. Additionally, the overlying coal pillars of the No. 3 and No. 4 coal seam rooms and pillar goaf were affected by the mining process of the longwall face, resulting in these two being successively damaged and failing. This caused the large-scale collapse of the two layers of coal pillars, resulting in the occurrence of strong ground pressure on the longwall face.

Both the No. 3 and No. 4 coal seam pillars, which consist of 11 coal pillars from No. 1 to No. 11, are numbered from right to left, as shown in Figure 9a. The first failure occurred at No. 4, No. 5, and No. 11 coal pillars of the No. 4 coal seam. Moreover, the No. 6 coal seam immediate, the controlled strata broke and revolved, and the No. 4 coal seam roof strata bent and sank. Fractures developed directly on the coal seam roof at the position of the setup entry and the No. 6 coal seam longwall face, as shown in Figure 9b. Eventually, the No. 4 and No. 5 coal pillars of the No. 3 coal seam failed, and fractures developed directly on the boundary of the coal pillar of the No. 4 coal seam, as shown in Figure 9c. Additionally, the No. 8 and No. 9 coal pillars of the No. 4 coal seam and No. 3 coal pillar of the No. 3 coal seam successively failed, the immediate roof of the No. 4 coal seam inferior key stratum and its controlled strata broke and rotated, and the No. 3 coal seam overlying strata was bent and subsided, as shown in Figure 9d. Overall, the coal pillars of the No. 4 coal seam above the goaf are all unstable, and the adjacent coal pillars that failed in the No. 3 coal seam have a large number of fractures in the roof, as shown in Figure 9e. The coal pillar of the No. 3 coal seam suffered a large-scale collapse, and the overlying roof strata broke and turned, causing the interburden of No. 3 and No. 4 coal seams to be entirely cut off. Moreover, the immediate roof of the No. 4 coal seam acted on the interburden of the No. 4 and No. 6 coal seams to cut instantaneously and fall down, causing the longwall face support crushing, as shown in Figure 9f.

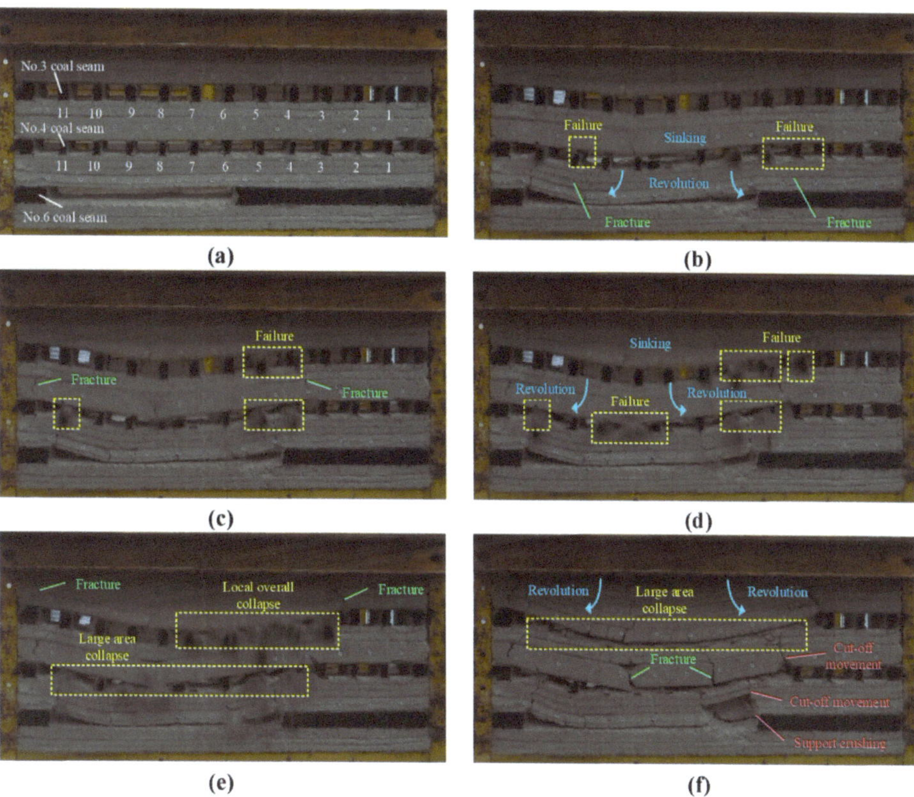



**Figure 9.** Dynamic collapse process of room-and-pillar mining pillars. (**a**) Before the coal pillar loses stability; (**b**) No. 4, No. 5, and No. 11 coal pillars fail first; (**c**) No. 4 and No. 5 coal pillars fail successively; (**d**) No. 8 and No. 9 coal pillars fail; (**e**) Large area of coal pillar collapse; (**f**) Support crushing under strong ground pressure at longwall face.

### 4.3. Analysis of the Disaster Mechanism of the Strong Ground Pressure

We obtained a model of the dynamic collapse of the coal pillar disaster mechanism based on the aforementioned process, as shown in Figure 10.

The immediate roof of the longwall face is a hard, inferior key stratum. The longwall face is affected by the advancing abutment pressure [45–47] when it advances to form a larger span. The diagonal failure of the coal pillars of the No. 4 coal seam above the setup entry and longwall face occurs first because the peak value of the advancing abutment pressure is larger than the compressive strength of the coal pillar, as shown in Figure 10a. The stress from the diagonal failure was instantly transferred to the pillars above the goaf, resulting in the breakage and rotation of the immediate roof that was exposed in the goaf of the longwall face of the No. 6 coal seam due to the sudden increase in stress. Additionally, the space of the immediate roof rotation was limited by the mining height, which forced the rotation to stop. Specifically, the No. 4 coal seam pillars overlying the mined-out area were subjected to excessive overburden load and collapsed. After this large-scale failure, the immediate roof also stopped moving in the limited space. On the other hand, the No. 3 coal seam pillars overlying the setup entry and longwall face were also diagonally damaged during the movement of the strata below, as shown in Figure 10b. Moreover, the advancing abutment pressure was continuously being transferred to the undamaged coal pillars of the No. 3 coal seam, resulting in the advanced coal pillar failure. The load of the No. 3 coal seam was concentrated on the coal pillars that had not failed in the middle. The coal pillars were destroyed and failed in a larger area due to the significant pressure from the overload.

The inferior key stratum of each coal seam and the controlled strata developed through fractures. The No. 3 coal seam roof strata also tended to move instantaneously because of the loss of the coal pillar support, as shown in Figure 10c. Moreover, the overlying strata of the No. 3 coal seam were entirely cut off along the through fractures of the overlying strata after the pillars of the No. 3 coal seam suffered a large area of collapse because there was no coal pillar support. This directly impacted the underlying strata of the No. 3 coal seam, and eventually, the interburden of the No. 3 and No. 4 coal seams were entirely cut off along its through fractures. The two huge impacts formed by the overall cutting of the two parts of the No. 4 coal seam overlying strata were superimposed on each other, acting on the interburden of the No. 4 and No. 6 coal seams. The violent impact made it instantly cut down along the entire intersecting fractures, directly acting on the longwall face that caused a strong ground pressure and caused the longwall face support crushing, as shown in Figure 10d.

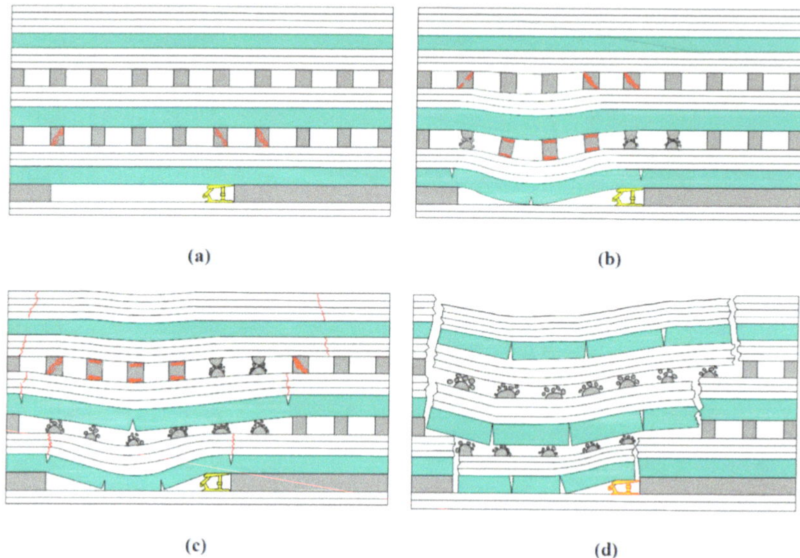

**Figure 10.** Diagram of disaster mechanism caused by dynamic collapse of coal pillars. (**a**) Diagonal failure of pillars of the No. 4 coal seam first occurs; (**b**) Continuous shear failure of adjacent coal pillars; (**c**) Local overall failure of pillars of No. 3 coal seam; (**d**) Whole-rock strata cut to strong ground.

## 5. Conclusions

In this study, the collapse and fracture of the coal pillar group and the movement characteristics of the overlying strata in the undermining process employed at the Yuanbao Bay 6107 longwall face are determined and investigated. The principles caused by the dynamic collapse of the coal pillar and the disaster mechanism of the strong ground pressure are identified and investigated using strata movement monitoring, borehole peeping, similar simulation experiments, and simplified model deepening analysis. The main conclusions of this study are as follows:

(1)    Results show that the overlying strata movement presents the movement characteristics of the entire step subsidence, while the staggered hole position of each borehole shows the characteristics of the stage step rise. Particularly, the strata movement data of each measuring point in boreholes No. 6 and No. 10 significantly increased when the strong ground pressure occurred. Simultaneously, the staggered position of each borehole moves upward, which is consistent with the time and position of the occurrence of the strong ground pressure on the 6107 longwall face. Moreover, the

collapse of the coal pillar left by the No. 4 coal seam leads to a strong ground pressure phenomenon on the 6107 longwall face.

(2) Without undermining, the No. 3 and No. 4 coal pillars, which are affected by their sizes, can maintain long-term stability. Meanwhile, in undermining conditions, the No. 4 coal pillars were found to be unable to maintain stability caused by the initial coal pillar collapse due to the influence of the advancing abutment pressure, while the No. 3 coal pillars can maintain stability.

(3) The overlying coal pillars of the No. 4 coal seam were first diagonally damaged at both ends above the goaf, and then the coal pillar above the middle of the goaf was instantaneously invalidated because of the effect of mining of the longwall face. Moreover, the coal pillars of the No. 4 coal seam had a wide range of overall collapse. On the other hand, the coal pillars of the No. 3 coal seam at the corresponding position of the goaf were diagonally damaged. Additionally, the pillars of the No. 3 coal seam were destabilized on a large scale, wherein failure on both sides and in the middle was apparent. Eventually, both pillars of the No. 3 and No. 4 coal seam experienced large-scale collapse.

(4) The causes of the strong ground pressure on the longwall face are the overall large cutting movement of the overburden and the large area collapse of the coal pillar, which is influenced by the advancing abutment pressure. After the large-scale collapse of the coal pillar, the overlying strata has a large subsidence movement without the support of the coal pillar. This may cause the strata to be cut down entirely, directly or indirectly impacting the longwall face, which causes a strong ground pressure on the longwall face and support crushing.

**Author Contributions:** For this paper, W.Z. put forward study ideas; D.Y. designed the article structure and wrote the paper; Z.L. revised the whole English writing style and discussed the design flow with W.Z.; X.Y. collected and analyzed the data from mine site; J.L. conducted the similar simulation experiment and analyzed the data. All authors have read and agreed to the published version of the manuscript.

**Funding:** This research was funded by the National Natural Science Foundation of China (Grant No. 52074265, No. 51874175).

**Acknowledgments:** The authors would like to thank the reviewers for their constructive comments, which improved the paper and anonymous colleagues for their kind efforts and valuable comments, which have improved this work.

**Conflicts of Interest:** The authors declare no conflict of interest.

## References

1. Zhu, W.B. Study on the Instability Mechanism of Key Strata Structure in Repeating Mining of Shallow Close Distance Seams. Ph.D. Thesis, China University of Mining and Technology, Xuzhou, China, 2010.
2. Mark, C.; Tuchman, R.J. *Multiple-Seam Longwall Mining in the United States: Lessons for Ground Control. New Technology for Ground Control in Multiple-Seam Mining*; Department of Health and Human Services, Public Health Service, Centers for Disease Control and Prevention, National Institute for Occupational Safety and Health: Pittsburgh, PA, USA, 2007; pp. 45–53.
3. Suchowerska, A.M. Geomechanics of Single Seam and Multi Seam Longwall Coal Mining. Ph.D. Thesis, University of Newcastle, Callaghan, Australia, 2014.
4. Porathur, J.L.; Srikrishnan, S.; Verma, C.P.; Jhanwar, J.C. Slope stability assessment approach for multiple seams highwall mining extractions. *Int. J. Rock Mech. Min. Sci.* **2014**, *70*, 444–449. [CrossRef]
5. Ghabraie, B.; Ren, G.; Barbato, J.; Smith, J. A predictive methodology for multi-seam mining induced subsidence. *Int. J. Rock Mech. Min. Sci.* **2017**, *93*, 280–294. [CrossRef]
6. Qian, M.G.; Shi, P.W.; Xu, J.L. *Ground Pressure and Strata Control*; China University of Mining and Technology Press: Xuzhou, China, 2010.
7. Chen, D.D.; He, F.L.; Xie, S.R.; Gao, M.M.; Song, H.Z. First fracture of the thin plate of main roof with three sides elastic foundation boundary and one side coal pillar. *J. Coal Sci. Eng. (China)* **2017**, *42*, 2528–2536.
8. Djamaluddin, I.; Mitani, Y.; Ikemi, H. GIS-Based Computational Method for Simulating the Components of 3D Dynamic Ground Subsidence during the Process of Undermining. *Int. J. Geomech.* **2012**, *12*, 43–53. [CrossRef]
9. Adelsohn, E.; Iannacchione, A.; Winn, R. Investigations on longwall mining subsidence impacts on Pennsylvania highway alignments. *Int. J. Min. Sci. Technol.* **2020**, *30*, 85–92. [CrossRef]

10. Sidorenko, A.A.; Sirenko, Y.G.; Sidorenko, S.A. An assessment of multiple seam stress conditions using a 3-D numerical modelling approach. *J. Phys. Conf. Ser. IOP Publ.* **2019**, *1333*, 032078. [CrossRef]

11. Kellich, W.; Lee, J.W.; Aziz, N.; Baafi, E. Numerical Modelling of Undermined River Valleys—A Case Study. In Proceedings of the 2005 Coal Operators' Conference, Wollongong, Australia, 26–28 April 2005; Aziz, N., Kininmonth, B., Eds.;

12. Coulthard, M.A. Undermining of an unlined tunnel in rock. In *FLAC and Numerical Modeling in Geomechanics*; CRC Press: Boca Raton, FL, USA, 2020; pp. 265–272.

13. Reed, G.; Kent, M.; Russell, F. An assessment of coal pillar system stability criteria based on a mechanistic evaluation of the interaction between coal pillars and the overburden. *Int. J. Min. Sci. Technol.* **2017**, *27*, 9–15. [CrossRef]

14. Poulsen, B.A.; Adhikary, D.P. A numerical study of the scale effect in coal strength. *Int. J. Rock Mech. Min. Sci.* **2013**, *63*, 62–71. [CrossRef]

15. Poulsen, B.A. Coal pillar load calculation by pressure arch theory and near field extraction ratio. *Int. J. Rock Mech. Min. Sci.* **2010**, *47*, 1158–1165. [CrossRef]

16. Singh, R.; Kumar, A.; Singh, A.K.; Coggan, J.; Ram, S. Rib/snook design in mechanised depillaring of rectangular/square pillars. *Int. J. Rock Mech. Min. Sci.* **2016**, *84*, 119–129. [CrossRef]

17. Atsushi, S.; Mitri, H.S. Numerical investigation into pillar failure induced by time-dependent skin degradation. *Int. J. Min. Sci. Technol.* **2017**, *27*, 591–597.

18. Maleki, H. Coal pillar mechanics of violent failure in U.S. Mines. *Int. J. Min. Sci. Technol.* **2017**, *27*, 387–392. [CrossRef]

19. Poulsen, B.A.; Shen, B. Subsidence risk assessment of decommissioned bord-and-pillar collieries. *Int. J. Rock Mech. Min. Sci.* **2013**, *60*, 312–320. [CrossRef]

20. Sinha, S.; Walton, G. Modeling behaviors of a coal pillar rib using the progressive S-shaped yield criterion. *J. Rock Mech. Geotech. Eng.* **2020**, *12*, 484–492. [CrossRef]

21. Kushwaha, A.; Singh, S.K.; Tewari, S.; Sinha, A. Empirical approach for designing of support system in mechanized coal pillar mining. *Int. J. Rock Mech. Min. Sci.* **2010**, *47*, 1063–1078. [CrossRef]

22. Xu, J.M.; Zhu, W.B.; Xu, J.L.; Wu, J.Y.; Li, Y.C. High-intensity longwall mining-induced ground subsidence in Shendong coalfield, China. *Int. J. Rock Mech. Min. Sci.* **2021**, *141*, 104730. [CrossRef]

23. Ju, J.F.; Xu, J.L. Prevention measures for support crushing while mining out the upper coal pillar in close distance shallow seams. *J. Min. Saf. Eng.* **2013**, *30*, 323–330.

24. Zhu, C.; He, M.C.; Zhang, X.H.; Tao, Z.G.; Yin, Q.; Li, L.F. Nonlinear mechanical model of constant resistance and large deformation bolt and influence parameters analysis of constant resistance behavior. *Rock Soil Mech.* **2017**, *42*, 1911–1924.

25. Wang, X.Z.; Zhu, W.B.; Xu, J.L.; Han, H.K.; Fu, X. Mechanism of overlying strata structure instability during mining below unconsolidated confined aquifer and disaster prevention. *Appl. Sci.* **2021**, *11*, 1778. [CrossRef]

26. Yu, S.C.; Xu, J.M.; Zhu, W.B.; Wang, S.H.; Liu, W.B. Development of a combined mining technique to protect the underground workspace above confined aquifer from water inrush disaster. *Bull. Eng. Geol. Environ.* **2020**, *79*, 3649–3666. [CrossRef]

27. Huang, Q.X.; Zhao, M.Y.; Huang, K.J. Study of roof double key strata structure and support resistance of shallow coal seams group mining. *Int. J. Min. Sci. Technol.* **2019**, *48*, 71–86.

28. Zhu, W.B.; Chen, L.; Zhou, Z.L.; Shen, B.T.; Xu, Y. Failure propagation of pillars and roof in a room and pillar mine induced by longwall mining in the lower seam. *Rock Mech. Rock Eng.* **2019**, *52*, 1193–1209. [CrossRef]

29. Feng, G.; Kang, Y.; Wang, X.C.; Hu, Y.Q.; Li, X.H. Investigation on the failure characteristics and fracture classification of shale under brazilian test conditions. *Rock Mech. Rock Eng.* **2020**, *53*, 3325–3340. [CrossRef]

30. Wang, Q.; He, M.C.; Li, S.C.; Jiang, Z.H.; Wang, Y.; Qin, Q.; Jiang, B. Comparative study of model tests on automatically formed roadway and gob-side entry driving in deep coal mines. *Int. J. Min. Sci. Technol.* **2021**, *31*, 591–601. [CrossRef]

31. Wang, Q.; Wang, Y.; He, M.C.; Jiang, B.; Li, S.C.; Jiang, Z.H.; Wang, Y.J.; Xu, S. Experimental research and application of automatically formed roadway without advance tunneling. *Tunn. Undergr. Space Technol.* **2021**, *114*, 103999. [CrossRef]

32. Feng, G.R.; Zhang, X.Y.; Li, J.J.; Yang, S.S.; Kang, L.X. Feasibility on the upward mining of the left-over coal above goaf with pillar supporting method. *J. China Coal Soc.* **2009**, *34*, 726–730.

33. Newman, C.; Newman, D. Numerical analysis for the prediction of bump prone conditions: A southern Appalachian pillar coal bump case study. *Int. J. Min. Sci. Technol.* **2021**, *31*, 75–81. [CrossRef]

34. Kozyreva, E.N.; Shinkevich, M.V. Evaluation of safe (by gas criterion) process parameters for longwalls and gateways based on operational gas content measurement in coal seams. In *IOP Conference Series: Earth and Environmental Science*; IOP Publishing: Bristol, UK, 2019; Volume 377, p. 012054.

35. Meshkov, A.; Kazanin, O.; Sidorenko, A. Methane Emission Control at the High-Productive Longwall Panels of the Yalevsky Coal Mine. In *E3S Web Conferences*; EDP Sciences: Evry, France, 2020; Volume 174, p. 01040.

36. Wang, M.L.; Zhang, H.X.; Zhang, G.Y. Safety assessment of ascending longwall mining over old rib pillar gobs. *J. Min. Saf. Eng.* **2008**, *25*, 87–90.

37. Dychkovskyi, R.; Shavarskyi, I.; Saik, P.; Lozynskyi, V.; Falshtynskyi, V.; Cabana, E. Research into stress-strain state of the rock mass condition in the process of the operation of double-unit longwalls. *Min. Miner. Depos.* **2020**, *14*, 85–94. [CrossRef]

38. Du, W.G.; Chai, J.; Zhang, D.D.; Lei, W.L. The study of water-resistant key strata stability detected by optic fiber sensing in shallow-buried coal seam. *Int. J. Rock Mech. Min. Sci.* **2021**, *141*, 104604. [CrossRef]

39. Ren, Y.W.; Yuan, Q.; Chai, J.; Liu, Y.L.; Zhang, D.D.; Liu, X.W.; Liu, Y.X. Study on the clay weakening characteristics in deep unconsolidated layer using the multi-point monitoring system of FBG sensor arrays. *Opt. Fiber Technol.* **2021**, *61*, 102432. [CrossRef]
40. Zhang, D.D.; Du, W.G.; Chai, J.; Lei, W.L. Strain Test Performance of Brillouin Optical Time Domain Analysis and Fiber Bragg Grating Based on Calibration Test. *Sens. Mater.* **2021**, *33*, 1387–1404.
41. Peng, S.S.; Chiang, H.S. *Longwall Mining*; Wiley: New York, NY, USA, 1984; p. 708.
42. Małkowski, P.; Niedbalski, Z.; Majcherczyk, T.; Bednarek, Ł. Underground monitoring as the best way of roadways support design validation in a long time period. *Min. Miner. Depos.* **2020**, *14*, 1–14. [CrossRef]
43. Yin, Q.; Wu, J.Y.; Zhu, C.; He, M.C.; Meng, Q.X. Shear mechanical responses of sandstone exposed to high temperature under constant normal stiffness boundary conditions. *Geomech. Geophys. Geo-Energy Geo-Resour.* **2021**, *7*, 35. [CrossRef]
44. Lou, J.F.; Gao, F.Q.; Yang, J.H.; Ren, Y.F.; Li, J.Z.; Wang, X.Q.; Yang, L. Characteristics of evolution of mining-induced stress field in the longwall panel: Insights from physical modeling. *Int. J. Coal Sci. Technol.* **2021**, *27*, 1–18.
45. Ren, Y.F.; Ning, Y. Changing feature of advancing abutment pressure in shallow long wall working face. *J. China Coal Soc.* **2014**, *39*, 38–42.
46. Jin, Z.Y.; Ma, L.Q.; Sun, Q. Impact factors of the distribution of advancing abutment pressure at comprehensive mechanized mining face. *Electron. J. Geotech. Eng.* **2015**, *20*, 1740–1741.
47. Cheng, Z.H.; Qin, H.Y.; Chen, L.; Zhao, X.D.; Yin, S.F. Relationship between mining thickness and peak position of advancing abutment pressure: A case study in Helin coal mine in China. *Arab. J. Geosci.* **2021**, *14*, 1–8. [CrossRef]

*Article*

# Stabilization of Rock Roadway under Obliquely Straddle Working Face

Peng Wang [1], Nong Zhang [1], Jiaguang Kan [1,*], Bin Wang [2] and Xingliang Xu [1]

[1] Key Laboratory of Deep Coal Resource Mining of the Ministry of Education, School of Mines, China University of Mining and Technology, Xuzhou 221116, China; wangpeng19@cumt.edu.cn (P.W.); zhangnong@cumt.edu.cn (N.Z.); 4645@cumt.edu.cn (X.X.)

[2] Admissions and employment Division, Wenzhou University, Wenzhou 325035, China; 20200035@wzu.edu.cn

* Correspondence: jgkan@cumt.edu.cn

**Abstract:** A floor rock roadway under an oblique straddle working face is a typical dynamic pressure roadway. Under the complex disturbance of excavation engineering works, the roadway often undergoes stress concentration and severe deformation and damage. To solve the problem of surrounding rock stability control for this roadway type, this study considered the East Forth main transport roadway in the floor strata of the 1762(3) working face of the Pansan coal mine. In situ ground pressure monitoring and numerical simulation calculation using the FLAC2D software were carried out. The influence laws of the surrounding rock lithology, the vertical and horizontal distance between the roadway and overlying working face, the positional relationship between the roadway and the overlying working face, and the support form and strength of the rock surrounding an oblique straddle roadway were obtained. Within the range of mining influence, the properties of the rock surrounding the roof and floor were very different, and the deformation of the rock surrounding the two sides exhibited regional difference. The influence range of the mining working face on the rock floor of the roadway was approximately 30–40 m, and that of horizontal mining was approximately 50–60 m. The mining influence on the rock surrounding the side roadway of the working face is large, but the mining influence on the roadway below is small. Using FLAC2D, the stress and displacement characteristics of the rock surrounding the obliquely straddle roadway were compared and analyzed when the bolt support, combined bolt and shed support, and bolt–shotcreting–grouting support were adopted, the proposed support scheme of bolting and shotcreting was successfully applied. The deformation of the rock surrounding the roadway was satisfactorily controlled, and the results were useful as a reference for similar roadway maintenance projects.

**Keywords:** obliquely straddle roadway; floor rock roadway; abutment pressure; stress evolution of surrounding rock; stability of surrounding rock

**Citation:** Wang, P.; Zhang, N.; Kan, J.; Wang, B.; Xu, X. Stabilization of Rock Roadway under Obliquely Straddle Working Face. *Energies* **2021**, *14*, 5759. https://doi.org/10.3390/en14185759

Academic Editors: Adam Smoliński, Manoj Khandelwal, Junlong Shang, Chun Zhu and Manchao He

Received: 27 July 2021
Accepted: 8 September 2021
Published: 13 September 2021



## 1. Introduction

Underground mining is dominant in China's coal mines. As the lifeblood of underground coal mine production, roadways account for a large proportion of mine engineering [1–4]. Dynamic pressure roadways account for 70–80% of coal mine roadways, among which the coal seam rock floor roadway is a typical dynamic pressure roadway [5–8]. After a floor rock roadway under an obliquely straddle working face is affected by the dynamic pressure of the overlying working face, owing to the change of the spatial relationship between the different areas of the roadway and the working face, the distribution of the stress field and the displacement field of the rock surrounding the roadway becomes more complex under the influence of the mining face, and the support difficulty increases [9–11]. Therefore, stricter requirements are proposed for the stability control of the rock surrounding a floor rock roadway under an obliquely straddle working face.

According to the positional relationship between the axial projection of the roadway floor to the plane of the overlying coal seam and the advancing vertical, parallel, and oblique direction of the upper coal seam working face, the cross-mining dynamic pressure roadway can be divided into three basic forms: the transverse straddle roadway, longitudinal straddle roadway, and oblique straddle roadway [12–14], as shown in Figure 1. During the mining of the overlying mining face, the bearing pressure generated by the mining space spreads to the floor. Under the complicated disturbance of the excavation engineering, the stress field and displacement field of the floor are redistributed. Hence, stress concentration occurs during driving or in the already driven cross-mining floor rock roadway. The original support system cannot guarantee the stability of the roadway, and the deformation and damage of the cross-mining roadway are severe, which restricts the normal production of the coal mine [15–18]. Analyzing the deformation and failure laws of the various rocks is an important method for investigating the stability of diagonal roadways. Chang and Haimson [19–22] obtained the strength and deformation characteristics of the rock under different loading conditions by carrying out true triaxial compression tests on different rocks. Many studies have extensively investigated the stability control of the cross-mining roadway [23–30]. Jiang [31] analyzed the characteristics of the surrounding rock deformation of the cross-span and longitudinal-span roadway, and put forward corresponding control countermeasures. Zhang and Liu [32,33] investigated the evolution of the stress field and displacement field of the floor under the influence of mining. The additional stress change of the floor can be divided into three areas along the coal seam direction, namely, the stress increase area, stress reduction zone, and stress recovery zone, to obtain the depth and shape for the obvious change zone of the rock floor stress. Aiming at the current situation that the Huaibei mining area has crossed the roadway for many times and cannot be used normally after repeated maintenance, Li [34] summarized 4 types of typical damage characteristics, and proposed control methods and technologies. Yuan [35] analyzed the mechanical characteristics of the surrounding rock of the longitudinal-span roadway based on the rheological characteristics, and pointed out that the rheological effect under the action of lateral stress is an important factor affecting the stability of the longitudinal-span roadway. Guo [36] conducted research on the stability of the crossing roadway in Inner Mongolia Chengyi Coal Mine and established a floor stress increment model. Based on elastic mechanics, Zhang [37] established the mechanical stress distribution model for the stope floor using the additional stress algorithm, and obtained the propagation law of the mining abutment pressure in the floor: The mining stress concentration coefficient of a point under the floor gradually decreased as the buried depth increased, and the degree of pressure relief gradually weakened as the buried depth increased. Xie [38] considered the formation process of the supporting pressure and its influence on the stability of the surrounding rock at the floor of the roadway, and analyzed the relationship between the stress of the surrounding rock and the displacement of the cross-mining roadway. Zhu [39] established a mechanical model to analyze the rheological characteristics of the rock surrounding the roadway under dynamic pressure in straddle mining, and proposed an expression for estimating the stress and deformation of the roadway with time. Zhang [40] used FLAC to analyze the influence range of mining on a crossing mining pressure roadway, and obtained the relationship between the vertical distance separating the roadway and the overlying working face and the horizontal distance of the stope boundary. Considering the high ground stress, large cross-section, and close distance characteristics, Ma [41] proposed a high-rigidity coupling support technology plan for a dynamic pressure roadway to improve the strength of the rock surrounding the roadway and the rigidity and stability of the supporting structure. The use of bolt, steel belt, anchor cable, metal mesh, and shotcrete coupling support technology is feasible. In summary, research on the distribution of the stress field and the displacement field and stability of the surrounding rock in cross-mining roadways has achieved good results. However, existing studies have mainly focused on transverse straddle roadways and longitudinal straddle roadways, while research on the stability of obliquely straddle roadways has been scarce. For the obliquely straddled roadway, there are two situations of the horizontal distance and vertical distance between each section of the roadway and the edge of the overlying working face: the horizontal distance is different, the vertical distance is different, or the horizontal and vertical distance are different. The influence factors of surrounding rock stability of crossing and longitudinal span roadways are relatively less and more controllable. The stress distribution of surrounding rock in different positions of obliquely straddle roadway is more different, so the stability of surrounding rock is relatively difficult to control. To date, the stress field and displacement field of an obliquely straddle floor rock roadway have not been elucidated, and their investigation is relatively difficult.

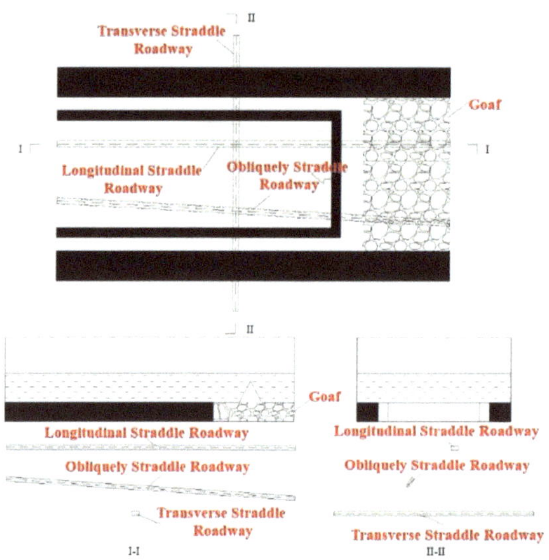


**Figure 1.** Layout diagram of rock roadway under overhead mining.

Based on previous studies and practical experience, and on in situ ground pressure monitoring, this study investigated the stability of the rock surrounding the roadway under different lithology conditions, different vertical distance between the roadway and the working face, different horizontal distance between the roadway and the working face, different positional relationship between the roadway and the working face, and different support form and strength. The Flac2D numerical simulation software was used to compare and analyze the control effect of the surrounding rock under different support schemes. Finally, an optimized support scheme was designed and evaluated. The results obtained by the industrial test revealed that the deformation control effect was satisfactory.

## 2. Engineering Background and Methods

### 2.1. Engineering Background

The Pansan coal mine is located in the northwest part of Huainan City, Anhui Province, China, and has a design production capacity of 5 MT/A. The Dongsi main transportation roadway of the Pansan coal mine is located in the floor rock of the 1762(3) working face. The elevation of the 1762(3) working face is −590 to −640 m, the mining height is 4.2 m, and the elevation of the Dongsi main transportation roadway is −637 m. The vertical distance between the Dongsi main transportation roadway and the 1762(3) working face is 30–43 m, and the horizontal projection distance between the Dongsi main transportation roadway and the 1762(3) working face is approximately 50 m. The direct roof of the roadway is 2.2 m thick mudstone, the main roof is 4.6 m thick sandy mudstone, the direct bottom is 2.8 m thick mudstone, and the main bottom is 3.5 m thick siltstone. Considering that the distance between the Dongsi transportation roadway close to the cut off of the 1762(3) working face is too close to the working face, and that the surrounding rock has poor lithology, the maintenance of the roadway is difficult and thus a new transportation roadway has been excavated to bypass this section. The layout of the roadway and working face is shown in Figure 2.

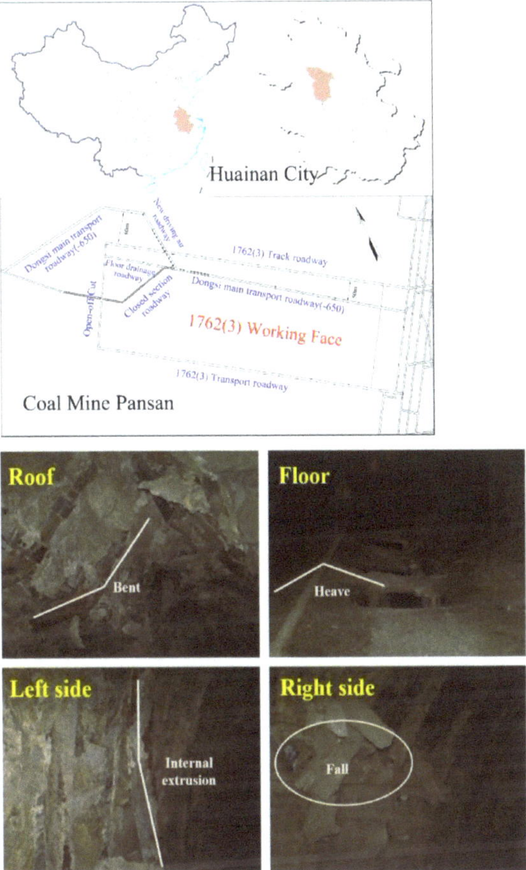



**Figure 2.** Mine general situation and roadway deformation.

Bolt support is used in a few sections of the Dongsi transportation roadway, and the remaining sections are simply supported by a U29 steel shed. The roadway section of the shed supporting section is arched, the section size is net-width $\times$ medium-height = 5.6 $\times$ 4.3 m, the shed distance is 600 mm. Two rows of bracing bars have been constructed, and the overall stability of the supporting structure is poor. Some U-shaped roof beams are bent and severely inclined, and most of the back plates behind the shed are not solid. From when the roadway starts being affected by the overlying working face mining until the influence tends toward stability, the surrounding rock is damaged in different degrees. Based on data obtained from the continuous observation of the roadway, the roadway deformation and failure situation are shown in Figure 2. When the distance between the roadway and the overlying working face is small, a part of the roof is bent down or even broken by the action of the two sides of the roadway, and a certain portion of the roof is bent upward by extrusion. Additionally, the shed legs of the side part are squeezed and even the back plate falls off.

## 2.2. Methods

This study mainly used the method of in situ monitoring and numerical simulation; the specific scheme is described below.

### 2.2.1. On-Site In Situ Mine Pressure Monitoring

Based on the influence of different factors on the stability of the rock surrounding the roadway, six types of measuring points were arranged: different lithology (2#, 1#), different vertical distance between the roadway and the overlying working face (3#, 4#), different horizontal distance between the roadway and the overlying working face (5#, 6#), different positional relationship between the roadway and the overlying working face (7#, 8#), different support form and strength (9#, 10#), and optimized supporting scheme (11#, 12#). The monitoring contents include the deformation of the surrounding rock, deformation velocity, roof separation layer, development of surrounding rock fracture, and so on. The monitoring methods and instruments are the cross-section method, roof separation instrument, and borehole peeping instrument. The specific layout of the measuring points is shown in Figure 3.

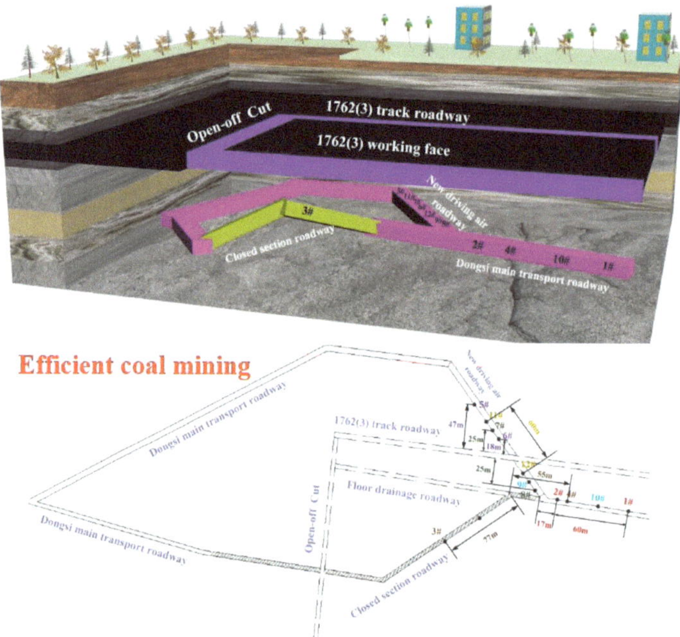


**Figure 3.** Layout of measuring points.

### 2.2.2. Numerical Simulation Software and Model

According to geological data for the Pansan coal mine, the FLAC2D numerical simulation software was used to establish the corresponding analysis model. The plane strain calculation model shown in Figure 4 was used to simulate the deformation process of the rock surrounding the roadway, and the surrounding rock was considered as a layered isotropic elastic medium. The calculation model size is length × width = 200 × 71 m, divided into 25,680 units, the rock mass element grid is divided into hexahedral elements. In order to balance the calculation accuracy and speed, the grid division around the roadway is dense and the grid division away from the roadway is sparse, which can effectively reduce the sawtooth shape in the stress nephogram. The left, right, and lower boundaries of the model were displacement fixed–constraint boundaries, and the upper boundary was the stress boundary. The uniformly distributed load was applied according to the thickness of the overlying strata. The buried depth of the roadway was about 600 m, the overlying load was 15 MPa, the lateral pressure coefficient was 1, and the gravity was set to 10 m/s$^2$. At the same time, the structural plane was applied at the upper and lower boundaries of the coal seam to avoid the mutual embedding of the overlying roof collapse and the floor strata after the excavation of the working face, the constitutive relationship of the surrounding rock is the Mohr Coulomb model. The roadway was located in the bottom sandstone layer of the working face, and the vertical distance from the overlying working face was 30.3 m, and the cross-section of the roadway was arched, the cross-section size is net-width × middle-height = 5.6 × 4.3 m. The mechanical parameters of the rock were selected as presented in Table 1. The simulation process is as follows: calculation of original rock stress balance; calculation of model not affected by mining; calculation of affected by mining of 13–1 coal seam.

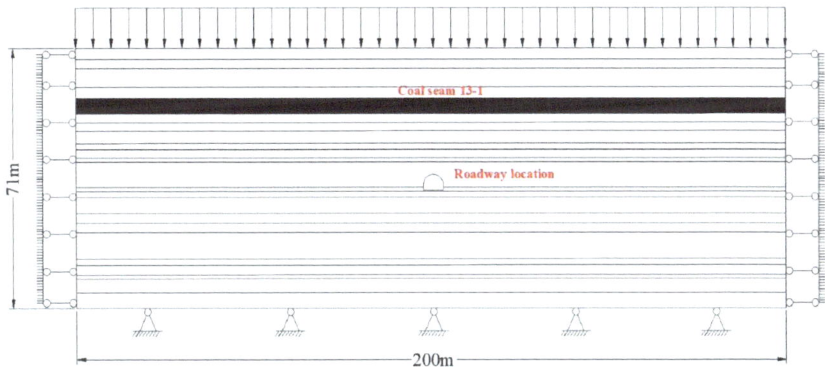


**Figure 4.** Numerical simulation model.

**Table 1.** Reference table of rock mechanics parameters.

| Name | Bulk Modulus (GPa) | Shear Modulus (GPa) | Density (kg/m$^3$) | Cohesion (MPa) | Friction Angle (°) |
|---|---|---|---|---|---|
| Sandstone | 7.3 | 4.2 | 2400 | 2 | 35 |
| Siltstone | 8.1 | 5.1 | 2500 | 2.1 | 29 |
| Mudstone | 3.2 | 2.74 | 2400 | 1.2 | 25 |
| Coal seam | 2.4 | 1.37 | 1350 | 0.8 | 23 |

## 3. Results and Analysis

### 3.1. In Situ Test for Surrounding Rock Stability of Obliquely Straddle Roadway under Different Influencing Factors

3.1.1. Influence of Different Lithology on Stability of Surrounding Rock

In the layout of the measuring points shown in Figure 3, the measuring points 2# and 1# were both arranged in the tunnel of the shed section. The distance between the two measuring points was 60 m, and the distance from the upper mining face was approximately 30 m. According to the borehole detection data, both sides of the roadway at the 2# and 1# survey points are medium sandstone with moderate hardness. The sandstone lithology in the 2# measuring point is relatively brittle and can easily fracture under the influence of mining. The roof is hard sandstone. The floor of the 1# measuring point contains approximately 7.2 m of sandy mudstone and clay rock, and the thickness of the sandy mudstone at the 2# measuring point is only approximately 3.3 m.

The deformation of the two sides, the deformation of the roof and floor, and the deformation velocity of the measuring points 2# and 1# are shown in Figure 5.

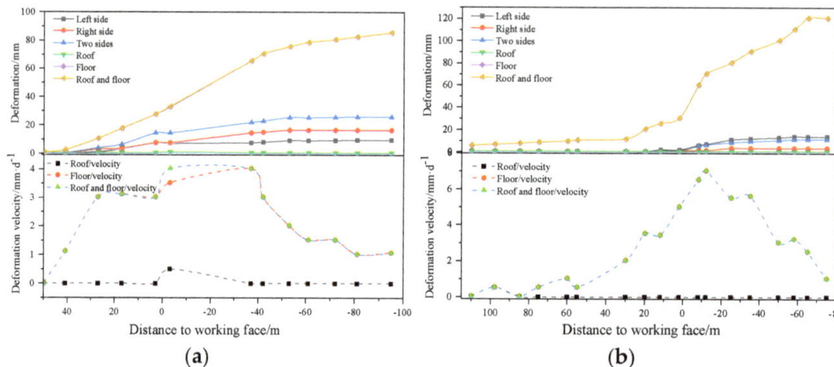

**Figure 5.** Comparison of deformation amount and deformation velocity of roadway surrounding rock (**a**) measuring point 2# (**b**) measuring point 1#.

As can be seen in Figure 5, the surface displacement of measuring point 2# starts being affected by the mining when it is approximately 40 m ahead of the working face, and 1# only starts being affected by advance mining when it is 20 m away from the working face. The accumulated deformation of the 2# measuring point is approximately 28 mm, and that of the 1# surface displacement measuring point is 14 mm. The deformation of the right side (leeward) of the roadway at the two measuring points is approximately 10 mm larger than that of the left side. After mining approximately 70 mm, the deformation of both sides gradually tends toward stability. After the working face is pushed through the measuring point, the accumulated deformation of the 2# and 1# measuring points is 90 and 120 mm, respectively, and the deformation of the roadway roof and floor at both sides continues. From the deformation velocity of the roof and floor, the maximum moving velocity for the roof and floor of the 2# and 1# measuring points is 4.25 and 7 mm/d, respectively. After the working face pushes over 70 m, the deformation velocity of the roadway at the two measuring points sharply decreases and gradually tends toward stability.

Through the deep detection of the surrounding rock near the 2# and 1# measuring points, the deep separation layer and fracture development of the surrounding rock at the side were observed as shown in Figure 6. From the drilling detection at different depths shown in Figure 6, it can be seen that the left and right sides of the roadway are relatively fractured within the range of 1.5 m near the 2# measuring point. In addition to the various cracks at approximately 2.5 m on the right side, there is no obvious fracture phenomenon in the surrounding rock at the depth of the measuring point. Moreover, from the perspective of the fragmentation degree, the right side is obviously larger than the left side, which leads to the displacement of the right side of the roadway and is greater than the displacement of the left side. The left side of the roadway near the 1# measuring point was relatively fractured in the range of 1.0 m, from 1.5 to 3.0 m, and crack development was not observed in the rock surrounding the roadway. However, when the depth was 3.5–4.5 m, cracks in the rock surrounding the roadway developed to a certain extent. This indicates that, in this section, the surrounding rock was relatively fractured, compared with other sections, and that the surrounding rock properties are poor. Therefore, when the mining stress is transferred to the side of the floor roadway, the surrounding rock in this section is prone to plastic deformation or even brittle failure. The fracture of the surrounding rock on the right side of the roadway is mainly concentrated in the range of 1.0 m, and there is essentially no crack development in the deep part of the surrounding rock. According to the shape of the hole's inner surface, there are circular ripples on the surface of the borehole when the hole is 1.5–2.5 m, while the surface of the borehole is spiral at the depth of the borehole, particularly when the hole is 3.5–4.5 m. This indicates that the hardness of the rock mass is quite different at different borehole depths. When the surrounding rock is soft, the surface of the drilled hole has a circular or non-obvious ripple. However, when the surrounding rock is hard, the ripple on the surface of the hole has spiral form.

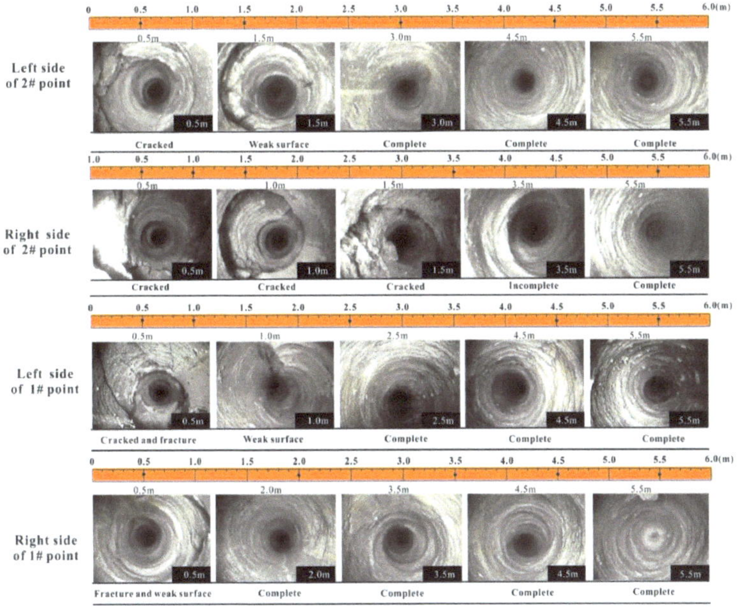

**Figure 6.** Borehole peep imaging.

### 3.1.2. Influence of Vertical Distance and Horizontal Distance between Roadway and Overlying Working Face End on Surrounding Rock Stability

To investigate the influence of the vertical distance and horizontal distance between the Dongsi main transportation roadway and the end of the overlying working face on the stability of the surrounding rock, four measuring points were arranged. In the vicinity of measuring point 3#, the vertical distance between the roadway and the overlying 1762(3) working face is only approximately 10 m, while in the vicinity of the 4# measuring point, the vertical distance between the roadway and the overlying working face is approximately 30 m. The horizontal distance between the measuring point 5# and the overlying 1762(3) working face is approximately 47 m, while that of 6# is approximately 18 m.

The vertical distance between the 3# and the 4# measuring point and the working face is different, and the deformation amount and deformation velocity curve of the upper part are shown in Figure 7.

As can be seen in Figure 7, the cumulative deformation of the two sides near the 3# measuring point is 90 mm, the maximum deformation speed reaches 3 mm/d, and the average deformation speed is 1 mm/d during the mining influence. The deformation of the surrounding rock slope tends toward stability when the working face pushes through the measuring point at approximately 100 m. The cumulative deformation of the two sides of the 4# measuring point is approximately 25 mm, and the maximum deformation velocity is 2 mm/d. The average deformation velocity during mining is only approximately 0.7 mm/d. The deformation of the side begins to stabilize after the working face pushes over the measuring point at approximately 50 m. Therefore, as the vertical distance between the roadway and the working face becomes smaller, the degree of the mining influence increases, and the mining influence time after the working face is mined becomes longer. For the 3# measuring point, the maximum deformation velocity of the roof and the floor of the roadway under the mining of the overlying working face reaches 12 mm/d, and the average deformation velocity is still 2 mm/d when the working face passes the 3# measuring point by 100 m. However, for the 4# measuring point, the maximum deformation velocity of the roadway top and floor is 6 mm/d, and the average deformation velocity of the roadway is only 1 mm/d at 80 m after the measuring point.

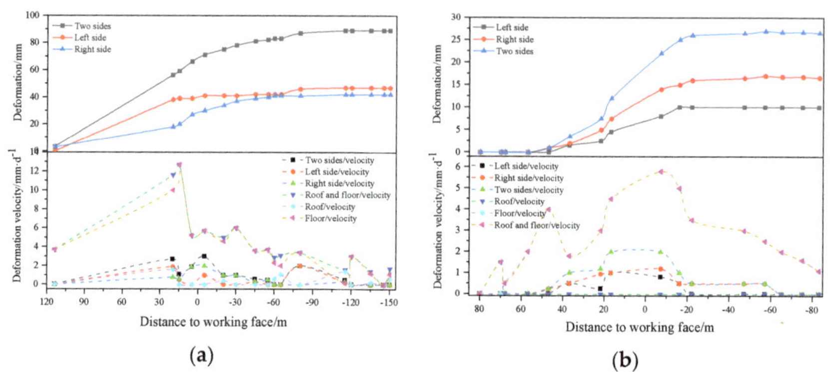

**Figure 7.** Comparison of deformation amount and deformation velocity of roadway surrounding rock (**a**) measuring point 3# (**b**) measuring point 4#.

Measuring points 5# and 6# are both located in the new combined roadway, and the supporting mode and surrounding rock properties are essentially the same. Before and after mining, owing to the difference of the horizontal distance from the working face, the rock surrounding the roadway also has different deformation and failure degrees. One of the most obvious characteristics is the difference of the roadway surface displacement; details are shown in Figure 8.

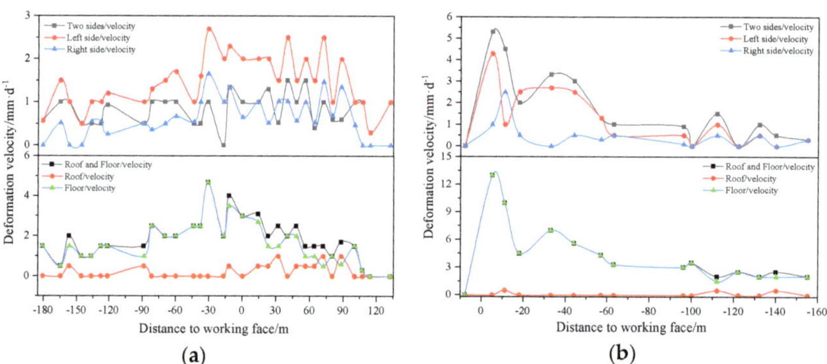

**Figure 8.** Comparison of deformation velocity of roadway surrounding rock (**a**) measuring point 5# (**b**) measuring point 6#.

As can be seen in Figure 8, during the mining of the overlying working face, the maximum deformation velocity on the two sides of measuring point 5# was approximately 2.5 mm/d. As the working face moved forward, the deformation fluctuated on both sides, but tended toward stability overall. Under the influence of mining, the maximum deformation velocity on both sides of the 6# measuring point was 5.5 mm/d, which is twice that of measuring point 5#. The maximum approaching velocity of the roof and floor near measuring point 5# was 4.5 mm/d, and its average moving speed was approximately 2 mm/d under the mining influence. The maximum moving speed of the roof and floor near measuring point 6# was approximately 14 mm/d, while the average deformation velocity of the working face before and after mining was 4 mm/d, which is twice that of measuring point 5#.

### 3.1.3. Influence of Position Relationship between Roadway and Overlying Working Face on Surrounding Rock Stability

In view of the particular position of the Dongsi transportation roadway, the working face and roadway in the joint roadway section are inclined across. Owing to the stability of the surrounding rock in the shed section, the deformation of the surrounding rock is small. In this study, the influence of the roadway on the surrounding rock deformation and failure was only analyzed under the lateral and lower conditions of the working face.

The horizontal distance between the 7# and 8# surface displacement measuring points and the track gateway is approximately 25 m, and the vertical distance is essentially the same as that of the overlying working face, wherein 7# is located at the lateral position of the overlying working face and 8# is located directly below the overlying working face. Because 7# and 8# are located in the newly excavated combined roadway section, the deformation law of the rock surrounding the roadway is quite different owing to the different position of the working face under the conditions of the support mode, same normal distance with the track gateway, and similar surrounding rock properties.

It can be seen from Figure 9 that the maximum deformation of the two sides of 7# measuring point is 70 mm, the maximum deformation velocity is 6 mm/D, and the average deformation velocity during mining is 2 mm/d. Moreover, the surrounding rock of the roadway is still not stable, and the deformation is still continuing after the working face passes through the measuring point about 160 m. The maximum approaching amount and velocity of roof and floor are 160 mm and 17 mm/d. After the working face passes this measuring point, the lateral abutment pressure still has a great influence on the deformation and failure of the surrounding rock of the roadway, and the average moving speed of the two sides can still reach 4 mm/d. The maximum deformation of both sides of the roadway at 8# measuring point is 9 mm, the maximum deformation velocity is about 1.5 mm/d, and the average deformation velocity affected by mining is 0.6 mm/d. When the working face passes the measuring point about 60 m, the surrounding rock begins to become stable, and the maximum approaching amount and velocity of roof and floor are 17 mm and 1.5 mm/d respectively. After the overlying working face crosses this point, the residual abutment pressure of goaf has little influence on the surrounding rock deformation, and when the working face has been mined for about 70 m, the surrounding rock of roadway has been stable.

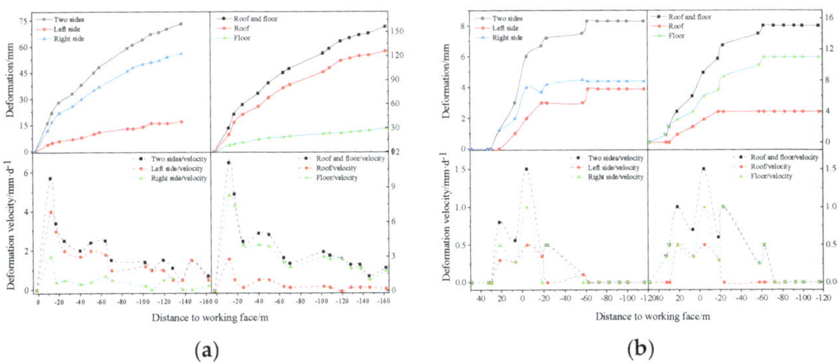

**Figure 9.** Comparison of deformation amount and deformation velocity of roadway surrounding rock (**a**) measuring point 7# (**b**) measuring point 8#.

### 3.1.4. Influence of Support Form and Strength on Surrounding Rock Stability

According to the support mode of the Dongsi main transportation roadway, the original support at the 10# measuring point was scaffolding support with a shed distance of 600 mm, and two rows of bracing bars were constructed. The roadway at the 9# measuring point is supported by bolt and anchor cable. The bolt specification is $\Phi 20 \times 2.2$ m, the spacing between the rows is $700 \times 700$ mm, the specification of the anchor cable is

18 mm × 6.3 m, the spacing between the rows is 1400 × 1400 mm, and the arrangement form of the anchor cable is 3–0-3. The measuring points 9# and 10# are located under the working face, and the surface displacement and deformation law are shown in Figure 10.

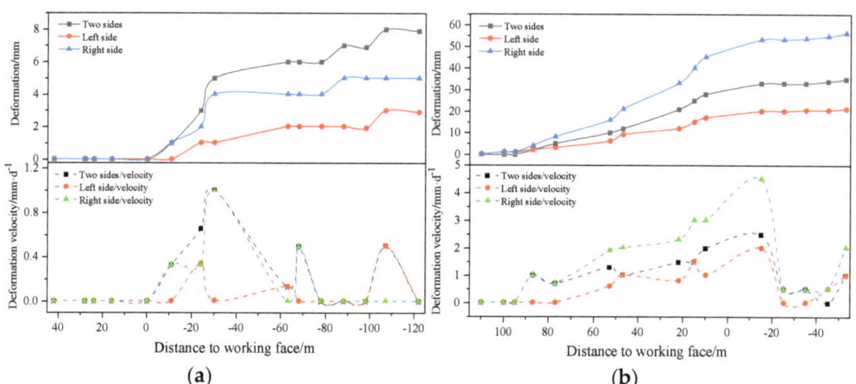

**Figure 10.** Comparison of deformation amount and deformation velocity of roadway surrounding rock (**a**) measuring point 9# (**b**) measuring point 10#.

As can be seen in Figure 10, when the roadway is supported by bolt and cable and the overlying working face is directly above the 9# measuring point, the slope starts being affected by the mining stress and the maximum deformation velocity is only 1 mm/d. When the working face passes the measuring point at approximately 70 m, the surrounding rock deformation velocity is 0.5 mm/d, which indicates that the roadway gradually becomes stable. The distance between the 10# measuring point and the working face is 90 m, owing to the influence of mining. When the working face crosses the measuring point at 50 m, its deformation speed can still reach 2 mm/d, which indicates that the rock surrounding the roadway is still affected by mining.

*3.2. Simulation of Surrounding Rock Stability for Inclined Span Roadway under Different Support Conditions*

3.2.1. Stress and Deformation of Rock Surrounding the Roadway Supported by Original Bolt under Mining Influence

In the process of coal seam mining, the relative positions of the roadway and the working face can be divided into three types according to the different distances between the roadway and the working face: in front of the working face, directly below the working face, and behind the working face. Therefore, the mining influence can be divided into the advance mining influence, mining process influence, and lagging mining influence. To elucidate the influence degree of mining at different roadway positions, combined with the advance progress of the working face, the deformation failure and stress distribution characteristics of the rock surrounding the roadway were simulated with consideration to the roadway being 60 and 20 m away from the working face, and the positions of 20 and 60 m across the roadway. The stress field distribution and surrounding rock deformation are shown in Figures 11–13.

As can be seen in Figures 11 and 12, the horizontal stress around the roadway was approximately 5.0 MPa and the vertical stress was approximately 2.5 MPa when there was no mining influence. When the working face advanced, the horizontal stress around the roadway became 10 MPa and the vertical stress was 5 MPa. When the horizontal distance between the working face and the roadway was 60 m, the advanced stress started exerting its influence, and the effect was obvious. When the working face crossed the roadway at 20 m, the horizontal stress and vertical stress of the roadway were still very high, which indicates that the influence of mining was unstable. When the working face crossed the roadway at 60 m, the horizontal and vertical stress values remained essentially unchanged, which indicates that the influence of mining on the rock surrounding the roadway had become stable. As can be seen in Figure 13, when there was not mining influence, the roadway roof subsidence was approximately 140 mm and the floor heave was approximately 50 mm. When the working face advanced and crossed the roadway at 60 m, the roadway roof subsidence reached 300 mm and the floor heave reached approximately 500 mm. As can be seen, the original support strength of the roadway no longer satisfied the needs of upper working face mining.

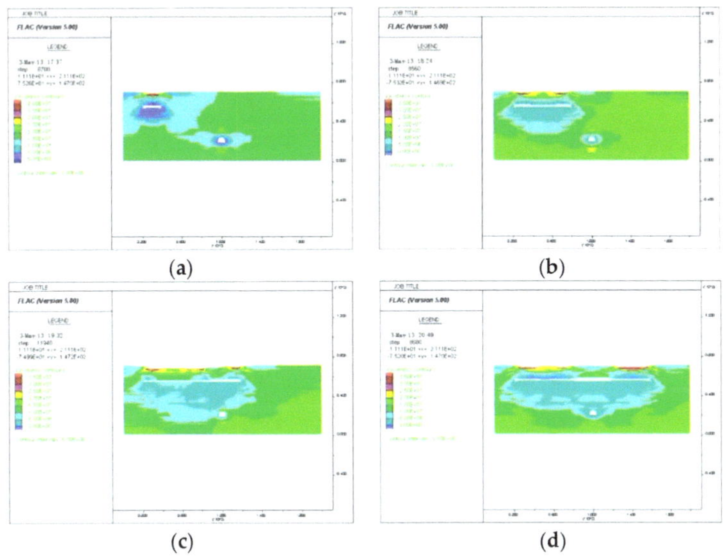

**Figure 11.** Horizontal stress distribution of roadway surrounding rock during bolt support. (**a**) Working face is 60 m away from the roadway. (**b**) Working face is 20 m away from the roadway. (**c**) Working face spans the roadway 20 m. (**d**) Working face spans the roadway 60 m.

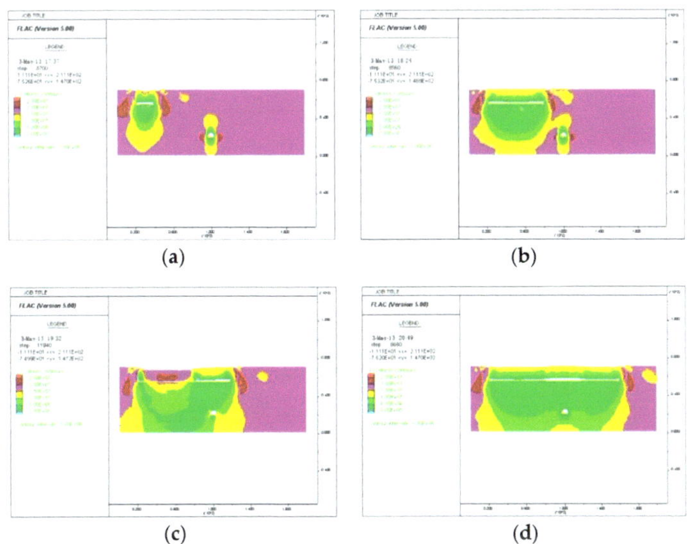

**Figure 12.** Vertical stress distribution of roadway surrounding rock during bolt support. (**a**) Working face is 60 m away from the roadway. (**b**) Working face is 20 m away from the roadway. (**c**) Working face spans the roadway 20 m. (**d**) Working face spans the roadway 60 m.

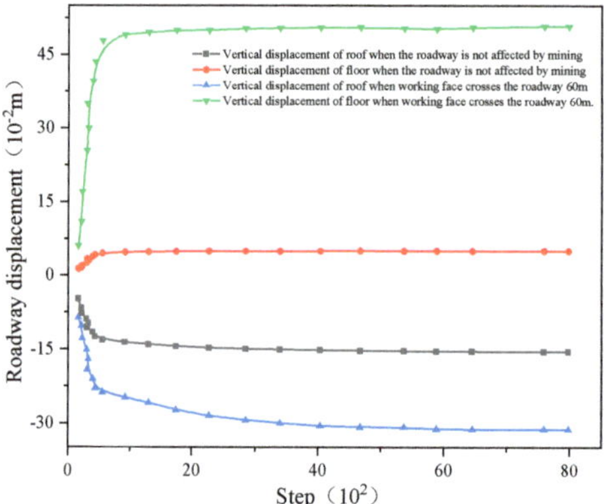

**Figure 13.** Roadway displacement.

### 3.2.2. Stress and Deformation of Rock Surrounding the Lower Roadway under Different Optimized Support Schemes

To ensure that the roadway can continue to be used while bearing enormous mining stress, in this study, the two schemes of combined bolt and shed support and anchor–shotcreting–grouting (Anchor Cable + shotcreting + grouting) were designed and compared. The deformation and failure law of the roadway and overlying strata, and the stress distribution characteristics of the surrounding rock brought about by the working face mining under different supporting forms, were analyzed with regard to the vertical distance of 30.3 m between the main roadway and the coal seam.

(1)   Simulation results and analysis of combined bolt and shed support.

The stress field distribution and deformation of the surrounding rock are shown in Figures 14–16.

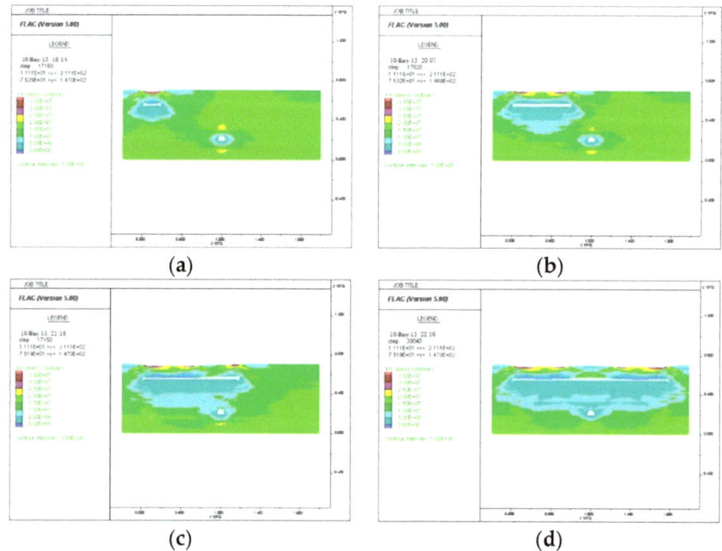

**Figure 14.** Horizontal stress distribution of surrounding rock of roadway supported by bolt and shed. (**a**) Working face is 60 m away from the roadway. (**b**) Working face is 20 m away from the roadway. (**c**) Working face spans the roadway 20 m. (**d**) Working face spans the roadway 60 m.

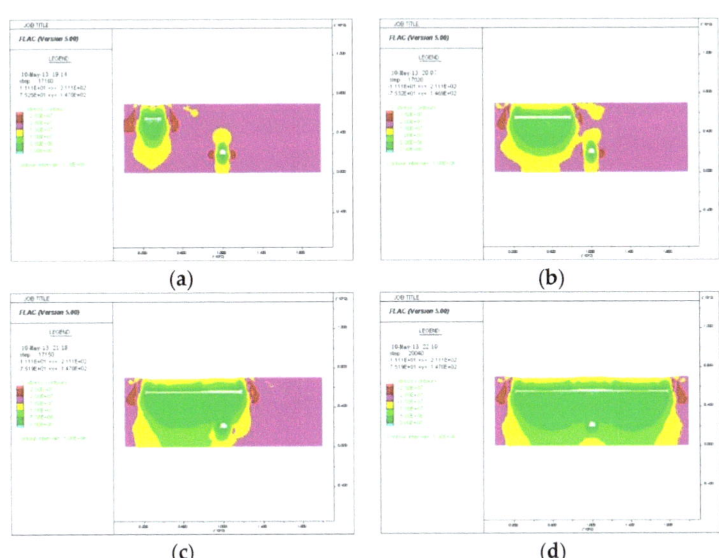

**Figure 15.** Vertical stress distribution of surrounding rock of roadway supported by bolt and shed. (**a**) Working face is 60 m away from the roadway. (**b**) Working face is 20 m away from the roadway. (**c**) Working face spans the roadway 20 m. (**d**) Working face spans the roadway 60 m.

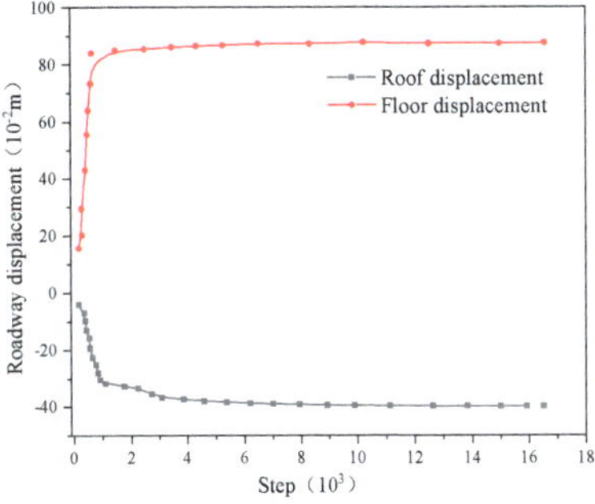


**Figure 16.** Vertical displacement of working face when crossing roadway 60 m.

According to the stress distribution in Figures 14 and 15, when the roadway was approximately 70 m away from the working face, the roadway was not affected by the mining of the overlying working face. When the working face was 60 m away from the roadway, the vertical stress of the roadway significantly increased, and the roadway started being affected by the mining of the overlying working face. Therefore, the maximum advance stress range was 60–70 m when the combined support of the bolt and shed was used. When there was no mining influence, the horizontal stress around the roadway was approximately 5.0 MPa and the vertical stress was approximately 2.5 MPa. After mining, the horizontal stress and vertical stress of the rock surrounding the roadway were approximately 10 and 5 MPa, respectively. Figure 16 shows the vertical displacement curve of the rock surrounding the roadway when the working face roadway crossed at 60 m. As can be clearly seen from the figure, the maximum roof subsidence and floor heave of the roadway were approximately 350 and 800 mm, respectively, as the upper working face advanced. Obviously, the combined support of the bolt and shed could not effectively control the deformation of the surrounding rock.

(2)    Simulation results and analysis for anchor–shotcreting–grouting support.

The stress field distribution and deformation of the surrounding rock are shown in Figures 17–19.

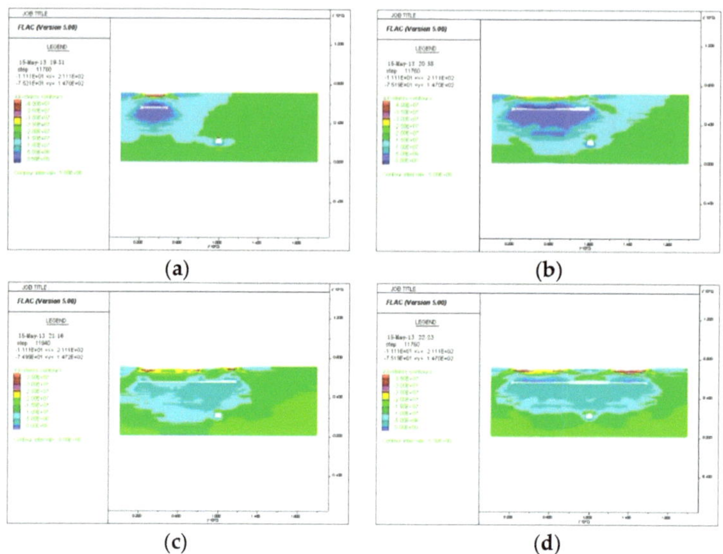

**Figure 17.** Horizontal stress distribution of surrounding rock of roadway supported by anchor–shotcreting–grouting support. (**a**) Working face is 60 m away from the roadway. (**b**) Working face is 0 m away from the roadway. (**c**) Working face spans the roadway 20 m. (**d**) Working face spans the roadway 60 m.

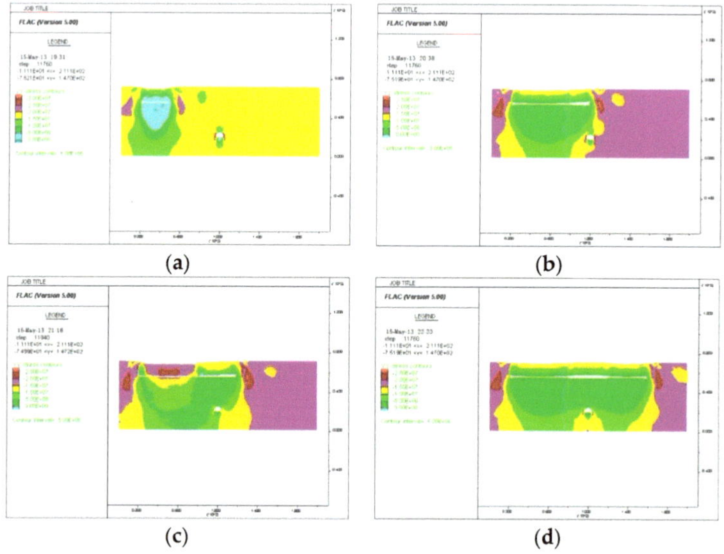

**Figure 18.** Vertical stress distribution of surrounding rock of roadway supported by anchor–shotcreting–grouting support. (**a**) Working face is 60 m away from the roadway. (**b**) Working face is 0 m away from the roadway. (**c**) Working face spans the roadway 20 m. (**d**) Working face spans the roadway 60 m.

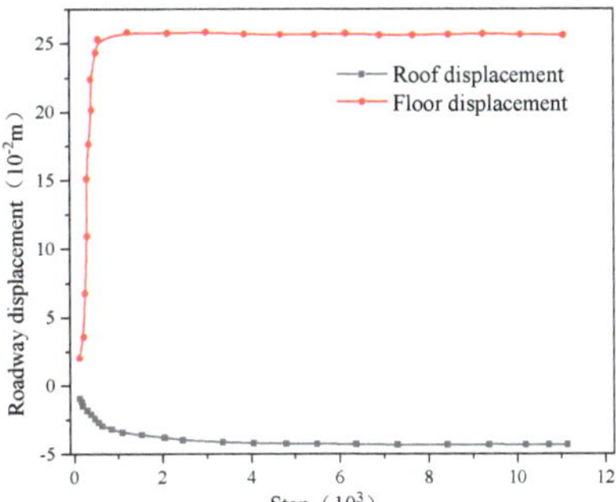

**Figure 19.** Vertical displacement of working face when crossing roadway 60 m.

As can be clearly seen in Figures 17 and 18, when the working face was 60 m away from the roadway, the roadway was not affected by mining. When the advancing distance was 50 m from the roadway, the roadway was already within the influence range of advanced mining in the upper coal seam. Therefore, it can be inferred that the maximum influence range of the mining advance stress was 50–60 m. When the working face crossed the roadway at 20 m, the roadway was still in the area influenced by mining. When the working face crossed the roadway at 40 m, the roadway was essentially not affected by the mining; when the working face crossed the roadway at 60 m, the stress did not change. Therefore, the influence area of the mining lag was approximately 20–40 m. The horizontal stress and vertical stress value of the rock surrounding the roadway did not obviously change compared with the combined bolt and shed support, which is approximately 5–10 MPa. The vertical stress distribution clearly shows that the bolt and shotcreting reinforcement can provide sufficient bearing capacity to the rock surrounding the roadway, and maintain its stability under large stress. Figure 19 shows the vertical displacement when the working face crossed the roadway at 60 m. As can be seen in the figure, when the upper working face was mined, the roof subsidence of the roadway was approximately 40 mm, that is, 1/8 of the combined bolt and shed support. Additionally, the floor heave was approximately 250 mm, which is 1/3 of the combined bolt and shed support. From the above analysis, it follows that the bolting and shotcreting support measures greatly improved the bearing capacity and stability of the rock surrounding the roadway.

### 3.3. Discussion

For the roof and floor, the surrounding rock properties greatly varied within the influence range of mining. The lithology of the rock surrounding the roadway was stress-transmission-oriented, and stress concentration could easily occur in parts with poor lithology, which is a key factor in roadway deformation and failure. The roof lithology of the Dongsi transportation roadway has hard characteristics, and the roof essentially did not sink under the mining influence of the 1762(3) working face. However, the surrounding rock properties at the floor are poor and the floor heave was severe after mining, accounting for 95% of the roof and floor displacement; the maximum floor heave speed was 16 mm/d. For the cross-mining roadway, when the distance between the roadway and the working face was large, the roof was not affected by mining owing to the rapid attenuation of mining stress. When the distance between the roadway and the working face was small, the mining stress attenuation was slow and still exerted great influence on the deformation and failure of the surrounding rock when it was transmitted to the floor roadway. Owing to the action of the roof, the stress was transferred to two sides and squeezed in the interior of the roadway, which lead to the inner extrusion of the shed legs and upward movement of the roof. Therefore, the vertical distance between the roadway and the working face was different, and the influence of the mining stress on the roof of the roadway floor was also different, particularly on the two sides. When the vertical distance from the overlying working face was the same, and the horizontal distance from the working face edge or coal pillar was different, the deformation and damage degree of the rock surrounding the roadway was also very different. For the 5# measuring point, the horizontal distance between the working face edge and the overlying working face was relatively large, which weakens the influence of the stress concentrated at the edge of the working face on the roadway. Moreover, the horizontal distance between the 6# measuring point and the overlying working face was only 18 m. The stress concentration of the surrounding rock was very high, owing to the severe deformation and damage of the rock surrounding the roadway. Therefore, the influence degree of the deformation of the rock surrounding the roadway decreased with the increase of the horizontal distance, and the influence degree on the roof and floor was much greater than that on the two sides. The data obtained from the monitoring points reveal that the vertical mining influence range of the working face was approximately 30–40 m, and the influence range of the horizontal mining was approximately 50–60 m.

In the lower and lateral part of the working face, when the horizontal distance between the roadway and the edge of the overlying working face or coal pillar was large, the influence of the stress concentrated around the working face edge or coal pillar on the floor roadway was weak, and the deformation of the rock surrounding the roadway was very small. When the roadway was close to the edge of the working face or the coal pillar, the floor roadway was in the strong mining influence range and the stress concentration degree of the rock surrounding the roadway was very high, which led to the severe deformation and failure of the roadway. When the relative position of the roadway and overlying working face was different, the influence degree of the mining stress caused by the working face on the rock surrounding the roadway was significant. In the lateral position of the working face, the stress could reach 2–3 times the stress of the original rock. Under the action of such high stress, the rock surrounding the roadway is prone to brittle failure. The rock surrounding the working face has a great impact on the surrounding rock for a long time after the working face is mined, which leads to the large deformation of the surrounding rock. Under the goal of the working face, the increase coefficient of the abutment pressure is typically less than 1. Because the abutment pressure is very small after the working face is mined, it does not influence the deformation of the rock surrounding the roadway, and tends to become stable soon thereafter. Hence, there is a big difference between 7# and 8#.

Based on the in situ ground pressure test and numerical simulation analysis, it was found that the bolt and cable support can effectively control the deformation of the roadway side, and that the shed support has no obvious control effect on the side. When the bolt and cable support is used, the mechanical properties of the surrounding rock are effectively improved, and the anti-deformation ability of the surrounding rock is enhanced. Generally, in the factors affecting the deformation and failure of the rock surrounding the cross-mining roadway, the internal structure of the surrounding rock and the nature of the rock mass determine the ability of the surrounding rock to bear the mining stress, and the vertical or horizontal distance between the roadway and the overlying working face determines the magnitude and degree of the mining influence on the roadway. The relationship between the position of the roadway and the working face determines the area affected by mining, and the supporting form and strength determine the bearing strength of the roadway affected by mining and its adaptability to the dynamic pressure roadway. However, under the original support conditions, the Dongsi main transportation roadway deformation was severe, and the combined effect of the bolt and shed support was not satisfactory. During the mining process of the upper coal seam, the vertical stress is concentrated on both sides of the roadway, and the horizontal stress is concentrated on the roof and floor of the roadway. The ground pressure behavior is severe, and the original support cannot effectively control the deformation of the rock surrounding the roadway. Hence, it is necessary to adopt high-strength support to control the surrounding rock deformation in time. Based on in situ test and theoretical analysis, the reasonable support scheme should be considered from two aspects of surrounding rock modification and support strengthening. First of all, bolt and cable support is used to improve the prophase support strength, control the deformation of shallow surrounding rock and eliminate the separation phenomenon as far as possible, further grouting to improve the internal structure of surrounding rock, improve the bearing strength of surrounding rock, and weaken the development degree of internal cracks in surrounding rock. Finally, surface shotcreting can improve the surface fragmentation of roadway, further improve the overall bearing capacity of support structure, and effectively control the surrounding rock deformation. When using "bolt shotcreting" to strengthen the support, the anchor cable can play an active role in controlling the deformation of the surrounding rock. Moreover, the anchoring depth is large, the bearing capacity is high, and a great pre-tightening force can be exerted. Grouting can consolidate the broken rock mass, improve the mechanical properties of the surrounding rock structure, and improve the overall strength of the surrounding rock to increase its bearing capacity, which plays an important role in ensuring the stability of the rock surrounding the Dongsi transportation roadway. At the same time, passive support should be added in time to ensure the safety of production.

## 4. Industrial Tests

### 4.1. Support Scheme Design

Based on the above analysis, the optimized support scheme design adopts bolt-shotcreting-grouting to reinforce the main roadway. For the bolt support section and shed support section, the support scheme and parameters are the same, but the construction sequence of each support measure is different. The bolt-supported roadway adopts roof anchor cable→shotcreting→shallow grouting, while the shed supporting the roadway adopts shotcreting→shallow grouting→roof anchor cable. The specific supporting parameters are as follows.

For the key parts of the roadway, nine sets of high-prestressed single anchor cables were arranged at the roof and side of the roadway, and a $\Phi$300-mm-disc anchor cable tray was equipped. The roof is made of a $\Phi$21.8 mm $\times$ 6.3 m steel strand with hole depth of 6.0 m. The two sides are made of a steel strand with specification of $\Phi$21.8 mm $\times$ 4.3 m with a hole depth of 4.0 m. One roll of k2360 and three rolls of z2360 resin cartridge were used for each hole of the roof anchor cable, and one roll of k2360 and two rolls of z2360 resin cartridge were used for the side anchor cable. The pre-tightening force was 80–100 kN, and the anchoring force was not less than 200 kN. The row spacing between the anchor cables was 1300 $\times$ 1000 mm, as shown in Figure 20a.

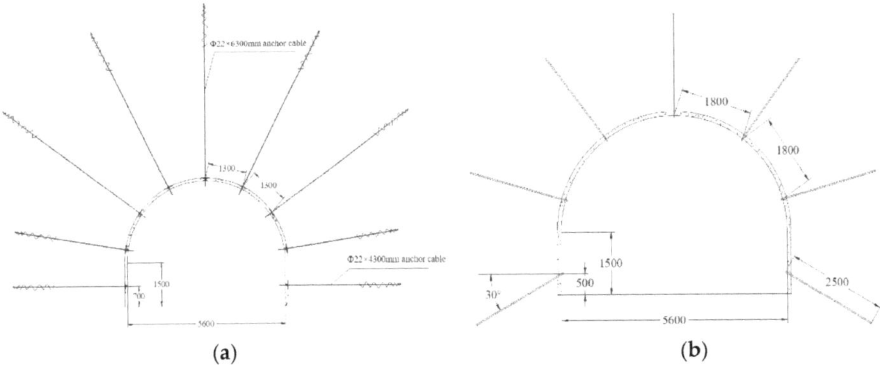


**Figure 20.** Schematic diagram of support scheme (mm). (**a**) Arrangement of side and roof anchor cables (**b**) Grouting reinforcement parameter.

The U-shaped steel shed and rock surrounding the roadway were closed by shotcrete. The thickness of the shotcrete should be 70–100 mm. The concrete ratio was cement with sand:gravel = 1:2:2. To ensure the grouting effect, the U-shaped steel shed must be completely closed. After spraying, the concrete should be watered and maintained to improve the strength of the shotcrete layer.

The grouting material was sulfoaluminate cement with a water cement ratio of 0.85–1.0, and was used to seal large cracks and reinforce the shallow fractured surrounding rock mass. The length of the grouting pipe was 2.5 m, and the grouting pressure was generally not more than 3.0 MPa. The grouting amount was not subjected to a large amount of slurry leakage. The grouting sequence was as follows: low pressure grouting was successively carried out from the grouting hole at the bottom corner of the roadway until the space behind the U-shaped steel shed wall was filled. As shown in Figure 20b, the details of the parameters are as follows:

(1) The grouting section and the anchor cable construction section were arranged at intervals. An air hammer was used to drill holes. Each section was arranged with seven holes, and the spacing between the rows was 1800 × 1000 mm. The grouting pipe at the side was 500 mm away from the roadway floor, and the construction was carried out with a downward inclination of 30°. The drill diameter was $\Phi = 42$ mm, and the grouting hole depth was 3.0 m;

(2) The length of the grouting bolt was 2.5 m, the front hole diameter was 8 mm, and the back-end hole diameter was 4 mm. The sealing material was hollow quick setting cement roll, and the sealing depth was 0.3 m.

*4.2. Monitoring Results*

According to the layout diagram of the measuring points shown in Figure 3, the surface displacement measuring points 11# and 12# were both arranged in the roadway of the air intake section. The distance between the two measuring points was 60 m and the points were approximately 30 m away from the upper mining face. After the reinforcement of the Dongsi transportation roadway with the combined bolt–shotcrete–grouting support, the deformation amount and deformation velocity curve of the roadway side, roof, and floor are shown in Figure 21. Furthermore, the deep surrounding rock near the 11# measuring point was monitored to master the deep layer separation of the surrounding rock. The separation value of the side and roof with the distance of the working face under different depths is shown in Figure 22.

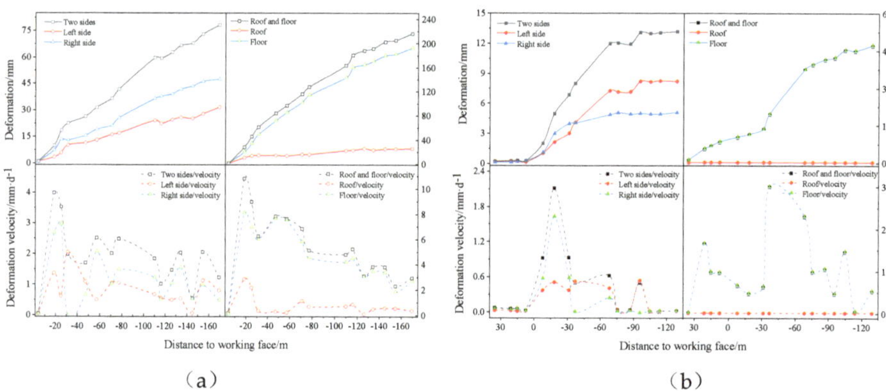


**Figure 21.** Deformation amount and deformation velocity of roadway surrounding rock (**a**) measuring point 11# (**b**) measuring point 12#.

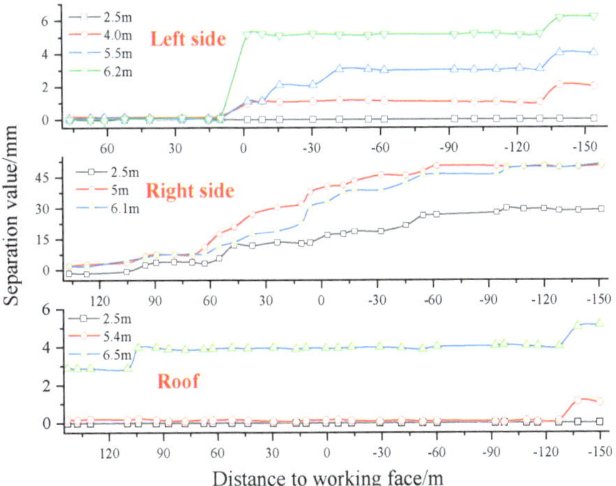



**Figure 22.** Surrounding rock separation near measuring point 11#.

As can be seen in Figure 21, the cumulative deformation of the 11# surface displacement measuring point was 80 mm, and after the working face passed the point at approximately 170 m, the deformation speed of the side was approximately 2 mm/d and the roadway deformation was still unstable. The accumulated deformation of the 12# measuring point was approximately 13 mm only when the point was 10 m away from the working face. When the working face passed 70 m, the rock surrounding the roadway began to stabilize. The accumulated deformation of the 11# and 12# measuring points was 200 and 50 mm, respectively, after the working face was pushed, and the deformation of the roadway roof and floor at the two measuring points was still ongoing. From the deformation velocity of the roof and floor, the maximum moving velocity of the 11# and 12# measuring points was 11 and 3mm/d, respectively. After the 12# measuring point was pushed over 100 m, the average moving speed of the top and bottom plate was only 0.5 mm/d. However, 170 m after measuring point 11#, the deformation speed could still reach 4 mm/d. After the working face was pushed over the range of 130 m, from the two measuring points to the end of the observation, the maximum cumulative deformation on both sides of the roadway was 80 mm, the maximum cumulative deformation of the roof and floor was 200 mm, the maximum deformation velocity of the two sides and the roof and floor was 2 and 4 mm/d, respectively. The working face continued to advance, the mining stress effect was not obvious, and the surrounding rock deformation tended toward stability, which satisfied the needs of production.

As can be seen from the two sides of the roadway and roof separation in Figure 22, the left side was within the range of 0–6.2 m, the layer separation value was 6 mm, and there was no layer separation within the range of 0–2.5 m. In the range of 0–5 m, the separation value of the right side was 51 mm. In the range of 0–2.5 m, the separation value was 29 mm and accounted for 57% of the total separation value. In the range of 5–6.1 m, there was no separation for the right side of the surrounding rock. Additionally, there was no layer separation in the range of 0–5.4 m near the roof, but there was separation of 5 mm in the range of 5.4–6.5 m. As can be seen, the layer separation of the right side of the roadway is much larger than that of the left side, which confirms that the deformation of the right side of the roadway was larger than that of the left side, to a certain extent. The rock surrounding the roadway roof is hard and does not get affected by mining. Therefore, the separation layer of the roadway roof was small. The surrounding rock at the floor of the roadway was fractured and had poor integrity, owing to the great effect exerted by mining, which can easily cause heave to the surrounding rock of the floor.

Based on a large amount of on-site monitoring data combined with numerical simulation methods, this paper has carried out research on the stability of the surrounding rock of the floor roadway of the diagonally spanning working face. There are few studies on this type of roadway. In the process of advancing the overlying working face, the influence range of bearing pressure is large and the time is long. Compared with the transverse straddle roadway and longitudinal straddle roadway, the distance between the floor roadway and the working face of the obliquely straddle working face is changing, and the deformation affected by mining is more complex. In the support design of this kind of roadway, more reference may be made to transverse straddle roadways and longitudinal straddle roadways. In this paper, based on the field monitoring and numerical simulation research, the laws of mining disturbance, surrounding rock deformation, stress release and transfer, and fracture expansion of this kind of roadway in the process of advancing the working face were described in more detail, which is more accurate. We carried out support scheme design and optimization according to local conditions to lay a foundation, and, compared with the original support method, the improved support scheme carried out targeted optimization design (based on mining influence and deformation characteristics), and more effectively combined the mutual coupling between the surrounding rock and anchor cable, rather than separate passive support, so as to provide reference and reference for the design of surrounding rock support scheme of floor roadway in similar obliquely straddle working face in this mine and even other mining areas.

## 5. Conclusions

The main conclusions drawn from this study are as follows:

(1) According to the geological conditions of the Dongsi main transportation roadway and its spatial relationship with the overlying 1762(3) working face, the laws of four key factors influencing the stability of the rock surrounding the roadway were obtained through theoretical and practical analysis. The results reveal that the surrounding rock properties of the roof and floor vary greatly within the mining influence range, and the rocks surrounding the two sides exhibit regional differences. The range of the vertical mining influence on the floor rock roadway is approximately 3040 m, and that of horizontal mining is approximately 50–60 m. The results reveal that the rock surrounding the lateral roadway is greatly affected by mining, while the lower roadway is less affected. The bolt cable support effectively controls the deformation of the roadway side, and the shed support has no obvious control effect on the side;

(2) Using the FLAC2D numerical software, the deformation and failure of the rock surrounding the inclined span working face and the dynamic pressure of the rock floor of the roadway were simulated and analyzed using different support forms. Before and after the working face mining, the influence range of the roadway mining was 60–110 m. The original support conditions of the roadway deformation were severe, and the combined effect of the bolt and shed support was not satisfactory. The use of anchor-shotcreting-grouting (the strength grade of shotcrete is C20, and the cement used for grouting is 425# grade ordinary portland cement) to strengthen the support significantly improved the surrounding rock conditions and strengthened the surrounding rock;

(3) The anchor–shotcreting–grouting support scheme is proposed and was successfully applied in practice. The ground pressure observation of the test roadway revealed the following: The cumulative deformation of the two sides of the roadway was 80 mm; the cumulative deformation of the roof and floor was 200 mm; the working face continued to advance; the mining stress effect was not obvious; the surrounding rock separation value was small; and the surrounding rock integrity was good. Finally, the deformation tended toward stability, which satisfied the production requirements.

**Author Contributions:** Data curation, J.K. and N.Z.; formal analysis, J.K. and P.W.; funding acquisition, J.K. and X.X.; investigation, B.W.; methodology, J.K. and B.W.; project administration, J.K., N.Z. and X.X.; software, P.W. and B.W.; writing—original draft, P.W.; writing—review and editing, J.K., P.W. and N.Z. All authors have read and agreed to the published version of the manuscript.

**Funding:** This work is supported by the National Natural Science Foundation of China (52074263,52034007), the Fundamental Research Funds for the Central Universities (2014QNA47) and the Postgraduate Research and Practice Innovation Program of Jiangsu Province (KYCX21_2332).

**Institutional Review Board Statement:** Not applicable.

**Informed Consent Statement:** Not applicable.

**Data Availability Statement:** Data is contained within the article.

**Conflicts of Interest:** The authors declare no conflict of interest.

# References


1. Kang, H.P. Support technologies for deep and complex roadways in underground coal mines: A review. *Int. J. Coal Sci. Technol.* **2014**, *1*, 261–277. [CrossRef]
2. Zhang, N.; Xue, F.; Zhang, N.C.; Feng, X.W. Patterns and security technologies for co-extraction of coal and gas in deep mines without entry pillars. *Int. J. Coal Sci. Technol.* **2015**, *2*, 66–75. [CrossRef]
3. Wang, G.F.; Pang, Y.H. Surrounding rock control theory and longwall mining technology innovation. *Energy Sci. Eng.* **2017**, *4*, 301–309. [CrossRef]
4. Palchik, V. Formation of fractured zones in overburden due to longwall mining. *Environ. Geol.* **2003**, *1*, 28–38. [CrossRef]
5. Wang, H.W.; Jiang, Y.D.; Xue, S.; Shen, B.T.; Wang, C.; Lv, J.G.; Yang, T. Assessment of excavation damaged zone around roadways under dynamic pressure induced by an active mining process. *Int. J. Rock Mech. Min. Sci.* **2015**, *77*, 265–277. [CrossRef]
6. Qin, D.D.; Wang, X.F.; Zhang, D.S.; Chen, X.Y. Study on Surrounding Rock-Bearing Structure and Associated Control Mechanism of Deep Soft Rock Roadway Under Dynamic Pressure. *Sustainability* **2019**, *11*, 1892. [CrossRef]
7. Prusek, S.; Masny, W. Analysis of damage to underground workings and their supports caused by dynamic phenomena. *J. Min. Sci.* **2015**, *1*, 63–72. [CrossRef]
8. Yao, Q.L.; Zhou, J.; Li, Y.N.; Tan, Y.M.; Jiang, Z.G. Distribution of Side Abutment Stress in Roadway Subjected to Dynamic Pressure and Its Engineering Application. *Shock Vib.* **2015**, *2015*, 929836.
9. Lu, S.L.; Sun, Y.L.; Jiang, Y.D. Position of floor rock roadway and adjacent coal seam roadway and cross mining pressure behavior law. *Coal Sci. Technol.* **1994**, *22*, 27–31.
10. Mo, S.; Sheffield, P.; Corbett, P.; Ramandi, H.L.; Oh, J.; Canbulat, I.; Saydam, S. A numerical investigation into floor buckling mechanisms in underground coal mine roadways. *Tunn. Undergr. Space Technol.* **2020**, *103*, 103497. [CrossRef]
11. Zhang, X.C.; Li, D.Y.; Chen, S.H.; Li, D.S.; Fan, D.W. Stability Prediction and Control of Surrounding Rocks for Roadway Affected by Overhead Mining. *J. Min. Eng.* **2008**, *25*, 361–365.
12. Zhang, P.S.; Lin, D.C.; Yang, J.; Wang, M.H. Numerical Simulation of Deep Varied Interval Riding Mining Roadway Stability Based on Time-effect of Rock. *J. Shandong Univ. Sci. Technol. (Nat. Sci.)* **2012**, *31*, 10–14.
13. Hebblewhite, B.K.; Lu, T. Geomechanical behaviour of laminated, weak coal mine roof strata and the implications for a ground reinforcement strategy. *Int. J. Rock Mech. Min. Sci.* **2004**, *1*, 147–157. [CrossRef]
14. Lu, X.; Zheng, Y.S.; Yan, M.C.; Li, W.; Cao, S.L. Study on ground pressure behavior and support technology of short distance over mining roadway. *Coal Technol.* **2019**, *38*, 47–50.
15. Malkowski, P.; Ostrowski, L.; Bachanek, P. Modelling the small throw fault effect on the stability of a mining roadway and its verification by in situ investigation. *Energies* **2017**, *12*, 2082.
16. Sun, Z.H.; Xie, W.B.; Zhang, J.W.; Cheng, X.J. Surrounding rock deformation regularity and control technology of soft rock roadway in floor during across mining. *Coal Technol.* **2016**, *35*, 65–67.
17. Gabet, T.; Malecot, Y.; Daudeville, L. Triaxial behaviour of concrete under high stresses: Influence of the loading path on compaction and limit states. *Cem. Concr. Res.* **2008**, *38*, 403–412. [CrossRef]
18. Okubo, S.; Fukui, K.; Hashiba, K. Development of a transparent triaxial cell and observation of rock deformation in compression and creep tests. *Int. J. Rock Mech. Min. Sci.* **2008**, *45*, 351–361. [CrossRef]
19. Chang, C.; Haimson, B. Waveform analysis in mitigation of blast-induced vibrations. *J. Geophys. Res.-Solid Earth* **2000**, *105*, 18999–19013. [CrossRef]
20. Haimson, B.C.; Chang, C.D. True triaxial strength of the KTB amphibolite under borehole wall conditions and its use to estimate the maximum horizontal in situ stress. *J. Geophys. Res.-Solid Earth* **2002**, *107*, 2257. [CrossRef]
21. Ning, S. Research on Surrounding Rock Deformation and Sectional Control Technology of Roadway Crossing through Passage. Master's Thesis, Shandong University of Science and Technology, Qingdao, China, 2019.
22. Zha, W.H.; Fu, X.M.; Yu, J.Y. Dynamic stepping and segmenting control and reinforcement technology of deep longitudinal spanning roadway. *Saf. Coal Mines* **2013**, *44*, 100–103.
23. Xu, B.G.; Wang, K. Study on pre—reinforcement technology of roadway affected by contugous overhead mining. *Coal Sci. Technol.* **2020**, *48*, 194–199.
24. Zhang, Z.Y. Study on the Support Law of the Right-Angle Trapezoidal Plate Breakage and Stress Propagation to the Floor. Master's Thesis, Shandong University of Science and Technology, Qingdao, China, 2017.

25. Konicek, P.; Soucek, K.; Stas, L.; Singh, R. Long-hole destress blasting for rockburst control during deep underground coal mining. *Int. J. Rock Mech. Min. Sci.* **2013**, *61*, 141–153. [CrossRef]
26. Xiong, L.J. Control Technical Research on the Stability of Wall Rock under the Deep Lose Continuous Roadway Passed by Working Face. Master's Thesis, AnHui University of Science and Technology, Huainan, China, 2013.
27. Li, X.H.; Yao, Q.L.; Zhang, N.; Wang, D.Y.; Zheng, X.G.; Ding, X.L. Numerical Simulation of Stability of Surrounding Rock in High Horizontal Stress Roadw ay Under Overhead Mining. *J. Min. Saf. Eng.* **2008**, *25*, 420–425.
28. Majdi, A.; Hassani, F.P.; Nasiri, M.Y. Prediction of the height of destressed zone above the mined panel roof in longwall coal mining. *Int. J. Coal Geol.* **2012**, *98*, 62–72. [CrossRef]
29. Hu, X.Y.; Zhang, H.L.; Zhang, L.G. Analysis on Stress Concentration Factors of Roadway Surrounding Rock Affected by Cross Mining. *Chin. J. Undergr. Space Eng.* **2015**, *11*, 658–664.
30. Li, Y.Y. Study on the Technology of Surrounding Rock Control in Floor Roadway under Repeated Overhead Mining. Master's Thesis, China University of Mining and Technology, Xuzhou, China, 2015.
31. Jiang, J.Q.; Han, J.S.; Feng, Z.Q. Sub classification of surrounding rock structure stability of cross mining roadway and its application. *J. Eng. Geol.* **1999**, *7*, 321–326.
32. Zhang, J.C.; Liu, T.Q. On the depth and distribution characteristics of coal seam floor mining fracture zone. *J. China Coal Soc.* **1990**, *15*, 46–55.
33. Liu, T.Q. Influence and control engineering of mining rock mass and its application. *J. China Coal Soc.* **1995**, *20*, 1–5.
34. Li, G.C.; Ma, C.Q.; Zhang, N.; Wang, P.P.; Ma, R. Research on failure characteristics and control measures of roadways affected by multiple overhead mining in Huaibei mining area. *J. Min. Saf. Eng.* **2013**, *30*, 181–187.
35. Yuan, A.Y.; Yang, Z.Y.; Yang, Y.M. Across Mechanical Analysis and Control Technology of Surrounding Rock of Roadway Under the Influence of Dynamic Pressure. *Met. Mine* **2016**, *2*, 47–50.
36. Guo, Y.; Zheng, X.G.; Guo, G.Y.; Zhao, Q.F.; Zhou, W.; An, T.L. Study on deformation failure and control of surrounding rock in soft rock roadway in close range coal seam with overhead mining. *J. Min. Saf. Eng.* **2018**, *35*, 56–63.
37. Zhang, H.L.; Wang, L.G. Computation of Mining Induced Floor Additional Stress and Its Application. *J. Min. Saf. Eng.* **2011**, *28*, 288–292.
38. Xie, W.B.; Shi, Z.F.; Yin, S.J. Stability Analysis of Surrounding Rock Masses of Roadway Under Overhead Mining. *Chin. J. Rock Mech. Eng.* **2004**, *23*, 1986–1991.
39. Zhu, Q.H. Mechanics Analysis of Surrounding Rock Deformation of Roadway Affected by Deep Riding Mining and Its Stability Control. Ph.D. Thesis, China University of Mining and Technology, Xuzhou, China, 2012.
40. Zhang, C.; Zhang, L.Y.; Li, K.; Li, M. Numerical Simulation of Rock Mass Stability in Mining Dynamic Pressure Straddling. *J. Xuzhou Inst. Technol. (Nat. Sci. Ed.)* **2011**, *26*, 40–46.
41. Ma, X.Z. Application of large rigidity coupling support technology in high stress cross mining roadway. *Energy Technol. Manag.* **2011**, *2*, 62–64.

*Article*

# Experimental Investigation on Uniaxial Compression Mechanical Behavior and Damage Evolution of Pre-Damaged Granite after Cyclic Loading

Jianhua Hu, Pingping Zeng, Dongjie Yang *, Guanping Wen, Xiao Xu, Shaowei Ma, Fengwen Zhao and Rui Xiang

School of Resources and Safety Engineering, Central South University, Changsha 410083, China; hujh21@csu.edu.cn (J.H.); 195512084@csu.edu.cn (P.Z.); Guanping-Wen@csu.edu.cn (G.W.); xiaoxu@csu.edu.cn (X.X.); mashaowei@csu.edu.cn (S.M.); zhaofengwen@csu.edu.cn (F.Z.); ruitorres@csu.edu.cn (R.X.)
* Correspondence: yangdjxx@csu.edu.cn

**Abstract:** Failure behavior of pillars in deep mines is affected by various cyclic loads that cause initial pre-damage. Pillars will be further damaged and developed in the long-term compressive stress until they are destroyed. To reveal the strength characteristics and crack damage fracture laws after rock pre-damage, uniaxial compression tests were carried out on granite specimens damaged by cyclic loading using the digital speckle correlation method. The experimental results indicate that the mechanical properties of pre-damaged specimens show large damage differences for different cycles. The damage variable of the pre-damaged specimens increases with the increase of cycle number and confining pressure. The damage of specimens is primarily due to the strength weakening effect caused by cycle numbers, and the confining pressure restriction effect is not obvious. The evolution laws of uniaxial compression damage propagation in the pre-damaged specimens show differences and obvious localization phenomenon. Pre-damaged specimens experienced three failure modes in the uniaxial compression test, namely tensile shear failure (Mode I), quasi-coplanar shear failure (Mode II), and stepped path failure (Mode III), and under different pre-damage stress environments with high confining pressures, the failure modes are dominated by Mode II and Mode III, respectively.

**Keywords:** digital speckle correlation method; cyclic load; high confining pressure; damage evolution; failure mode

**Citation:** Hu, J.; Zeng, P.; Yang, D.; Wen, G.; Xu, X.; Ma, S.; Zhao, F.; Xiang, R. Experimental Investigation on Uniaxial Compression Mechanical Behavior and Damage Evolution of Pre-Damaged Granite after Cyclic Loading. *Energies* **2021**, *14*, 6179. https://doi.org/10.3390/en14196179

Academic Editor: Kamel Hooman

Received: 13 August 2021
Accepted: 23 September 2021
Published: 28 September 2021

**Publisher's Note:** MDPI stays neutral with regard to jurisdictional claims in published maps and institutional affiliations.



## 1. Introduction

Instability in rock engineering is primarily due to the further expansion of joints, faults, and cracks [1]. There are studies that show the effect of mechanical excavation as a cyclic load in deep mining, especially with roadheader excavation [2], which will lead to stress redistribution on pillars or surrounding rocks. At the same time, the rock masses are subjected to cyclic loading, causing meso-damage of the surrounding rock or pillars [3]. Mechanical properties for permanently supported pillars are further damaged and deteriorated under compressive stress after excavation, which introduces safety hazards to the engineering structure [4].

In recent years, rock mass damage evolution research has made significant achievements. Fan et al. [5] conducted uniaxial compression tests on specimens containing multi-non-persistent joints with $PFC^{3D}$, and they found that the joint specimens exhibited four failure modes. Hu et al. [6] proposed an interval value for the damage initial stress range of intact rocks under uniaxial compression. Xiao et al. [7] presented the fatigue damage evolution laws of rocks under cyclic loading and believed that it primarily depended on the maximum stress, amplitude, and initial fatigue damage. Bagde et al. [8] studied the effects of amplitude and frequency on the mechanical properties of rock under uniaxial cyclic loading, and the average Young's modulus and dynamic axial stiffness of the rock was found to decrease with loading frequency and amplitude. Chen et al. [9] analyzed the crack propagation laws of the granite in three fatigue stages during uniaxial cyclic loading, and the results showed that distinguishing crack growth was identified in quartz grains at the initial degradation stage. At the second stage, no significant crack growth in quartz grains was identified. On the other hand, in feldspar grains, development of cracks was observed. At the final accelerated stage, many intergranular cracks were identified. Zhang et al. [10] carried out uniaxial cyclic loading tests on cracked rock. The results showed that the damage mechanism was caused by the initiation, propagation, and penetration of cracks. Fuenkajorn et al. [11] found that the compressive strength of salt rocks under cyclic loading decreased as the cycle numbers increased. Taheri et al. [12] studied the crack evolution laws in the Hawkesbury sandstone under different cyclic loading and found that unstable cracks propagation occurred at around 65% of the cumulative axial strain. Jia et al. [13] studied the relationship of deformation parameters to the cycle numbers of fine-grained sandstone under triaxial cyclic loading. The experimental results showed that the axial and lateral irreversible strain initially increased with a decreasing rate, followed by steady state increases, before failure within a given stress level. When the specimens approached failure, an accelerated increase in the irreversible strain occurred. Wang et al. [14] studied the effects of volumetric change on the granite fatigue performances under triaxial cyclic loading (confining pressure was 3–10 MPa), and they found that the stress level corresponding to the transition from volumetric compaction to volumetric dilation could be considered as the threshold for fatigue failure. Liu et al. [15] studied the effects of the confining pressure (2–50 MPa) and frequencies on the sandstone damage evolution during cyclic loading and unloading. The results showed that with an increase in the frequency, the axial strain and the number of cycles at failure increased at the same confining pressure. On research methods, the digital speckle correlation method (DSCM) is helpful for analyzing the local deformation and instability of the intact rock mechanics experiment process and has become an important method for studying the local deformation of geotechnical materials [16]. Xie et al. [17] observed the strain field on coal–rock composites with various height ratios using DSCM under uniaxial compression, and they determined that the strain cloud gradient and local strain area move upward and increase with increasing coal rock height. Tan et al. [18] discussed the influences of crack numbers on the local deformation of layer-crack rock models by using the DSCM technique in uniaxial compression tests. Based on speckle images, they found that with the increase of fissure number of layer-crack specimens, the bearing capacity of specimens decreased, and the appearing time of nonuniform deformation phenomenon in the specimen surface decreased. Sun et al. [19] combined uniaxial compression tests with DSCM to research deformation field evolution laws of backfill specimens. The results showed that the peak strain of the GCGFB specimen was large, and the failure characteristics of the GCGFB sample showed multiple shear deformation zones, which is indicative of a typical cylindrical splitting failure. Cao et al. [20] used DSCM and the numerical methods in $PFC^{2D}$ to analysis the failure behavior of brittle jointed specimens during uniaxial loading. They divided the failure modes into three categories: stepped path failure, failure through parallel plane, and failure through cross plane.

However, these research results rarely involve the cyclic damage law of rock under high confining pressure. As for the damaged rock in deep high stress field under cyclic loading, after excavation, the three-dimensional stress changes to two-dimensional stress, and the research on its mechanical behavior and damage law is rare. On the other hand, the granite has the characteristics of good stability and low permeability, and in deep mining, some orebodies occur near granite, so the damage research of granite under cyclic load is particularly important. This paper primarily studied the damage state of rocks under the long-term action of in situ stress in the depth of 1852–4444 m (in situ stress between 50–120 MPa) [21]. Uniaxial stress-resistance tests were conducted on granite specimens with various cyclic loading numbers, and the strength characteristics, damage evolution, and failure modes of pre-damaged granite specimens were analyzed in combination with DSCM, providing the theoretical basis for a mechanical analysis in deep mines.

## 2. Experiment

### *2.1. Equipment*

The main equipment used was the American MTS815 triaxial electro-hydraulic testing machine, the vacuum water-saturation equipment, the uniaxial compression rigidity testing machine, and the DSCM acquisition system (Figure 1). The camera type of the DSCM system is WP-UT500M, with a resolution of 5 million pixels, a rate of 20 frames/s, an image size of 2448 pixels × 2048 pixels, and a fluorescent light source.

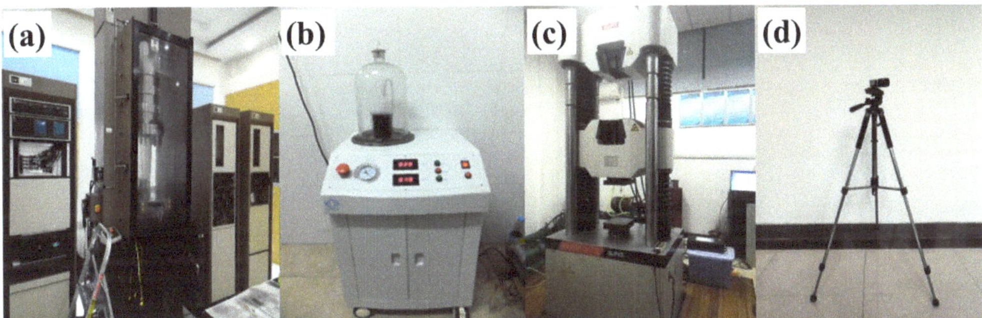


**Figure 1.** Experimental equipment. (**a**) MTS815 testing machine for 2000 KN; (**b**) rock water-saturation equipment; (**c**) uniaxial compression rigidity testing machine; (**d**) DSCM acquisition system.

### *2.2. Experimental Scheme*

#### 2.2.1. Pre-Damage Treatment of Specimens

The granite specimen was a standard cylinder of 50 mm × 100 mm, with uniform dense texture and no damage. To obtain the pre-damaged granite specimens, the intact specimens were first subjected to a triaxial compression test. The cyclic loading scheme is shown in Figure 2. The granites were classified into four groups (A, B, C, and D), and the basic physical parameters of group A3, B3, C3, and D are shown in Table 1. The procedures are mainly as follows:

1.  Four groups of rock specimens were vacuumed and saturated with water using the vacuum water saturation equipment, first for 240 min dry pumping and then for 120 min wet pumping to completely saturate the specimens;
2.  Uniaxial compressive strength of specimens in Group D were measured with the MTS815 testing machine loaded at a rate of 2 kN/s (Group D was the control group);
3.  The remaining rock specimens of Group A, B, and C were cyclically loaded and unloaded for 10, 20, and 30 times under confining pressures of 50 MPa, 80 MPa, 100 MPa, and 120 MPa, respectively, without destroying the specimens, to realize the pre-damage treatment inside the granite specimens (see Table 2 for the specific scheme). The load cycle path used constant confining pressure and unloading axial pressure; both loading and unloading rates were 2 kN/s, loading stress point was 153 MPa (about 93% of the average uniaxial compressive strength), and unloading stress point was 5 MPa (about 3% of the average uniaxial compressive strength).

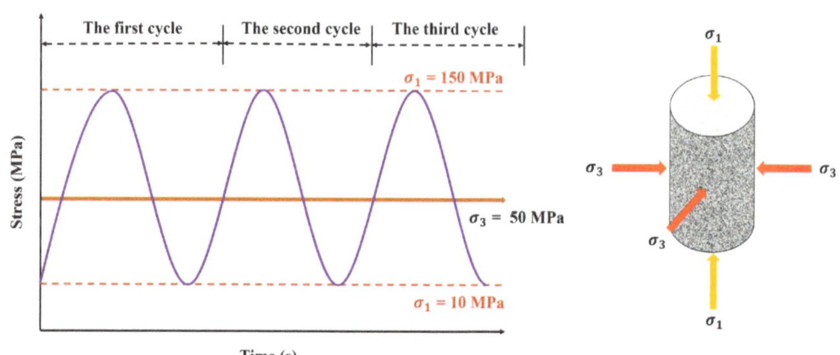


**Figure 2.** Cyclic loading scheme.

**Table 1.** Physical parameters of granite specimens.

| Group | Number of Cycles | Height (mm) | Diameter (mm) | Dry Mass (g) | Natural Mass (g) | Saturated Mass (g) | Saturated Mass after Load (g) |
|---|---|---|---|---|---|---|---|
| A3-1 |  | 100.67 | 49.11 | 495.25 | 495.79 | 496.58 | 497.20 |
| A3-2 | 10 | 101.01 | 49.12 | 492.53 | 493.02 | 493.91 | 499.00 |
| A3-3 |  | 100.30 | 48.91 | 491.73 | 492.17 | 493.03 | 497.62 |
| B3-1 |  | 100.17 | 49.34 | 493.82 | 494.49 | 495.35 | 497.42 |
| B3-2 | 20 | 100.45 | 49.05 | 494.89 | 495.54 | 496.51 | 497.02 |
| B3-3 |  | 100.43 | 49.45 | 498.07 | 498.61 | 499.54 | 502.21 |
| C3-1 |  | 100.95 | 49.04 | 494.66 | 495.20 | 496.02 | 496.68 |
| C3-2 | 30 | 100.30 | 49.59 | 502.68 | 503.24 | 504.73 | 504.77 |
| C3-3 |  | 100.75 | 49.16 | 496.48 | 497.02 | 497.96 | 498.56 |
| D-1 |  | 100.50 | 49.12 | 492.24 | 492.74 | 493.65 | —— |
| D-2 | 0 | 100.77 | 49.04 | 495.15 | 495.73 | 496.61 | —— |
| D-3 |  | 99.45 | 49.17 | 491.03 | 491.57 | 492.40 | —— |

**Table 2.** Specific scheme of pre-damage tests.

| Confining Pressure (MPa) | Cyclic Loading and Unloading of Specimens | | | Uniaxial Compression Specimens |
| --- | --- | --- | --- | --- |
|  | 10 Cycles | 20 Cycles | 30 Cycles |  |
| 50 | A1-1 | B1-1 | C1-1 |  |
|  | A1-2 | B1-2 | C1-2 |  |
|  | A1-3 | B1-3 | C1-3 |  |
| 80 | A2-1 | B2-1 | C2-1 |  |
|  | A2-2 | B2-2 | C2-2 |  |
|  | A2-3 | B2-3 | C2-3 | D-1 |
| 100 | A3-1 | B3-1 | C3-1 | D-2 |
|  | A3-2 | B3-2 | C3-2 | D-3 |
|  | A3-3 | B3-3 | C3-3 |  |
| 120 | A4-1 | B4-1 | C4-1 |  |
|  | A4-2 | B4-2 | C4-2 |  |
|  | A4-3 | B4-3 | C4-3 |  |

## 2.2.2. Uniaxial Compression Test with DSCM

The fundamentals of DSCM is to match the same pixel areas (sub-images) between two digital speckle images of the specimen's surface before deformation (reference image) and after deformation (deformed image) [22]. Its main purpose is to search the displacement component of the pixels in the speckle sub-image (Figure 3). To identify the sub-images in the two images, the properties of the sub-images are represented by the gray function. Thus, by tracking the grayscale value of each pixel in the reference image and the deformation image and evaluating their image correlation, the displacement field on the specimen's surface can be calculated, and then the strain field can be obtained by further calculation. The correlation coefficient is defined as follows [23]:

$$C(u,v) = \frac{\sum_{i=1}^{m}\left[(x_i,y_i) - \overline{f}\right]\left[g(x_i',y_i') - \overline{g}\right]}{\sqrt{\sum_{i=1}^{m}\left[f(x_i,y_i) - \overline{f}\right]^2}\sqrt{\sum_{i=1}^{m}\left[g(x_i',y_i') - \overline{g}\right]^2}} \tag{1}$$

where $f(x_i,y_i)$ is the grayscale value of the reference image at the point $(x_i,y_i)$ and $g(x_i',y_i')$ is the grayscale value of the deformed image at the point $(x_i',y_i')$; $\overline{f}$, $\overline{g}$ are the average grayscale values of the sub-images in the reference image and the deformed image, respectively.

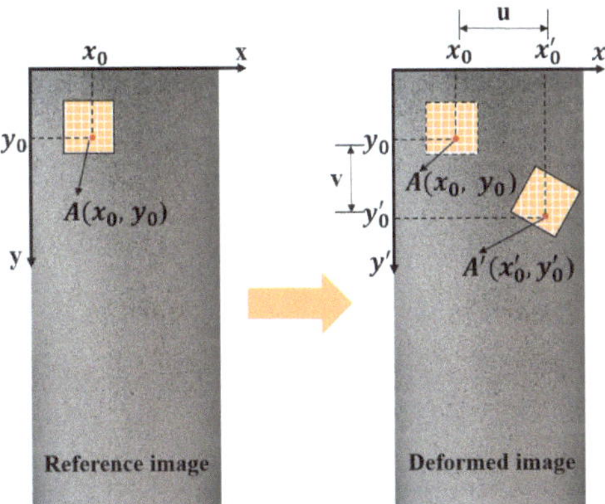


**Figure 3.** Schematic of the DSCM.

Combined with the DSCM method, secondary uniaxial compression tests were performed on the pre-damaged granite to explore the damage evolution and failure laws. The observation surface of each specimen was uniformly sprayed with a layer of matte white paint, and they were then randomly sprayed with matte black paint. After air drying, a uniaxial compression rigidity testing machine was used for displacement loading at 0.02 mm/min, and a CCD camera was used for shooting until the specimen was destroyed. The experimental scheme is shown in Figure 4.

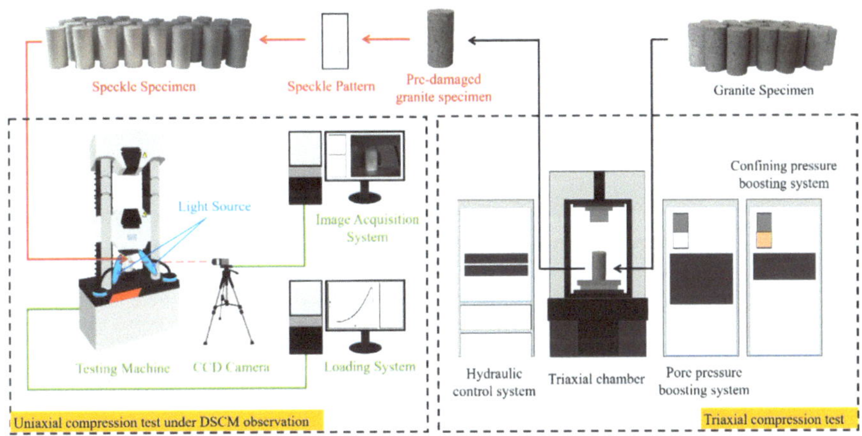


**Figure 4.** Schematic of experimental scheme.

## 3. Results

### 3.1. Uniaxial Compressive Strength Results for Intact Specimens

The stress–strain curves of specimens (control group) in group D are illustrated in Figure 5, and the uniaxial compressive strength is 164.20, 162.57, and 167.87 MPa, respectively. To study the damage laws of the other groups of pre-damaged granite specimens during uniaxial compression, the uniaxial compressive strength of intact specimens is taken as the average value of 164.88 MPa. In addition, it can be seen from the stress–strain curve of uniaxial compression that intact granite belongs to brittle rock, and there is almost no plastic yield stage.

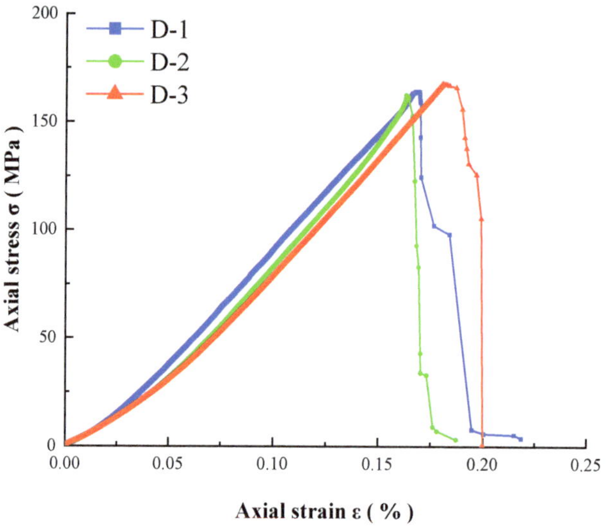


**Figure 5.** Stress–strain curve of intact granite.

### 3.2. Uniaxial Compressive Strength Results for Pre-Damaged Specimens

The stress–strain curves for pre-damaged specimens under 50 MPa confining pressure are illustrated in Figure 6. There are significant differences in the axial stress–strain curves after different cycles. The uniaxial compressive strength of pre-damaged specimens with 10 cycles is the largest, and the stress–strain curves showed obvious brittleness, which is very similar to that of the intact granite specimens. With the increase of cycle numbers, the uniaxial compressive strength tended to decrease, and there was strain softening behavior after the peak, which exhibited significant ductility. This indicates that the mechanical properties of pre-damaged specimens in high cycle numbers (20 cycles and 30 cycles) were significantly different from those in low cycle numbers (10 cycles); especially under high cycle numbers, the mechanical properties of pre-damaged specimens decreased significantly. In addition, the stress–strain curve of the pre-damaged specimens has many micro-stress drops, which show pre-peak and post-peak fluctuations, indicating that the pre-damaged specimens were not destroyed once, but destroyed step by step. Because the pre-damage specimen was cyclically loaded and unloaded during triaxial compression pre-damage, it resulted in the repeated initiation and closure of cracks or micro-holes inside the specimen. When cycle numbers increased, the interaction of defects inside the specimen and its damage accumulation effect lead to the expansion of the local deformation area of the specimen under uniaxial compression, greatly reducing the mechanical properties of the specimen. Moreover, when the pre-damaged specimen had good bearing capacity, it is possible that yield failure could have first occurred locally in the specimen. Therefore, a slight stress drop could occur before the specimen fails completely. This result was observed in many studies [24–26].

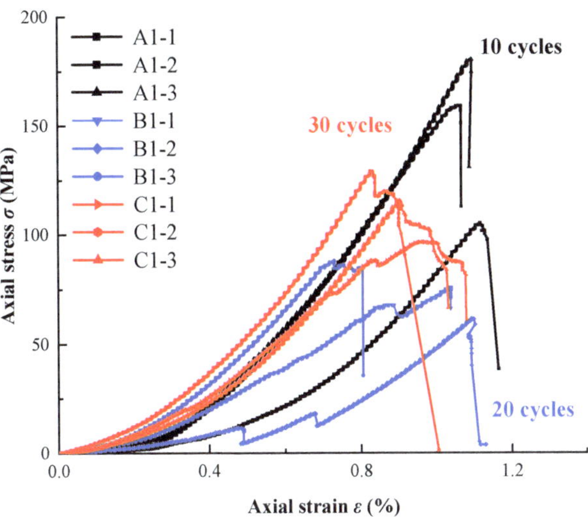



**Figure 6.** Axial stress–strain curves of pre-damaged specimens with $\sigma_3$ = 50 MPa.

### 3.3. Strength Characteristics of Pre-Damaged Specimens

Table 3 lists the average uniaxial compressive strength of pre-damaged specimens under different damage conditions, and its relationship with the cycle numbers and confining pressures is shown in Figure 7.

**Table 3.** The average uniaxial compressive strength of pre-damaged specimens under different damage conditions.

| N [1] | $\sigma_c$ (MPa) | | | |
|---|---|---|---|---|
| | $\sigma_3$=50 MPa | $\sigma_3$=80 MPa | $\sigma_3$=100 MPa | $\sigma_3$=120 MPa |
| 10 | 148.13 | 118.86 | 114.60 | 105.32 |
| 20 | 109.21 | 101.03 | 100.87 | 93.04 |
| 30 | 113.87 | 107.12 | 85.65 | 53.59 |

[1] $\sigma_3$ is the confining pressure, N is cycle numbers, and $\sigma_c$ is the average uniaxial compressive strength.

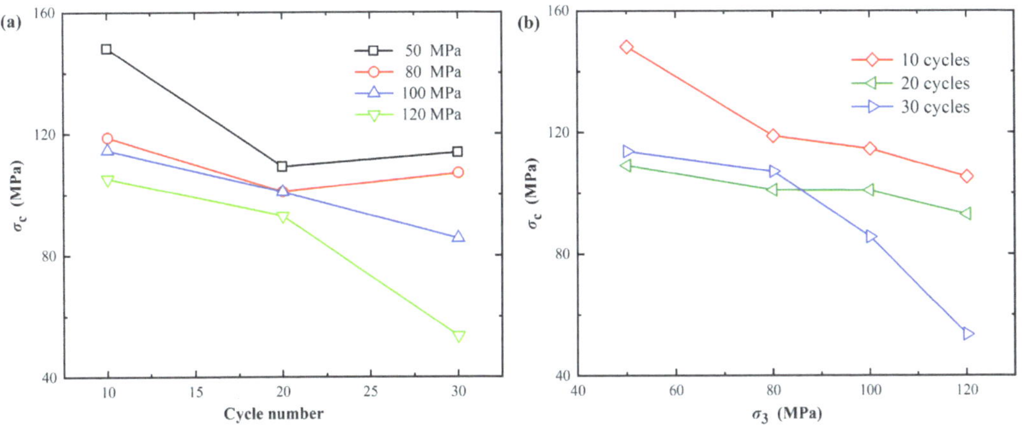



**Figure 7.** Strength characteristics of pre-damaged specimens. (**a**) Relationship between $\sigma_c$ and the cycle number; (**b**) relationship between $\sigma_c$ and $\sigma_3$.

The uniaxial compressive strength of pre-damaged specimens with less than 20 cycles decreased with the increase of cycle numbers under different confining pressures (Figure 7a). Due to cyclic loading and unloading, cracks were generated inside the specimen, which reduced the bearing capacity of the specimen. Therefore, the uniaxial compressive strength of the specimen was weakened. However, the pre-damaged specimens with more than 20 cycles showed complex strength laws. Under the confining pressures of 50 MPa and 80 MPa, the uniaxial compressive strength of the pre-damaged specimens increased with the increase of cycle numbers and decreased with the increase of cycle numbers at 100 MPa and 120 MPa. In the high cyclic loading and unloading numbers (more than 20 cycles) of pre-damaged specimen, a large number of cracks inside the specimen was initiated and accumulated. Under the confining pressures of 50 MPa and 80 MPa, it is shown that the uniaxial compressive strength of the pre-damaged granite specimen did not always decrease with the increase of the number of cycles. This result differs from that revealed in He et al. [27]. This difference may be due to Griffith cracks, which affect the strength of the specimen according to its size. It has also been shown that if the deviatoric stress of cyclic loading is lower than the fatigue threshold (generally 94% of rock peak strength), the peak strength obtained by post-monotonic loading may increase [28]. In this test, the upper limit of deviatoric stress cycle was about 93% of the peak strength of granite, so a slight increase in strength was also possible. Under the confining pressure of 100 MPa and 120 MPa, due to the excessive confining pressure, the closed damage cracks inside the specimen started and propagated again, so the uniaxial compressive strength of pre-damaged specimen would continue to weaken. In general, the uniaxial compressive strength of pre-damaged granite specimens is much lower than that of intact granite specimens, which further proves that the bearing capacity of rock decreases after fatigue damage.

The relationship between uniaxial compressive strength and confining pressure is nonlinear (Figure 7b). Under different cycle numbers (10 cycles, 20 cycles, and 30 cycles), the uniaxial compressive strength of pre-damaged specimens decreased with the increase of confining pressure, which decreased by 28.9%, 14.8%, and 52.9%, respectively. This shows that the restriction effect of confining pressure on the internal microcrack evolution of the specimen was weakened with the increase of cycle numbers, and the damage of the specimen was mainly dominated by cycle numbers.

## 4. Rock Damage Mechanisms and Failure Modes Analysis

### 4.1. Relationship of Damage Variables to Cycle Numbers and Confining Pressures

Cyclic damage of rocks refers to the existence of internal cracks, micro-holes, and other defects in rocks in a certain pressure field when subjected to a certain cyclic load, resulting in micro-stress concentrations at their ends, and the interaction between micro-defects results in the activation of primary cracks or holes and dislocation between rock particles [29]. The meso-damage evolution within the rock is related to cycle numbers and confining pressure, among which the existence of initial confining pressure restricts the generation of micro-defects and particle dislocation. The damage variable D is the description of the meso-damage evolution within the rock [30]. Therefore, the damage variable D for pre-damaged specimens in this paper is defined as:

$$D = \frac{\sigma_{c0} - \sigma_c}{\sigma_{c0}} \times 100\% \tag{2}$$

where $\sigma_c$ is the uniaxial compressive strength of the pre-damaged specimen, $\sigma_{c0}$ is the uniaxial compressive strength of the intact specimen, and D is the damage variable of the pre-damaged specimen. It is generally believed that when D = 0, the rock is in a nondestructive state, at which time $\sigma_{c0} = \sigma_c$; when $0 < D < 100$, the rock is in the process of deformation failure and is damaged to varying degrees. When D = 100, it reaches rock material failure, corresponding to a total loss condition. According to Equation (2), the damage law of pre-damaged specimens under different cycle numbers and confining pressures is obtained (Figure 8).

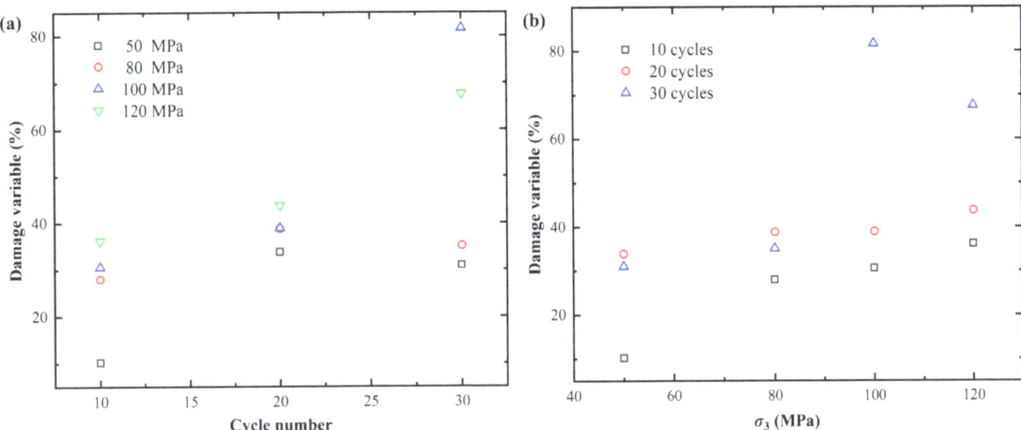

**Figure 8.** Damage variables of pre-damaged specimens. (**a**) Relationship between D and cycle numbers; (**b**) relationship between D and $\sigma_3$.

According to Section 3.1, the uniaxial compressive strength of the intact specimen $\sigma_{c0}$ was 164.88 MPa. In Figure 8, as cycle numbers increased, the damage variables for the pre-damaged specimens increased, and the magnitude of the damage changes varied with different confining pressure. At 50 MPa and 80 MPa, the increasing rate of the specimen's damage gradually slows down, but there is a certain degree of damage in each cycle. At 100 MPa and 120 MPa, the increasing rate of the specimen's damage accelerated with the increase of cycle numbers, with an increase of 167% and 87%, respectively. The damage characteristics of pre-damaged specimens were not only related to cycle numbers, but also related to the confining pressure. With the increase of confining pressure, the damage variable of pre-damaged specimens after the same number of cycles became higher. When the specimen was subjected to a high number of cyclic loads, the confining pressure effect was not obvious.

### 4.2. Damage Evolution Mechanism of Pre-Damaged Specimens

To further analyze the evolution mechanism (initiation, propagation, and coalescence) of surface cracks in pre-damaged granites, representative specimens were selected and analyzed in detail with the digital speckle correlation method. The surface principal strain cloud maps of the pre-damaged specimen under uniaxial compression are illustrated in Figures 9–11, in which different colors represent different principal strain regions; blue represents low principal strain regions and red represents high principal strain regions. From the principal strain cloud map corresponding to the marks on the stress–strain curve, it can be intuitively seen that there is a one-to-one correspondence between the macroscopic mechanical behavior and the principal strain field during the loading process of pre-damaged specimens.

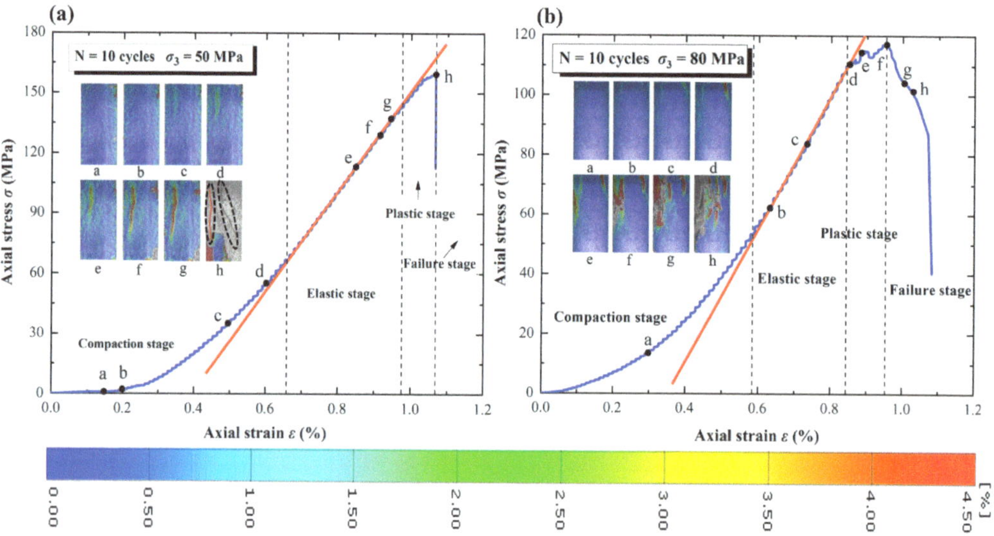

**Figure 9.** Crack evolution process of pre-damaged specimen after 10 cycles. (**a**) $\sigma_3 = 50$ MPa; (**b**) $\sigma_3 = 80$ MPa.

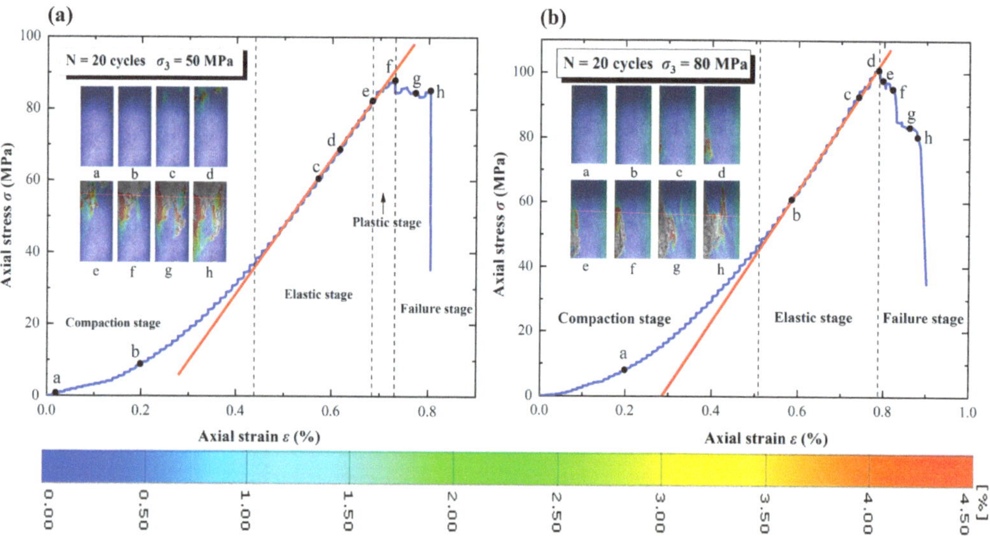

**Figure 10.** Crack evolution process of pre-damaged specimen after 20 cycles. (**a**) $\sigma_3 = 50$ MPa; (**b**) $\sigma_3 = 80$ MPa.

The deformation characteristics of the granite specimen after pre-damage show four stages. First, the compaction stage, which has an upward–concave stress–strain curve, is more significant compared to the undamaged granite specimens of group D (Figure 5). This is followed by the elastic stage; the stress–strain curve in this stage is basically straight, as shown in the red line part in Figure 9a. Next is the plastic stage, when the curve is concave, obviously showing the phenomenon of strain increase. Entering the plastic phase means that the rock will have irreversible plastic deformation. Finally, there is the failure stage of the rock. It should be noted that the plastic stage of the granite specimens after pre-damage in this test was very short, and in some cases almost non-existent, as shown in Figures 10b and 11a, which exhibit the characteristics of plastic-brittle failure [31]. In order to combine the processed speckle images, we took the typical representative images of each stage of the granite specimens for analysis.

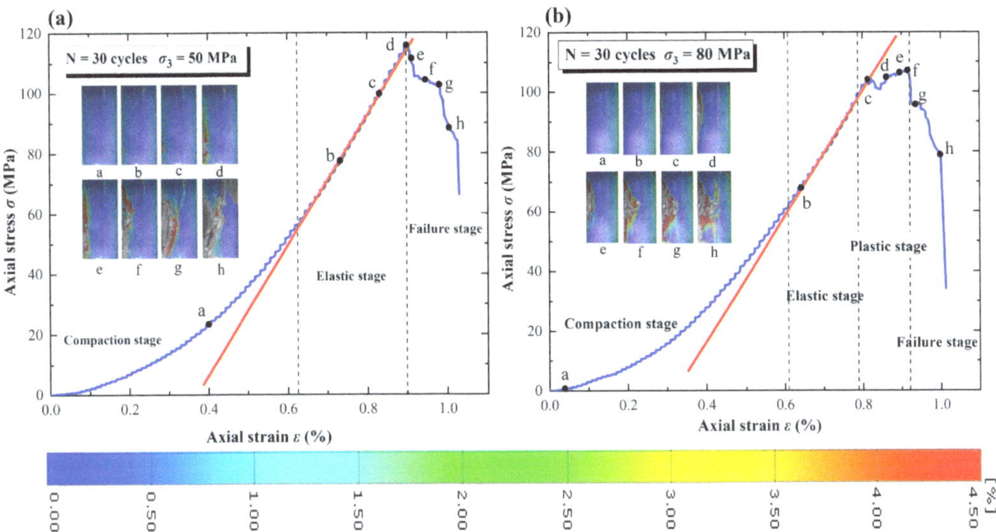

**Figure 11.** Crack evolution process of pre-damaged specimen after 30 cycles. (**a**) $\sigma_3$ = 50 MPa; (**b**) $\sigma_3$ = 80 MPa.

At 50 MPa and 10 cycles, the stress at points a–h were 0.6%$\sigma_c$, 1.6%$\sigma_c$, 23.0%$\sigma_c$, 37.8% $\sigma_c$, 72.6%$\sigma_c$, 85.7%$\sigma_c$, 88.6%$\sigma_c$, and 100%$\sigma_c$, respectively (Figure 9a). Points a, b, c, and d correspond to the compaction stage of the curve; the specimen deformation was small at this time. When loaded to point c, an obvious tensile zone was formed in the upper left corner of the specimen, and crack initiation began. Points e, f, and g correspond to the elastic stage of the curve. When loaded to point d–g, the tensile band in the upper left corner extended further along the direction of compressive stress until it reached both ends of the specimen. Point h corresponds to the failure stage of the curve, and it is the peak stress point. When loaded to point h, a penetrating shear band rapidly formed in the upper right corner of the specimen. At this time, the tensile crack zone on the left side of the specimen was not obvious. Because the surface expanded by the tensile crack was peeled off, the speckle surface could not be captured by the system during image processing, which lead to the failure of recognition and calculation. In the whole process of crack evolution, tensile crack was generated first, then shear crack was generated. In the final failure of the specimen, the tensile crack and shear crack were independent of each other, and there was no coalescence between the cracks. This damage evolution was concentrated in the pre-peak stage, which started from the initial elastic stage to generate cracks, and then propagated crack in the elastic stage and plastic stage, until the specimen was destroyed at the peak, suggesting that the overall evolution was relatively slow.

At 80 MPa and 10 cycles, the stress at points a–h were 11.8%$\sigma_c$, 52.8%$\sigma_c$, 70.5%$\sigma_c$, 93.8% $\sigma_c$, 95.6%$\sigma_c$, 100%$\sigma_c$, 89.1%$\sigma_c$, and 86.7%$\sigma_c$, respectively (Figure 9b). Point a corresponds to the compaction stage of the curve, at which time there was a small strain concentration in the upper left corner of the specimen. Points b and c correspond to the elastic stage of the curve, and when loaded to point c, a tensile band began to form at the upper right side of the specimen. Points d and e correspond to the plastic stage of the curve; when loaded to point d, the stretch band in the upper right corner deepened further, while the shear band in the upper left corner did not change significantly. When loaded to point e, a wide range of shear bands were formed in the upper left corner of the specimen. Points f–h correspond to the failure stage of the curve. When loaded to point f, cracks on the upper left corner of the specimen propagated and penetrate, leading to fracture. After point f, the stress drops rapidly. When the stress drops to points g and h, the specimen cracks propagated rapidly and coalesced into and penetrated each other, eventually destroying a large area in the upper left corner of the specimen. During the whole evolution, numerous tensile cracks and partial shear cracks were initiated, all of which propagated along the direction of compressive stress, and the various cracks interacted with each other, eventually penetrating to form stepped shear cracks, on one side of which the stepped shear cracks generally rupture. This is similar at 100 MPa and 120 MPa confining pressures.

At 50 MPa and 20 cycles, the stress at points a–h were 2.8%$\sigma_c$, 12.2%$\sigma_c$, 70.8%$\sigma_c$, 81.2%$\sigma_c$, 95.8%$\sigma_c$, 100%$\sigma_c$, 97.5%$\sigma_c$, and 98.9%$\sigma_c$, respectively (Figure 10a). Point a and point b correspond to the compaction stage of the curve, at which time the deformation was relatively uniform. Points c–d correspond to the elastic stage of the curve, and when loaded to point c, a strain concentration occurred in the upper left corner of the specimen and the specimen began to deform unevenly. Point e corresponds to the plastic stage of the curve. When loaded to point e, multiple tensile bands initiated above the specimen, all of which propagated along the direction of compressive stress. Points f–h correspond to the failure stage of the curve, and when loaded to points f–h, the tensile crack terminals connected with each other to form two crossed macroscopic shear cracks, namely quasi-coplanar shear cracks. This damage evolution was concentrated near the peak, and it started to initiate cracks in the middle elastic stage, then crack propagation and penetration occurred in the plastic failure stage, and finally destroyed the specimen. The evolution of cracks under other confining pressures (Figure 10b) is similar to Figure 9b in that tensile cracks were initiated first, and then the interactions between the tensile cracks merged to form stepped shear cracks.

At 50 MPa and 30 cycles, the stress at points a–h were 20.2%$\sigma_c$, 67.6%$\sigma_c$, 86.9%$\sigma_c$, 100%$\sigma_c$, 90.4%$\sigma_c$, 90.2%$\sigma_c$, 86.9%$\sigma_c$, and 72.6%$\sigma_c$, respectively (Figure 11a). Point a corresponds to the compaction stage of the curve, points b and c correspond to the elastic stage of the curve, and the deformation at points a and b is uniform. When loaded to point c, a strain concentration began to form in the lower left corner of the specimen. Points d–h correspond to the failure stage of the curve. When loaded to points d–f, an obvious tensile band formed on the left side of the specimen and began to deform unevenly. When loaded to points g–h, a shear crack through the specimen formed, which interacted with the tensile crack, forming a "Y" type quasi-coplanar shear crack. This damage evolution was concentrated in the post-peak stage, and the evolution speed was rapid.

At 80 MPa and 30 cycles, the stress at points a–h were 3.8%$\sigma_c$, 64.2%$\sigma_c$, 96.0%$\sigma_c$, 98.1%$\sigma_c$, 99.1%$\sigma_c$, 100%$\sigma_c$, 89.3%$\sigma_c$, and 73.6%$\sigma_c$, respectively (Figure 11b). Point a corresponds to the compaction stage of the curve, and the deformation was uniform. Point b corresponds to the initial elastic phase of the curve. When loaded to point b, the deformation was still uniform. Points c–e correspond to the plastic stage of the curve. When loaded to point c, a stress concentration began to appear on the left side of the specimen with a slight rupture. When loaded to points d–e, the strain area on the left side of the specimen gradually expanded to form multiple shear cracks, and the strain area was concentrated toward the center of the specimen. Points f–h correspond to the failure stage of the curve. Point f was the peak stress point, and when the stress dropped to point h, the specimen strain zone exhibited an "X" type quasi-coplanar shear crack. Under the confining pressure of 100 MPa and 30 cycles, the crack evolution process was similar to that of the stepped crack formed above. Under the confining pressure of 120 MPa and 30 cycles, the crack evolution process was a typical quasi-coplanar shear crack evolution process, but its evolution began very early.

DSCM results showed that the surface crack initiation period of the specimen was related to the cycle numbers. The larger the cycle numbers, the surface crack initiation period changed from the elastic stage to the plastic stage, respectively. During the uniaxial compression failure process of pre-damaged specimens, the final macroscopic failure of all specimens was the result of the continuous accumulation of early damage [32]. Due to the influence of the boundary, stress concentration phenomenon usually occurred at the end of the specimen, so the crack initiation was mostly concentrated at the end and always expanded along the direction of the maximum principal strain, leading to the occurrence of early damage. In addition, tensile crack was always the earliest crack in the pre-damaged granite specimen, and the initiation of secondary cracks around the tensile crack was mostly the result of the joint action of tension and shear. The shear gradually played a leading role in the crack propagation, and the crack evolution presented a localized phenomenon. These results are consistent with those of crack evolution in [16].

### 4.3. Failure Mode of Pre-Damaged Specimens

Tensile failure is caused by the transverse tensile stress of the failure surface exceeding the maximum tensile stress of the rock, and the crack is called tensile crack. The path of propagation of the tensile crack is smooth. Shear failure is caused by the shear stress of the failure surface exceeding the maximum shear stress of the rock, and the crack is called shear crack. The surface of a shear crack is considerably rougher than that of a tensile crack [33–35].

The pre-damaged specimens showed different failure modes under uniaxial compression. Based on the two basic failure modes of tensile crack and shear crack, the failure modes of the pre-damaged specimen can be divided into three categories: (1) tensile shear failure (mode I), (2) quasi-coplanar shear failure (mode II), and (3) stepped path failure (mode III) (Figure 12).

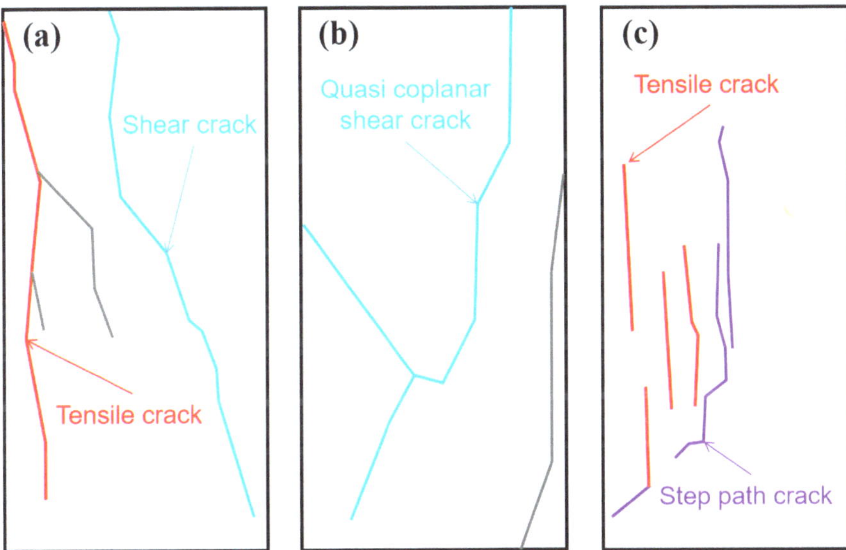


**Figure 12.** Failure modes of typical specimens. (**a**) Failure mode I; (**b**) failure mode II; (**c**) failure mode III.

(1)  Mode I: The main cracks in this failure mode are tensile cracks and shear cracks, and these two main cracks are independent of each other at the time of final failure. No crack coalescence occurs, so this failure mode is called tensile shear failure (Figure 12a). The red line is a tensile crack, the blue line is a shear crack, and the gray line is secondary shear cracks in Figure 12a. The final failure in this mode will cause spalling on the tensile crack or shear crack surface, and the overall degree of damage is small. This failure mode is also identified in [33,36], which is called mixed coalescence mode.

(2)  Mode II: The main cracks in this failure mode are two intersecting shear cracks, so this mode is called quasi-coplanar shear failure (Figure 12b). The blue line is a quasi-coplanar shear crack, and the gray line is a shear secondary crack in Figure 12b. The damage in this mode is generally marked by peeling from its conjugate surface or

collapse, and the degree of damage is large. Such quasi-conjugate or single shear plane failure results have been reported in [37].

(3)  Mode III: The main cracks in this failure mode are tensile cracks. Because multiple tensile cracks interact, propagate and coalescence to form a shear crack in a stepped path, this mode is called a stepped path failure (Figure 12c). The purple line is a step path crack, and the blue line is a tensile crack in Figure 12c. The main feature of this failure mode is that there are generally multiple groups of parallel tensile cracks above or below the step. Failure mode III in this mode generally shows layers of peeling in the tensile crack area, and the degree of damage is relatively large. This result is consistent with those of a granite specimen that contains multiple pre-existing holes in [33].

Both cycle numbers and confining pressure affect the failure mode of pre-damaged specimens to varying degrees. Table 4 summarizes failure modes of pre-damaged specimens under uniaxial compression, with red lines representing tensile cracks, blue lines representing shear cracks, and purple lines representing stepped cracks. When confining pressure is 50 MPa, pre-damaged specimens with 10 cycles belong to mode I, and specimens with 20 and 30 cycles belong to mode II. When confining pressure is 80 MPa, pre-damaged specimens with 10 cycles belong to the transition failure mode I–III, specimens with 20 cycles belong to failure mode III, and specimens with 30 cycles belong to failure mode II. When confining pressure is 100 MPa, all pre-damaged specimens belong to mode III. When confining pressure is 120 MPa, pre-damaged specimens with 10 and 20 cycles belong to failure mode III, and specimens with 30 cycles belong to failure mode II.

**Table 4.** Failure mode of pre-damaged specimens under uniaxial compression.

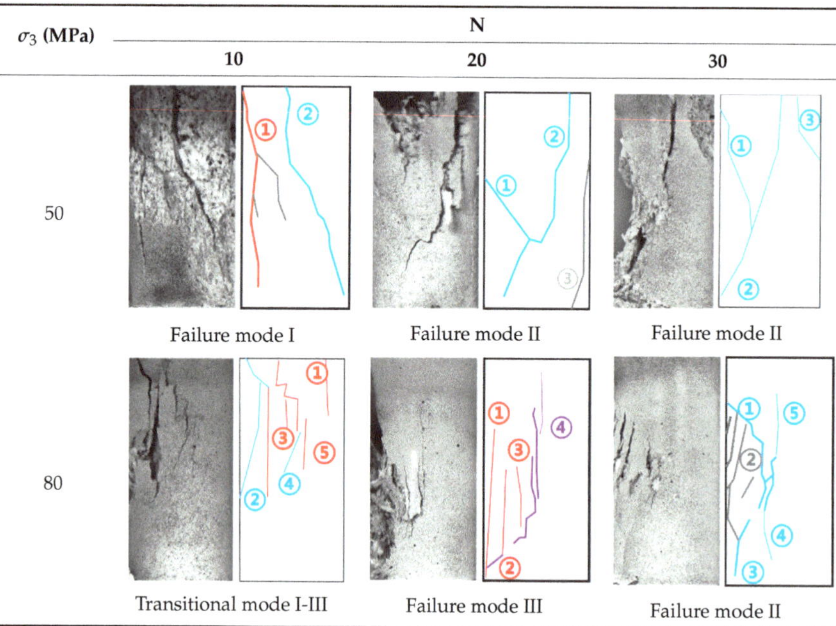

| $\sigma_3$ (MPa) | N | | |
|---|---|---|---|
| | **10** | **20** | **30** |
| 50 | Failure mode I | Failure mode II | Failure mode II |
| 80 | Transitional mode I-III | Failure mode III | Failure mode II |

**Table 4.** *Cont.*

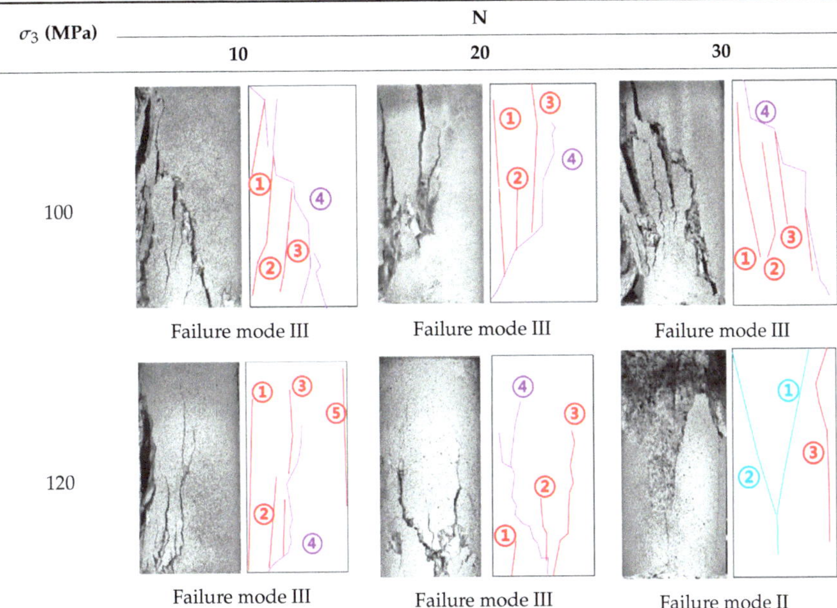



| $\sigma_3$ (MPa) | N | | |
|---|---|---|---|
| | 10 | 20 | 30 |

## 5. Conclusions

(1) The mechanical properties of granite specimens that undergo high cycle pre-damage are significantly different from those of their low cycle pre-damage specimens. The mechanical properties of the high cycle pre-damaged specimens decreased significantly. Under uniaxial compression, the pre-damaged specimens showed complex strength characteristics, and uniaxial compressive strength presents a non-linear relationship with cycle numbers and confining pressures. Damage to pre-damaged specimens was primarily due to the strength weakening effect caused by cycle numbers.

(2) The damage variables of pre-damaged specimens positively correlate with the cycle numbers and confining pressure. Damage to the pre-damaged specimens continues to increase with increasing cycle numbers, and the magnitude of damage varied with various confining pressure. The greater the confining pressure, the greater the damage variable in pre-damaged specimens. When specimens were subjected to high cycle numbers, the confining pressure restriction effect was not obvious.

(3) The evolution laws of uniaxial compression damage propagation in pre-damaged specimens showed differences and obvious localization phenomenon under different cycle numbers and confining pressures. The number of initial cracks was less at low cycle numbers. However, the number of initial cracks was large under high cycle numbers, and they coalesced with each other to form stepped shear cracks. Tensile cracks were always the earliest cracks to appear in pre-damaged granite specimens, and the greater the cycle numbers, the more concentrated the crack initiation was in the plastic deformation stage.

(4) There are three uniaxial compression failure modes for pre-damaged specimens, namely tensile shear failure (mode I), quasi-coplanar shear failure (mode II), and stepped path failure (mode III). The failure modes of pre-damaged specimens are related to confining pressure. Under different pre-damage stress environments with high confining pressures, the failure modes are dominated by Mode II and Mode III, respectively.

**Author Contributions:** Conceptualization, J.H. and D.Y.; methodology, D.Y.; validation, D.Y., G.W. and X.X.; data curation, P.Z. and F.Z.; writing-original draft preparation, P.Z.; writing-review and editing, J.H. and D.Y.; visualization, P.Z., S.M. and R.X.; supervision, J.H.; funding acquisition, J.H. All authors have read and agreed to the published version of the manuscript.

**Funding:** This research was funded by the National Key Research and Development Program of China (2017YFC0602901); the National Natural Science Foundation of China (41672298); and the Central South University Postgraduate Independent Exploration and Innovation Project Fund (2021zzts0872).

**Institutional Review Board Statement:** Not applicable.

**Informed Consent Statement:** Not applicable.

**Data Availability Statement:** All data, models, and code generated or used during the study appear in the submitted article.

**Conflicts of Interest:** The authors declare no conflict of interest.

## References


1. Wang, Y.T.; Zhou, X.P.; Xu, X. Numerical simulation of propagation and coalescence of flaws in rock materials under compressive loads using the extended non-ordinary state-based peridynamics. *Eng. Fract. Mech.* **2016**, 248–273. [CrossRef]
2. Singh, P.K. Blast vibration damage to underground coal mines from adjacent open-pit blasting. *Int. J. Rock Mech. Min. Sci.* **2002**, *39*, 959–973. [CrossRef]
3. Vaneghi, R.G.; Ferdosi, B.; Okoth, A.D.; Kuek, B. Strength degradation of sandstone and granodiorite under uniaxial cyclic loading. *J. Rock Mech. Geotech. Eng.* **2018**, *10*, 117–126. [CrossRef]
4. Zhou, Z.L.; Wang, H.Q.; Cai, X.; Zang, H.Z.; Chen, L.; Liu, F. Bearing characteristics and fatigue damage mechanism of multi-pillar system subjected to different cyclic loads. *J. Cent. South Univ.* **2020**, *27*, 542–553. [CrossRef]
5. Fan, X.; Kulatilake, P.H.S.W.; Chen, X. Mechanical behavior of rock-like jointed blocks with multi-non-persistent joints under uniaxial loading: A particle mechanics approach. *Eng. Geol.* **2015**, *190*, 17–32. [CrossRef]
6. Hu, J.H.; Yang, D.J. Meso-damage evolution and mechanical characteristics of low-porosity sedimentary rocks under uniaxial compression. *Trans. Nonferr. Met. Soc. China* **2020**, *30*, 1071–1077. [CrossRef]
7. Xiao, J.Q.; Ding, D.X.; Jiang, F.L.; Xu, G. Fatigue damage variable and evolution of rock subjected to cyclic loading. *Int. J. Rock Mech. Min. Sci.* **2010**, *47*, 461–468. [CrossRef]
8. Bagde, M.N.; Petroš, V. Fatigue and dynamic energy behaviour of rock subjected to cyclical loading. *Int. J. Rock Mech. Min. Sci.* **2009**, *46*, 200–209. [CrossRef]
9. Chen, Y.Q.; Watanabe, K.; Kusuda, H.; Kusaka, E.; Mabuchi, M. Crack growth in Westerly granite during a cyclic loading test. *Eng. Geol.* **2011**, *117*, 189–197. [CrossRef]
10. Zhang, P.; Xu, J.G.; Li, N. Fatigue properties analysis of cracked rock based on fracture evolution process. *J. Cent. South Univ. Technol.* **2008**, *15*, 95–99. [CrossRef]
11. Fuenkajorn, K.; Phueakphum, D. Effects of cyclic loading on mechanical properties of Maha Sarakham salt. *Eng. Geol.* **2010**, *112*, 43–52. [CrossRef]
12. Taheri, A.; Yfantidis, N.; Olivares, C.L.; Connelly, B.J.; Bastian, T.J. Experimental Study on Degradation of Mechanical Properties of Sandstone under Different Cyclic Loadings. *Geotech. Test. J.* **2016**, *39*, 673–687. [CrossRef]
13. Jia, C.J.; Xu, W.Y.; Wang, R.B.; Wang, W.; Zhang, J.C.; Yu, J. Characterization of the deformation behavior of fine-grained sandstone by triaxial cyclic loading. *Constr. Build. Mater.* **2018**, *162*, 113–123. [CrossRef]
14. Wang, Z.C.; Li, S.C.; Qiao, L.P.; Zhao, J.G. Fatigue Behavior of Granite Subjected to Cyclic Loading under Triaxial Compression Condition. *Rock Mech. Rock Eng.* **2013**, *46*, 1603–1615. [CrossRef]
15. Liu, E.L.; Huang, R.Q.; He, S.M. Effects of Frequency on the Dynamic Properties of Intact Rock Samples Subjected to Cyclic Loading under Confining Pressure Conditions. *Rock Mech. Rock Eng.* **2011**, *45*, 89–102. [CrossRef]
16. Hu, J.H.; Wen, G.P.; Lin, Q.B.; Cao, P.; Li, S. Mechanical properties and crack evolution of double-layer composite rock-like specimens with two parallel fissures under uniaxial compression. *Theor. Appl. Fract. Mech.* **2020**, *108*, 102610–102621. [CrossRef]
17. Xie, Z.Z.; Zhang, N.; Meng, F.F.; Han, C.L.; An, Y.P.; Zhu, R.J. Deformation Field Evolution and Failure Mechanisms of Coal–Rock Combination Based on the Digital Speckle Correlation Method. *Energies* **2019**, *12*, 2511. [CrossRef]
18. Tan, Y.L.; Guo, W.Y.; Zhao, T.B.; Yu, F.H.; Huang, B.; Huang, D.M. Influence of Fissure Number on the Mechanical Properties of Layer-Crack Rock Models under Uniaxial Compression. *Adv. Civ. Eng.* **2018**, *2018*, 1–12. [CrossRef]
19. Sun, Q.; Cai, C.; Zhang, S.K.; Tian, S.; Li, B.; Xia, Y.J.; Sun, Q.W. Study of localized deformation in geopolymer cemented coal gangue-fly ash backfill based on the digital speckle correlation method. *Constr. Build. Mater.* **2019**, *215*, 321–331. [CrossRef]
20. Cao, R.H.; Lin, H.; Cao, P. Strength and failure characteristics of brittle jointed rock-like specimens under uniaxial compression: Digital speckle technology and a particle mechanics approach. *Int. J. Min. Sci. Technol.* **2018**, *28*, 669–677. [CrossRef]

21. Li, X.P.; Wang, B. Research on Distribution Rule of Geostress in Deep Rock in Chinese Mainland. *Adv. Mater. Res.* **2011**, *243–249*, 2116–2122. [CrossRef]
22. Zhou, Z.L.; Cai, X.; Ma, D.; Cao, W.Z.; Chen, L.; Zhou, J. Effects of water content on fracture and mechanical behavior of sandstone with a low clay mineral content. *Eng. Fract. Mech.* **2018**, 47–65. [CrossRef]
23. Ma, S.P.; Wang, L.G.; Jin, G.C. Damage evolution inspection of rock using digital speckle correlation method (DSCM). *Exp. Mech. Nano Biotech.* **2006**, *326–328*, 1117–1120. [CrossRef]
24. Liu, L.; Xu, W.Y.; Zhao, L.Y.; Zhu, Q.Z.; Wang, R.B. An experimental and numerical investigation of the mechanical behavior of granite gneiss under compression. *Rock Mech. Rock Eng.* **2017**, *50*, 499–506. [CrossRef]
25. Huang, B.; Lu, W. Experimental investigation of the uniaxial compressive behavior of thin building granite. *Constr. Build. Mater.* **2021**, *267*. [CrossRef]
26. Yang, S.Q.; Huang, Y.H.; Tian, W.L.; Yin, P.F.; Jing, H.W. Effect of high temperature on deformation failure behavior of granite specimen containing a single fissure under uniaxial compression. *Rock Mech. Rock Eng.* **2019**, *52*, 2087–2107. [CrossRef]
27. He, C.; Yang, J. Laboratory study of dynamic mechanical characteristic of granite subjected to confining pressure and cyclic blast loading. *Lat. Am. J. Solids Struct.* **2018**, *15*, 1–16. [CrossRef]
28. Taheri, A.; Royle, A.; Yang, Z.; Zhao, Y. Study on variations of peak strength of a sandstone during cyclic loading. *Geomech. Geophys. Geo-Energy Geo-Resour.* **2016**, *2*, 1–10. [CrossRef]
29. Wu, F.; Chen, J.; Zou, Q.L. A nonlinear creep damage model for salt rock. *Int. J. Damage Mech.* **2018**, *28*, 758–771. [CrossRef]
30. Shang, J.L.; Hu, J.H.; Zhou, K.P.; Luo, X.W. Porosity increment and strength degradation of low-porosity sedimentary rocks under different loading conditions. *Int. J. Rock Mech. Min. Sci.* **2015**, *75*, 216–223. [CrossRef]
31. Liu, G.; Xiao, F.K.; Guo, Z.B.; Chi, X.H.; Jiang, Y.N.; Yu, H.; Hou, Z.Y.; Zhao, R.X. Plastic characteristics of rock under point load. *Sci. Technol. Eng.* **2018**, *18*, 217–222.
32. Amitrano, D. Rupture by damage accumulation in rocks. *Int. J. Fract.* **2006**, *139*. [CrossRef]
33. Huang, Y.H.; Yang, S.Q.; Tian, W.L. Cracking process of a granite specimen that contains multiple pre-existing holes under uniaxial compression. *Fatigue Fract. Eng. Mater. Struct.* **2019**, *42*, 1341–1356. [CrossRef]
34. Shi, G.C.; Yang, X.J.; Yu, H.C.; Zhu, C. Acoustic emission characteristics of creep fracture evolution in double-fracture fine sandstone under uniaxial compression. *Eng. Fract. Mech.* **2019**, *210*, 13–28. [CrossRef]
35. Lee, H.; Jeon, S. An experimental and numerical study of fracture coalescence in pre-cracked specimens under uniaxial compression. *Int. J. Solids Struct.* **2011**, *48*, 979–999. [CrossRef]
36. Liu, J.P.; Li, Y.H.; Xu, S.; Jin, C.Y.; Liu, Z.S. Moment tensor analysis of acoustic emission for cracking mechanisms in rock with a pre-cut circular hole under uniaxial compression. *Eng. Fract. Mech.* **2015**, *135*, 206–218. [CrossRef]
37. Lin, P.; Wong, R.H.C.; Tang, C.A. Experimental study of coalescence mechanisms and failure under uniaxial compression of granite containing multiple holes. *Int. J. Rock Mech. Min. Sci.* **2015**, *77*, 313–327. [CrossRef]

*Article*

# Two-Phase Flow Effects on Seismic Wave Anelasticity in Anisotropic Poroelastic Media

**Juan E. Santos** [1,2,3,†], **José M. Carcione** [4,†] **and Jing Ba** [1,*]

[1] School of Earth Sciences and Engineering, Hohai University, Nanjing 211100, China
[2] Department of Mathematics, Purdue University, West Lafayette, IN 47907, USA
[3] Facultad de Ingeniería, Universidad de Buenos Aires, IGPUBA, Av. Las Heras 2214, Buenos Aires C1127AAR, Argentina
[4] National Institute of Oceanography and Applied Geophysics—OGS, 34010 Trieste, Italy
[*] Correspondence: jingba@188.com
[†] These authors contributed equally to this work.

**Abstract:** We study the wave anelasticity (attenuation and velocity dispersion) of a periodic set of three flat porous layers saturated by two immiscible fluids. The fluids are very dissimilar in properties, namely gas, oil, and water, and, at most, three layers are required to study the problem from a general point of view. The sequence behaves as viscoelastic and transversely isotropic (VTI) at wavelengths much longer than the spatial period. Wave propagation causes fluid flow and slow P modes, inducing anelasticity. The fluids are characterized by capillary forces and relative permeabilities, which allow for the existence of two slow modes and the presence of dissipation, respectively. The methodology to study the physics is based on a finite-element uspcaling approach to compute the complex and frequency-dependent stiffnesses of the effective VTI medium. The results of the experiments indicate that there is higher dissipation and anisotropy compared to the widely used model based on an effective fluid that ignores the effects of surface tension (capillarity) and viscous flow interference between the two fluid phases.

**Keywords:** capillary pressure; two-phase fluids; porous medium; anisotropy, attenuation; finite elements

**Citation:** Santos, J.E.; Carcione, J.M.; Ba, J. Two-Phase Flow Effects on Seismic Wave Anelasticity in Anisotropic Poroelastic Media. *Energies* **2021**, *14*, 6528. https://doi.org/10.3390/en14206528

Academic Editor: Nikolaos Koukouzas

Received: 21 August 2021
Accepted: 28 September 2021
Published: 12 October 2021

**Publisher's Note:** MDPI stays neutral with regard to jurisdictional claims in published maps and institutional affiliations.

## 1. Introduction

Wave anelasticity in porous media is a topic with applications in many areas of geophysics and material science [1]. Waves generate fluid flow in the pores and energy losses that can be observed in field and laboratory experiments [2–4]. Recent laboratory experiments performed in the seismic range have shown the frequency dependence of anelasticity in sandstones with partial gas or oil saturations [5–7], while experiments conducted in Reference [8] show a significant attenuation in the extensional and bulk deformation modes, as well as numerical simulations in close agreement with laboratory data.

The theoretical pioneering work of Biot [9–11] describes waves propagation in porous media saturated by single-phase fluids. The theory predicts a shear wave and two compressional waves, a fast one, where the solid and fluid move together, and a slow one, where the displacement is out of phase. At low frequencies, the slow wave is diffusive and becomes a propagating wave at high frequencies. Significant anelasticity of the fast wave is due to mode conversion at mesoscopic scale heterogeneities (mesoscopic loss). White et al. [12] were the first to introduce this loss mechanism based on Biot's theory for a periodic sequence of two flat and thin porous layers saturated with a single-phase fluid. This mechanism is also termed WIFF or wave-induced fluid flow loss.

The study of propagation in partially saturated porous media has been presented in Reference [13–15]. Lo et al. [16] present an Eulerian model for waves propagating in porous media saturated by two fluids. In this model, the stress-strain relations are obtained by taking into account capillary pressure assumed as a unique function of saturation and ignoring hysteresis, while the dynamic equations are formulated from a mass balance equation for each solid and fluid phase. Three compressional waves are predicted to exist, one wavelike and two of them of diffusive type. The model is applied to analyze waves in soils saturated by water with air or oil. A macroscopic model based on a two-component Biot model and a mixture theory is presented in Reference [17]. Other models to analyze the wave behavior in porous rocks saturated my immiscible fluids appeared in [1,15,18–22].

Concerning studies on porous media by using numerical simulations, the work by Thovert et al. [23] analyzes wave propagation using an homogenization approach. The authors consider the existence of connected pores that may or may not percolate, obtain the macroscopic coefficients and apply the results to a synthetic porous medium. Hamzehpour et al. [24] study acoustic waves in two-dimensional fractured porous media using finite differences to discretize the differential equations. This analysis concludes that, near the source, waves amplitude decay exponentially. A generalization of the Biot theory, appearing in Reference [25], takes into account more than one fluid phase saturating the pores, such as brine-gas and oil-gas mixtures. This theory predicts three compressional waves and a shear wave. The stress-strain relations are derived from the complementary virtual work principle in order to include the capillary pressure relation, while the dissipation function is defined in terms of Darcy's law based on two fluids [26].

The presence of an additional slow wave is expected to induce more attenuation and velocity dispersion, as shown in Reference [27], for the case of vertical propagation across a periodic set of two thin layers saturated with two immiscible fluids. This work considers the more general case of a sequence of three layers, also leading to a VTI behavior at long wavelengths, but to a different frequency dependence of the velocity and attenuation. Qi et al. [28] consider capillary effects in random media of patchy saturation by using a membrane stiffness, but this approach leads to an increase in phase velocities and lower attenuation. An experimental study on how capillarity affects the compressional P-wave velocity for patchy saturation during an imbibition process is presented by Liu et al. [29], where the patch size depends on the fluid saturation. The effect of capillarity is related to the injection rate, changing with imbibition. The results exhibit an increase of the experimentally measured P-wave static velocity as capillary pressure increases.

This study is based on the theory of Cavallini et al. [30] valid for periodic $N$ layers saturated for single-phase fluids and a FE procedure to obtain an equivalent VTI medium to an horizontally layered medium consisting of a three periodic poroelastic solid saturated by immiscible fluids. An FE procedure is used to determine a VTI medium equivalent at long wavelengths to a periodic sequence of thin poroelastic layers. It consists of defining five independent boundary-value problems, each one associated with either a compressibility or shear time-harmonic test, whose solutions are obtained by the FE method. The results are validated against the theory by Cavallini et al. [30] for effective single-phase fluids. Then, several examples are given, where the velocities and attenuation are obtained by using effective-single phase and two-phase fluids.

## 2. The Differential Model

The poroelastic medium is saturated by non-wetting and wetting fluid phases, whose properties and variables are indicated by subscripts or superscripts "$n$" and "$w$", respectively. The corresponding saturations are denoted by $S_w = S_w(\mathbf{x})$ and $S_n = S_n(\mathbf{x})$, with $\mathbf{x} = (x, y, z)$, respectively, with associated residual saturations $S_{rw}$ and $S_{rn}$. It is assumed that both fluid phases occupy the whole pore space and that a continuous network of these phases exists (funicular regime) [31]. Thus,

$$S_w + S_n = 1, \quad S_{rn} < S_n < 1 - S_{rw}, \quad S_{rw} < S_w < 1 - S_{ro}.$$

The Fourier transforms of the particle displacement of the three phases composing the material, i.e., the solid, non-wettting and phases phases, are denoted as $\mathbf{u}^s = (u_i^s)$, $\tilde{\mathbf{u}}^n = (\tilde{u}_i^n)$, and $\tilde{\mathbf{u}}^w = (\tilde{u}_i^w)$, $i = 1, 2, 3$, respectively. Set $\mathbf{u} = (\mathbf{u}^s, \mathbf{u}^n, \mathbf{u}^w)$. Define $\phi = \phi(\mathbf{x})$

as the matrix effective porosity, with the relative displacement and variation in fluid content of each fluid phase defined as

$$\mathbf{u}^{\theta} = \phi(\tilde{\mathbf{u}}^{\theta} - \mathbf{u}^{s}), \quad \zeta^{\theta} = -\nabla \cdot \mathbf{u}^{\theta}, \; \theta = n, w.$$

Define $P_w$ and $P_n$ as the Fourier transforms of infinitesimal changes in the wetting and non-wetting fluid pressures, respectively, with respect to an initial equilibrium state of pressures $\bar{P}_w$, $\bar{P}_n$, porosity $\bar{\phi}$, and non-wetting saturation $\bar{S}_n$. The capillary relation is [26,31,32]

$$P_{ca} = P_{ca}(S_n + \bar{S}_n) = \bar{P}_n + P_n - (\bar{P}_w + P_w) = P_{ca}(\bar{S}_n) + P_n - P_w \geq 0. \tag{1}$$

Ignoring hysteresis, $P_{ca}$ is an increasing function of $S_n$ and positive.

Let $\varepsilon_{ij}(\mathbf{u}^{s})$ and $e^{s} = \varepsilon_{ii}(\mathbf{u}^{s})$ be the Fourier transforms of the solid strain tensor and corresponding linear invariant, respectively. Denoting with $\tau_{ij}(\mathbf{u})$ the stress-tensor components of the bulk material, the constitutive relations are:

$$\begin{aligned}
\tau_{ij}(\mathbf{u}) &= 2\mu\,\varepsilon_{ij} + \delta_{ij}(\lambda_u\,e^{s} - B_1\,\zeta^{n} - B_2\,\zeta^{w}), & (2)\\
\mathcal{T}_n(\mathbf{u}) &= (\bar{S}_n + \beta)P_n - \beta P_w = -B_1\,e^{s} + M_1\,\zeta^{n} + M_3\,\zeta^{w}, & (3)\\
\mathcal{T}_w(\mathbf{u}) &= \bar{S}_w\,P_w = -B_2\,e^{s} + M_3\,\zeta^{n} + M_2\,\zeta^{w}, & (4)
\end{aligned}$$

where

$$\beta = \frac{P_{ca}(\bar{S}_n)}{P'_{ca}(\bar{S}_n)}.$$

In (2), $\lambda_u = K_u - \frac{2}{3}\mu$, with $K_u$ and $\mu$ being the wet-rock bulk and dry-rock shear moduli, respectively. Thus,

$$E_u = \lambda_u + 2\mu \tag{5}$$

is the wet-rock P-wave modulus. The coefficients in (2)–(4) can be determined as indicated in Appendix A.

The model describing the response of a poroelastic medium saturated with two fluids in the diffusive range of frequencies consists of a partial differential equations imposing the stress equilibrium of the bulk material (Equation (6)), together with a generalization of the two-phase Darcy law [26,31,32] (Equations (7) and (8)). Thus, if $\omega$ denotes the angular frequency, these equations are

$$\nabla \cdot \tau(\mathbf{u}) = 0, \tag{6}$$
$$i\omega\,d_n\,\mathbf{u}^{n} - i\omega\,d_{nw}\,\mathbf{u}^{w} + \nabla\mathcal{T}_n(\mathbf{u}) = 0, \tag{7}$$
$$i\omega\,d_w\,\mathbf{u}^{w} - i\omega\,d_{nw}\,\mathbf{u}^{n} + \nabla\mathcal{T}_w(\mathbf{u}) = 0. \tag{8}$$

The coefficients in (7) and (8) depend on the absolute permeability $\kappa$, fluid viscosities $\eta_l$ and relative permeabilities $K_{rl}(S_l)$, $l = n, w$. They are defined as

$$d_l(\bar{S}_l) = (\bar{S}_l)^2\frac{\eta_l}{\kappa K_{rl}(\bar{S}_l)}, \quad l = n, w, \tag{9}$$
$$d_{nw}(\bar{S}_n, \bar{S}_w) = \epsilon(d_n(\bar{S}_n)d_w(\bar{S}_w)). \tag{10}$$

The coefficient $d_{nw}(S_n)$ in (10) is a dissipative function, where $\epsilon$ is small. It describes the viscous drag between the immiscible fluids. In the absence of experimental data, it has the form given in (10).

## 3. The Equivalent Viscoelastic Transversely-Isotropic Medium

For long wavelengths, compared to the layer thicknesses, a fluid-saturated porous medium is seen as an effective or equivalent VTI medium characterized by five frequency dependent and complex coefficients that can be determined by means of harmonic experiments performed on a representative sample of the medium. These experiments, given next for the 2D case, are defined as boundary value problems (BVPs), whose solutions are obtained by using an FE method. For single-phase fluids, the analytical solutions given in Appendix B are used to validate the numerical simulations.

Let us consider a representative squared sample $\Omega = (0, L)^2$ with boundary $\Gamma$ in the $(x, z)$-plane, where $x$ and $z$ denote the coordinates along the horizontal and vertical directions. Let $\Gamma^R, \Gamma^L, \Gamma^T$, and $\Gamma^B$ denote the right, left, top, and bottom boundaries of $\Omega$. In addition, let $\chi$ be a unit tangent on $\Gamma$ oriented counterclockwise, and $\nu$ the unit outer normal on $\Gamma$.

Let $\mathcal{E}(\tilde{\mathbf{u}}^s)$ and $\mathcal{T}(\tilde{\mathbf{u}}^s)$ be the Fourier transforms of the strain and stress tensors at the macroscale, with $\tilde{\mathbf{u}}^s$ denoting the macroscopic displacement vector of the solid. The stress-strain equations for a VTI medium can be stated as follows:

$$\mathcal{T}_{11}(\tilde{\mathbf{u}}_s) = p_{11}\,\mathcal{E}_{11}(\tilde{\mathbf{u}}_s) + p_{12}\,\mathcal{E}_{22}(\tilde{\mathbf{u}}_s) + p_{13}\,\mathcal{E}_{33}(\tilde{\mathbf{u}}_s), \tag{11}$$

$$\mathcal{T}_{22}(\tilde{\mathbf{u}}_s) = p_{12}\,\mathcal{E}_{11}(\tilde{\mathbf{u}}_s) + p_{11}\,\mathcal{E}_{22}(\tilde{\mathbf{u}}_s) + p_{13}\,\mathcal{E}_{33}(\tilde{\mathbf{u}}_s), \tag{12}$$

$$\mathcal{T}_{33}(\tilde{\mathbf{u}}_s) = p_{13}\,\mathcal{E}_{11}(\tilde{\mathbf{u}}_s) + p_{13}\,\mathcal{E}_{22}(\tilde{\mathbf{u}}_s) + p_{33}\,\mathcal{E}_{33}(\tilde{\mathbf{u}}_s), \tag{13}$$

$$\mathcal{T}_{23}(\tilde{\mathbf{u}}_s) = 2\,p_{55}\,\mathcal{E}_{23}(\tilde{\mathbf{u}}_s), \tag{14}$$

$$\mathcal{T}_{13}(\tilde{\mathbf{u}}_s) = 2\,p_{55}\,\mathcal{E}_{13}(\tilde{\mathbf{u}}_s), \tag{15}$$

$$\mathcal{T}_{12}(\tilde{\mathbf{u}}_s) = 2\,p_{66}\,\mathcal{E}_{12}(\tilde{\mathbf{u}}_s), \tag{16}$$

where $p_{12} = p_{11} - 2p_{66}$. To compute the five independent frequency dependent and complex stiffness coefficients in (11)–(16), we solve Equations (6)–(8) in $\Omega$ imposing no change in fluid content of both fluid phases, i.e., with the boundary conditions

$$\mathbf{u}^n \cdot \nu = 0, \quad \mathbf{u}^w \cdot \nu = 0, \quad (x, z) \in \Gamma, \tag{17}$$

and additional boundary conditions stated below for each coefficient $p_{IJ}$.

To determine $p_{33}$, the boundary conditions are imposed:

$$\tau(\mathbf{u})\nu \cdot \nu = -\Delta P, \quad (x, z) \in \Gamma^T, \tag{18}$$

$$\tau(\mathbf{u})\nu \cdot \chi = 0, \quad (x, z) \in \Gamma, \tag{19}$$

$$\mathbf{u}^s \cdot \nu = 0, \quad (x, z) \in \Gamma \setminus \Gamma^T. \tag{20}$$

The solution of this BVP satisfies the equations (cf. (2)) $\varepsilon_{11} = \varepsilon_{22} = 0, \varepsilon_{ij} = 0,$ $i \neq j, \xi^n = \xi^w = 0$. Thus, $\mathcal{E}_{11} = \mathcal{E}_{22} = 0$, and (13) reduces to

$$\mathcal{T}_{33} = p_{33}\mathcal{E}_{33}. \tag{21}$$

Now, $p_{33}$ can be computed from (21) by calculating $\mathcal{T}_{33}$ and $\mathcal{E}_{33}$ as averages over the sample $\Omega$ of $\tau_{33}$ and $\epsilon_{33}$, i.e.,

$$\mathcal{T}_{33} = \frac{1}{\Omega} \int_{\Omega} \tau_{33} d\Omega, \qquad \mathcal{E}_{33} = \frac{1}{\Omega} \int_{\Omega} \epsilon_{33} d\Omega. \tag{22}$$

To determine $p_{11}$, the following boundary conditions are implemented:

$$\tau(\mathbf{u})\nu \cdot \nu = -\Delta P, \quad (x, z) \in \Gamma^R, \tag{23}$$

$$\tau(\mathbf{u})\nu \cdot \chi = 0, \quad (x, z) \in \Gamma, \tag{24}$$

$$\mathbf{u}^s \cdot \nu = 0, \quad (x, z) \in \Gamma \setminus \Gamma^R. \tag{25}$$

The solution of this BVP for $p_{11}$ satisfies $\varepsilon_{33} = \varepsilon_{22} = 0, \varepsilon_{ij} = 0, i \neq j, \xi^n = \xi^w = 0$. Hence, $\mathcal{E}_{33} = \mathcal{E}_{22} = 0$ and (11) reduces to

$$\mathcal{T}_{11} = p_{11}\mathcal{E}_{11}. \tag{26}$$

Equation (26) determines $p_{11}$ since $\mathcal{T}_{11}$ and $\mathcal{E}_{11}$ can be computed as averages over $\Omega$ of $\tau_{11}$ and $\varepsilon_{11}$, as indicated in (22), for determining $p_{33}$.

The numerical experiment to determine $p_{13}$ is defined by applying the boundary conditions:

$$\boldsymbol{\tau}(\mathbf{u})\boldsymbol{\nu} \cdot \boldsymbol{\nu} = -\Delta P, \quad (x,z) \in \Gamma^R \cup \Gamma^T, \tag{27}$$

$$\boldsymbol{\tau}(\mathbf{u})\boldsymbol{\nu} \cdot \boldsymbol{\chi} = 0, \quad (x,z) \in \Gamma, \tag{28}$$

$$\mathbf{u}^s \cdot \boldsymbol{\nu} = 0, \quad (x,z) \in \Gamma^L \cup \Gamma^B. \tag{29}$$

In this experiment, $\varepsilon_{22} = 0, \varepsilon_{ij} = 0, i \neq j, \xi^n = \xi^w = 0$. Then, $\varepsilon_{11}$ and $\varepsilon_{33}$ do not vanish, and $\mathcal{E}_{11}$ and $\mathcal{E}_{33}$ can be obtained as

$$\mathcal{E}_{11} = \frac{1}{\Omega} \int_\Omega \varepsilon_{11} d\Omega, \qquad \mathcal{E}_{33} = \frac{1}{\Omega} \int_\Omega \varepsilon_{33} d\Omega. \tag{30}$$

Next, from Equations (11) and (13), it follows that

$$\mathcal{T}_{11} = p_{11}\mathcal{E}_{11} + p_{13}\mathcal{E}_{33}\varepsilon_{33}, \quad \mathcal{T}_{33} = p_{13}\mathcal{E}_{11} + p_{33}\mathcal{E}_{33}. \tag{31}$$

Note that it follows from (27) that $\mathcal{T}_{11} = \mathcal{T}_{33} = -\Delta P$, so that using (31) $p_{13}$ is determined as

$$p_{13} = \frac{p_{11}\mathcal{E}_{11} - p_{33}\mathcal{E}_{33}}{\mathcal{E}_{11} - \mathcal{E}_{33}}. \tag{32}$$

The following boundary conditions are used to determine $p_{55}$:

$$-\boldsymbol{\tau}(\mathbf{u})\boldsymbol{\nu} = \mathbf{g}, \quad (x,z) \in \Gamma^T \cup \Gamma^L \cup \Gamma^R, \tag{33}$$

$$\mathbf{u}^s = 0, \quad (x,z) \in \Gamma^B, \tag{34}$$

where

$$\mathbf{g} = \begin{cases} (0, \Delta G), & (x,z) \in \Gamma^L, \\ (0, -\Delta G), & (x,z) \in \Gamma^R, \\ (-\Delta G, 0), & (x,z) \in \Gamma^T. \end{cases}$$

This BVP satisfies the conditions $\varepsilon_{ij} \neq 0$ only for $i = 1, j = 3$, so that $\mathcal{E}_{ij} \neq 0$ only for $i = 1, j = 3$, and (15) reduces to

$$\mathcal{T}_{13} = 2\, p_{55}\, \mathcal{E}_{13}. \tag{35}$$

Next, by computing the average of the local strain $\varepsilon_{13}$ over the sample

$$\mathcal{E}_{13} = \frac{1}{\Omega} \int_\Omega \varepsilon_{13} d\Omega, \tag{36}$$

the stiffness $p_{55}$ can be determined from (35).

Finally, $p_{66}$ is computed by using the boundary conditions:

$$-\boldsymbol{\tau}(\mathbf{u})\boldsymbol{\nu} = \mathbf{g}_2, \quad (x_1, x_2) \in \Gamma^B \cup \Gamma^R \cup \Gamma^T, \tag{37}$$

$$\mathbf{u}_s = 0, \quad (x_1, x_2) \in \Gamma^L, \tag{38}$$

where

$$\mathbf{g}_2 = \begin{cases} (\Delta G, 0), & (x_1, x_2) \in \Gamma^B, \\ (-\Delta G, 0), & (x_1, x_2) \in \Gamma^T, \\ (0, -\Delta G), & (x_1, x_2) \in \Gamma^R. \end{cases}$$

Then, $p_{66}$ is determined as indicated for $p_{55}$.

The five BVPs are solved with the FE method. The components of the displacement vector of the solid are represented with locally bilinear polynomials, while global continuity is imposed on $\Omega$. Furthermore, only the normal components of the vector displacements of the two fluids need to be continuous across the common edges of adjacent computational cells. This condition is imposed by using polynomials linear in the $x$-direction and constant in the $z$-direction to represent the first component of each fluid phase, while polynomials constant in $x$ and linear in $z$ are used for the second component.

## 4. Results. Numerical Experiments

The solution of the five BVPs to determine the complex stiffnesses $p_{IJ}$ stated in Section 3 is obtained as a function of the propagation direction and frequency with the FE method. Appendix A gives analytical expressions for the phase and energy velocities and dissipation factors of the qP, qSV, and SH waves.

In all the experiments, the numerical samples are discretized with a $90 \times 90$ uniform mesh representing six periods, each one with three layers of equal 20 cm thickness, saturated with two fluids. The material properties of the solid and fluid phases are listed in Tables 1 and 2.

**Table 1.** Physical properties of the frame.

|  | L1 | L2 | L3 |
|---|---|---|---|
| $K_s$ (GPa) | 33.4 | 33.4 | 33.4 |
| $\rho_s$ (g/cm$^3$) | 2.65 | 2.65 | 2.65 |
| $\phi$ | 0.3 | 0.2 | 0.1 |
| $K_m$ (GPa) | 7.2 | 14.5 | 23.5 |
| $\mu$ (GPa) | 6.5 | 13 | 21.1 |
| $\kappa$ (darcy) | 1 | 0.24 | 0.02 |

**Table 2.** Physical properties of the fluids.

|  | Brine | Oil | Gas |
|---|---|---|---|
| bulk modulus (GPa) | 2.2 | 2 | 0.0096 |
| density (g/cm$^3$) | 975 | 870 | 70 |
| viscosity (Pa · s) | 0.001 | 0.3 | 0.00015 |

The behavior of the two fluids is described with relative permeabilities, $K_{rn}(S_n)$ and $K_{rw}(S_n)$, and capillary pressure function, $P_{ca}(S_n)$, defined by [33–35]:

$$K_{rn}(S_n) = (1 - (1 - S_n)/(1 - S_{rn}))^2, \tag{39}$$

$$K_{rw}(S_n) = ([1 - S_n - S_{rw}]/(1 - S_{rw}))^2, \tag{40}$$

$$P_{ca}(S_n) = A\left(1/(S_n + S_{rw} - 1)^2 - S_{rn}^2/[S_n(1 - S_{rn} - S_{rw})]^2\right), \tag{41}$$

with $A$ in (41) being the capillary pressure amplitude. The wetting phase is brine, and the non-wetting phase is oil or gas. In all the experiments, the residual saturations are $S_{rw} = 0.01$, $S_{rn} = 0$, the capillary-pressure amplitude is 30 kPa, and $\epsilon = 0.01$ in the definition of the coefficient $d_{nw}$ in (10). Other choices of $\epsilon$ yield similar results. The functions in (39)–(41) are an analytical representation of the experimental curves obtained in laboratory experiments and used in the simulation of two-phase fluid flow and wave propagation. The functions (39)–(41), of common use in reservoir engineering, are useful to model the two-phase funicular regime. (see Reference [26,31,32,36] for additional information on flow of two fluids in porous media). Capillary pressure and relative permeability functions may be obtained from log-well data, as indicated in Reference [37].

The experiments show dissipation factors and phase and energy velocities of waves computed by using two and one fluid phases. The properties of the effective fluid (viscosity $\eta^{(*)}$, density $\rho^{(*)}$, and bulk modulus $K^{(*)}$) are obtained with Reuss averages for the bulk moduli and arithmetic averages for densities and viscosities:

$$\eta^{(*)} = \eta_n S_n + \eta_w S_w,$$

$$\rho^{(*)} = \rho_n S_n + \rho_w S_w,$$

$$\frac{1}{K^{(*)}} = \frac{S_n}{K_n} + \frac{S_w}{K_w}. \tag{42}$$

The following cases are analyzed in the experiments

- Case 1: Six periods of three layers ($L_1$, $L_2$, $L_3$) of layer thickness 20 cm, where the non-wetting fluid saturations in each layer are $L_1$: 1% gas, $L_2$: 5% gas, and $L_3$: 5% oil. The curves labeled "single-phase" are obtained by using the classical Biot theory and effective single-phase fluids.
- Case 2: Six periods of three layers ($L_1$, $L_2$, $L_3$) of layer thickness 20 cm, where the non-wetting fluid saturations in each layer are $L_1$: 1% oil, $L_2$: 5% oil, and $L_3$: 5% gas.

The following notation is used in the figures: cpII and $Q_{\mathrm{II}} = \mathrm{Re}(p_{\mathrm{II}})/\mathrm{Im}(p_{\mathrm{II}})$ are the phase velocity and dissipation factor of the waves traveling parallel and perpendicular to the layering, "11-waves" and "33-waves", respectively.

Figures 1a,b and 2a,b show the phase velocities and corresponding dissipation factors of "33-waves" and "11-waves" as a function of frequency for Cases 1 and 2 computed with the FE method. The plots compare the results for two fluids and one (effective) fluid, where the latter is defined in Equations (42).

At low frequencies, the more realistic case of two-phase fluids has lower velocities than for single-phase fluids, with similar asymptotic values at higher frequencies. As expected, phase velocities for "33-waves" are lower than for "11-waves".

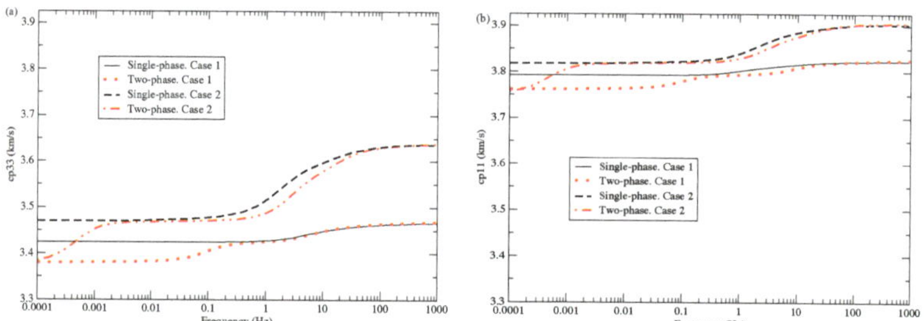

**Figure 1.** Phase velocity of "33-waves" (**a**) and "11-waves" (**b**) for Case 1 and 2 as a function of frequency.

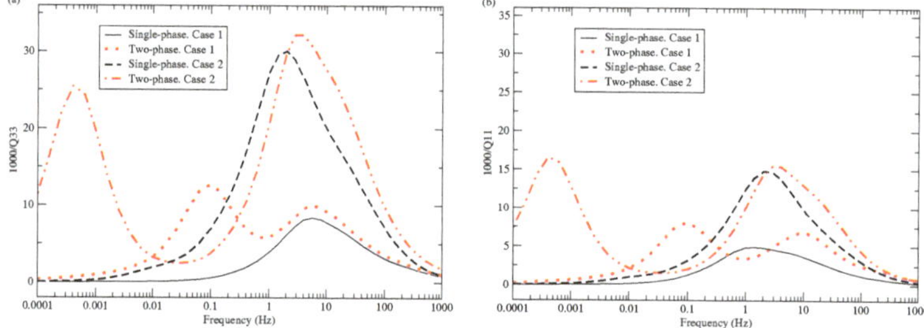

**Figure 2.** Dissipation factor of "33-waves" (**a**) and "11-waves" (**b**) for Cases 1 and 2 as a function of frequency.

The dissipation factors corresponding to two-phase fluids (Figure 2) exhibit a very different behavior as compared with the effective single-phase fluids, since there is an additional attenuation peak at lower frequencies, much stronger for Case 2. This new phenomenon can be explained as follows. Ignoring the cross-dissipative coefficient $d_{nw}$, the particle velocity of the $l$-phase $i\omega\mathbf{u}^l$, $l = o, g, b$ is equal to the gradient of the generalized fluid pressure $\mathcal{T}^l$ times the absolute permeability $\kappa$ and the l-phase mobility $\gamma_l(S_l)$, defined as $\gamma_l(S_l) = [K_{rl}(\bar{S}_l)]/\eta_l$ [26]. For the values of brine, oil and gas saturation is used in the examples, at 1% (5%) oil saturation, and $\gamma_o$ is five (three) orders of magnitude lower than $\gamma_b$ and $\gamma_g$. Thus, fluid flows differently within the pore space, with the fluids interfering each other and inducing additional energy losses, which are not taken into account by the single-phase effective-fluid approach. Furthermore, Case 2 has more oil content than Case 1, thus inducing higher attenuation (see Figure 2). Moreover, the attenuation peaks for "33-waves" have higher amplitudes than for "11-waves", a known effect in finely layered porous media. An additional result observed in the experiments is that, at a given saturation, varying the amplitude of the capillary pressure does not affect the attenuation.

Figures 3–7 exhibit polar plots of energy velocities and dissipation factors at 10 Hz versus the propagation angle. They compare analytical solution for single-phase effective fluids (Appendix B) to the FE solutions based on single-phase effective fluids and the two-phase fluids. Figures 3, 5, and 7 show that the energy velocities of the three waves have a perfect match, which validates the FE numerical scheme. On the other hand, Figures 4 and 6 show a perfect agreement between the single-phase analytical and FE dissipation factors of the qP and qSV waves (SH waves are lossless), but the attenuation corresponding to the two-phase fluids, including capillarity effects, is higher, in agreement with the previous remarks regarding Figure 2.

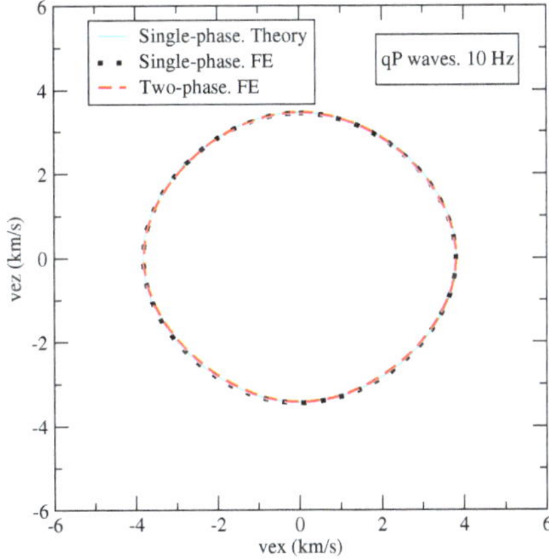


**Figure 3.** Polar representation at 10 Hz of the energy velocities of qP waves for two-phase fluids obtained with the FE method and single-phase effective fluids (analytical and numerical).

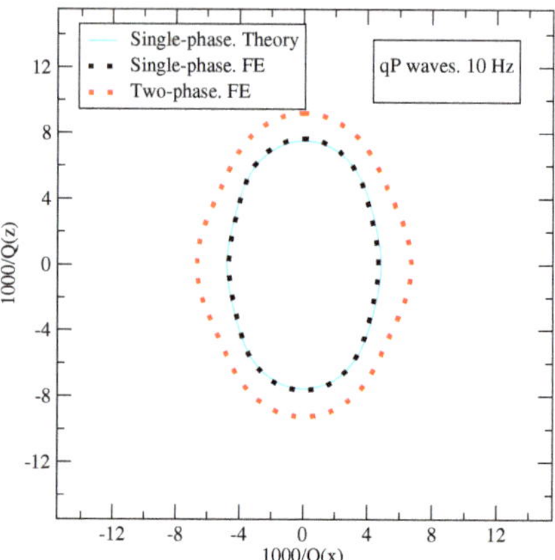

**Figure 4.** Polar representation at 10 Hz of the dissipation factor of the qP waves for two-phase fluids obtained with the FE method and single-phase effective fluids (analytical and numerical).

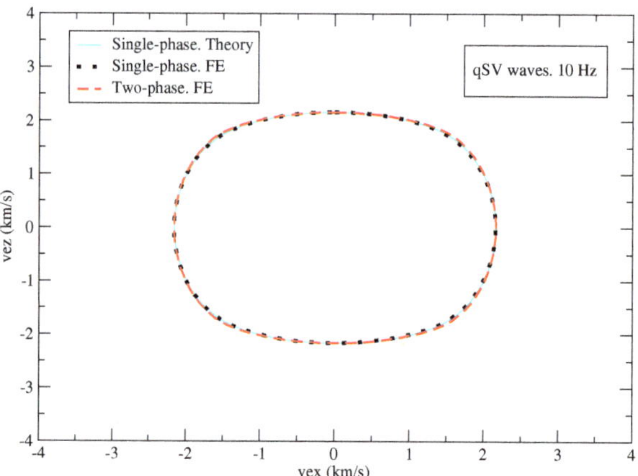

**Figure 5.** Polar representation at 10 Hz of the energy velocity of qSV waves for two-phase fluids obtained with the FE method and single-phase effective fluids (analytical and numerical).

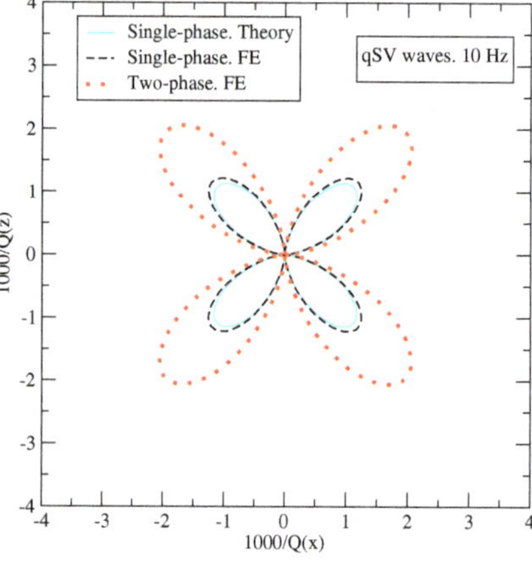

**Figure 6.** Polar representation at 10 Hz of the dissipation factor of qSV waves for two-phase fluids obtained with the FE method and single-phase effective fluids (analytical and numerical).

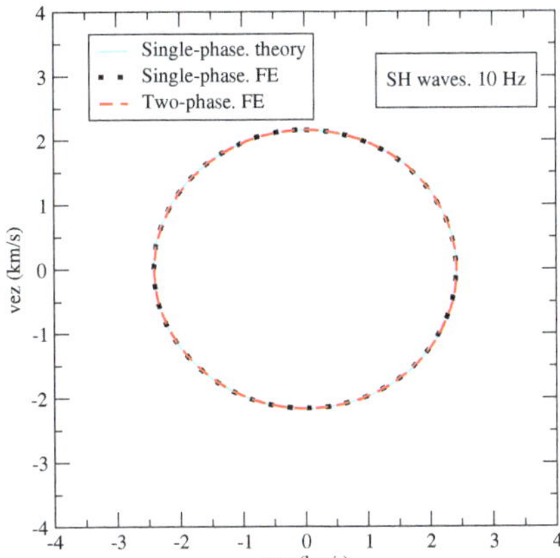



**Figure 7.** Polar representation at 10 Hz of the energy velocity of SH waves for two-phase fluids obtained with the FE method and single-phase effective fluids (analytical and numerical).

Figures 8 and 9 display the modulus of the vertical component of the particle velocities $\mathbf{v}^w = i\omega\mathbf{u}^w$ and $\mathbf{v}^n = i\omega\mathbf{u}^n$ of the wetting and non-wetting fluids at 10 Hz, respectively, that appear in the left-hand side of the two-phase Darcy law (7) and (8). These vertical components, corresponding to Case 2, are denoted by $v^{n,z}$ and $v^{w,z}$ in the plots. The higher mobility $\gamma_b$, as compared with the gas and oil mobilities $\gamma_g$ and $\gamma_o$, explains the higher values of the wetting particle velocity in Figure 9 when compared with those of the non-wetting phase in Figure 8. As stated above, the interference between the two fluid phases due to their different particle velocities induces energy losses and velocity dispersion.

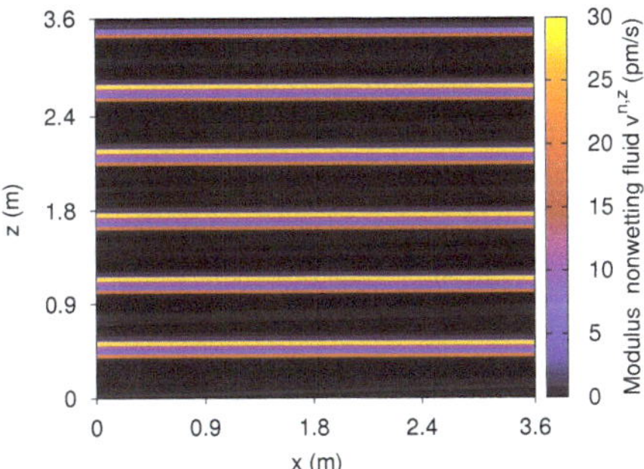

**Figure 8.** Case 2. Modulus of the vertical component of the particle velocity of non-wetting fluids $v^{n,z} = i\omega\mathbf{u}^{n,z}$ obtained with the FE method. Frequency is 10 Hz.

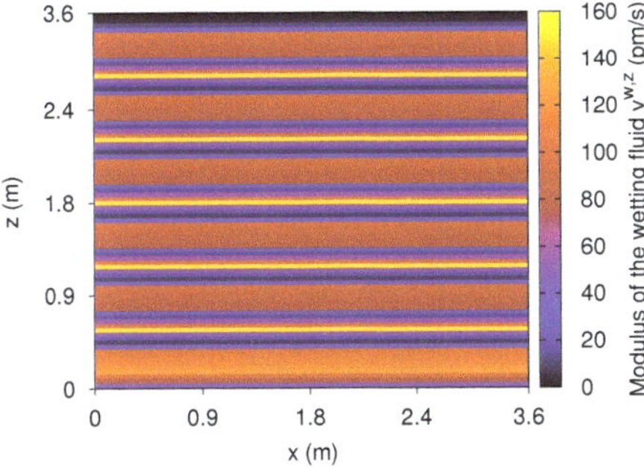

**Figure 9.** Case 2. Modulus of the vertical component of the particle velocity of wetting fluids $v^{w,z} = i\omega\mathbf{u}^{w,z}$ obtained with the FE method. Frequency is 10 Hz.

Finally, Figure 10 exhibits the modulus of the gradient of the total fluid pressure $\mathcal{T} = \mathcal{T}_n + \mathcal{T}_w$ at 10 Hz for Case 2, which is clearly observed at the layer interfaces. This pressure gradient is another useful illustration of the WIFF mechanism for the case of two-phase fluid saturation.

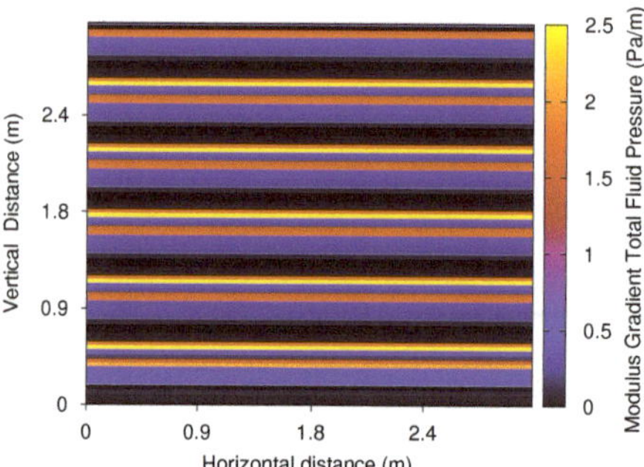

**Figure 10.** Case 2. Modulus of the gradient of the total fluid pressure $\mathcal{T} = \mathcal{T}_n + \mathcal{T}_w$ obtained with the FE method. Frequency is 10 Hz.

## 5. Discussion

Wave-induced fluid flow in porous media is known to be a major cause of attenuation and velocity dispersion. Concerning the existence of additional slow waves when multiphase fluids saturate a porous medium, Albers [17] developed a macroscopic linear model to study wave propagation in unsaturated soils. This model, which assumes that capillary pressure and relative permeabilities depend on saturation, predicts the existence of three compressional waves, one of them fast and two slow, and a shear wave, with phase velocities and attenuation depending on saturation, and the coefficients in the stress-strain relations derived as in Reference [25].

Furthermore, Lo and Sposito [16] present differential equations to model dilatational waves in porous media saturated with two immiscible fluids. Their analysis predicts three compressional waves (P1, P2, P3), with the attenuation of the P1 wave highly affected by the pore fluids. On the other hand, the attenuation of the P2 and P3 waves is found to be related to the inverse of the sum of the relative mobilities of the two fluids, thus dominated by the fluid with larger relative mobility. This model also assumes that capillary pressure is a unique function of saturation (ignoring hysteresis), with the dissipation coefficients are defined in terms of the absolute permeability and the saturation-dependent relative permeability functions using the form given in Reference [25]. Quoting these authors "the model equations by Brutsaert [13], Garg and Nayfeh [14], Berryman et al. [15] and Tuncay and Corapcioglu [19] can be considered as special cases (of Reference [16]) and Santos et al. [25] (model) as a version based on a Lagrangian framework".

## 6. Conclusions

Finite-element time-harmonic quasistatic experiments were used to study the wave anelasticity of a periodic set of three thin and flat layers saturated with two fluids, focusing on the effects of surface tension (capillarity forces). The results are compared with the case in which the two fluids are replaced by an effective single-phase one. This later case is validated against the analytical solution, whereas there is no analytical solution for the more general and realistic scenario involving capillary forces. The results show that, in this case, the loss is higher for both waves traveling normal and parallel to the layers, and there are two attenuation peaks, as opposed to one peak for the effective single-phase fluid.

The energy velocities and dissipation factors of the three waves, shown as polar plots at 10 Hz versus the propagation angle, exhibit a perfect agreement between the analytical and FE single-phase fluid solutions. While the energy velocities coincide in all cases, the dissipation factors are always higher for the case of two-phase fluids, as mentioned above. The physics can be explained by combined effects of capillary pressure and different mobilities of each fluid phase in the two-phase Darcy law.

The existence of an additional slow wave compared to the classic slow wave of the Biot theory is due to surface tension effects represented by a saturation-dependent capillary pressure. On the other hand, relative mobilities induce relative motions between the fluids, causing losses absent when an effective fluid is considered, as it is mostly the case in the literature regarding wave propagation in porous media. Future work includes a more detailed analysis of the influence of wettability, phase mobilities, and saturation on wave anelasticity of rocks saturated with immiscible fluids.

**Author Contributions:** J.E.S. contributed with the methodology, investigation, funding acquisition and writing. J.M.C. contributed with validation, investigation and formal analysis. J.B. contributed with formal analysis, Writing—review & editing and funding acquisition. All authors have read and agreed to the published version of the manuscript.

**Funding:** This work was partially funded by ANPCyT, Argentina (PICT 2015 1909) and Universidad de Buenos Aires (UBACyT 20020190100236BA).

**Acknowledgments:** The authors wish to thank the Purdue University's central Information Technology (ITaP) for their technical support when running the codes used in the numerical experiments. This work is supported by the National Natural Science Foundation of China (Grant no. 41974123), and the Jiangsu Province Outstanding Youth Fund Project (Grant no. BK20200021).

**Conflicts of Interest:** The authors declare no conflict of interest.

## Appendix A. Determination of the Coefficients in the Constitutive Relations

Following Santos et al. [25], $K_u$ is computed using the relations

$$
\begin{aligned}
K_u &= K_s(K_m + \Xi)/(K_s + \Xi), \\
\Xi &= K_f(K_m - K_s)/\bar{\phi}(K_f - K_s), \\
K_f &= \alpha(\gamma \bar{S}_n C_n + \bar{S}_w C_w)^{-1}, \quad \alpha = 1 + (\bar{S}_n + \beta)(\gamma - 1), \\
\gamma &= \left(1 + P'_{ca}(\bar{S}_n)\bar{S}_n \bar{S}_w C_w\right)\left(1 + P'_{ca}(\bar{S}_n)\bar{S}_n \bar{S}_w C_n\right)^{-1},
\end{aligned}
\tag{A1}
$$

where $K_m(\mathbf{x})$, $K_s(\mathbf{x})$, $K_w$, and $K_n$ are the bulk modulus of the empty matrix, of the grains, and the wetting and non-wetting fluid phases, respectively, with compressibilities $C_l = K_l^{-1}$, $l = m, s, n, w, c$.

The remaining coefficients can be obtained by using the following relations:

$$
\begin{aligned}
B_1 &= \chi K_c[(\bar{S}_n + \beta)\gamma - \beta], \\
B_2 &= \chi K_c[(\bar{S}_w], \\
M_1 &= -M_3 - B_1 C_m \delta^{-1}, \quad M_2 = (rB_2)q^{-1}, \\
M_3 &= -M_2 - B_2 C_m \delta^{-1},
\end{aligned}
\tag{A2}
$$

with

$$
\begin{aligned}
\chi &= [\delta + \bar{\phi}(C_m - C_c)]\left\{\alpha\left[\delta + \bar{\phi}\left(C_m - C_f\right)\right]\right\}^{-1}, \\
q &= \bar{\phi}(C_n + 1/P'_{ca}(\bar{S}_n)\bar{S}_n \bar{S}_w), \quad \delta = C_s - C_m, \\
r &= (\bar{S}_n + \beta)C_s + (C_c - C_m)\left[qB_2 + (\bar{S}_n + \beta)\left(1 - C_s C_c^{-1}\right)\right].
\end{aligned}
\tag{A3}
$$

## Appendix B. Analytical Solution

*Appendix B.1. Frequency-Dependent Stiffnesses*

Cavallini et al. [30] (Appendix A.4, specifically, Equation (61)) yield the frequency dependent and complex stiffness, $p_{33}$, of a set of $N$ porous layers, each of thickness $d_i$, such that the stratification period is $d = \sum_i^N d_i$. The assumptions of theory are that the stiffnesses matrix is symmetric and the flow is normal to the layering. Each layer is defined by the porosity: $\phi$, grain and fluid bulk moduli: $K_s$ and $K_f$, dry-rock bulk and shear moduli: $K_m$ and $\mu$, grain and fluid densities: $\rho_s$ and $\rho_f$, fluid viscosity: $\eta$ and permeability: $\kappa$.

The White model was generalized by Krzikalla and Müller [38] to anisotropic media, to obtain the five stiffnesses, $p_{IJ}$, of the symmetric $6 \times 6$ matrix of the effective transversely isotropic medium,

$$p_{IJ}(\omega) = c_{IJ} + \left( \frac{c_{IJ} - c_{IJ}^r}{c_{33} - c_{33}^r} \right) [p_{33}(\omega) - c_{33}], \tag{A4}$$

where $c_{IJ}^r$ and $c_{IJ}$ are the relaxed and unrelaxed values, i.e., the low- and high-frequency values. Gelinsky and Shapiro [39] (their Equations (14) and (15)) give the expressions of these stiffnesses,

$$
\begin{aligned}
c_{66}^r &= B_1^* = \langle \mu \rangle, \\
c_{11}^r - 2c_{66}^r &= c_{12}^r = B_2^* = 2\left\langle \frac{\lambda_m \mu}{E_m} \right\rangle + \left\langle \frac{\lambda_m}{E_m} \right\rangle^2 \left\langle \frac{1}{E_m} \right\rangle^{-1} + \frac{B_6^{*2}}{B_8^*}, \\
c_{13}^r &= B_3^* = \left\langle \frac{\lambda_m}{E_m} \right\rangle \left\langle \frac{1}{E_m} \right\rangle^{-1} + \frac{B_6^* B_7^*}{B_8^*}, \\
c_{33}^r &= B_4^* = \left\langle \frac{1}{E_m} \right\rangle^{-1} + \frac{B_7^{*2}}{B_8^*} = \left[ \left\langle \frac{1}{E_m} \right\rangle - \left\langle \frac{\alpha}{E_m} \right\rangle^2 \left\langle \frac{E_G}{ME_m} \right\rangle^{-1} \right]^{-1}, \\
c_{55}^r &= B_5^* = \langle \mu^{-1} \rangle^{-1}, \\
B_6^* &= -B_8^* \left( 2\left\langle \frac{\alpha \mu}{E_m} \right\rangle + \left\langle \frac{\alpha}{E_m} \right\rangle \left\langle \frac{\lambda_m}{E_m} \right\rangle \left\langle \frac{1}{E_m} \right\rangle^{-1} \right), \\
B_7^* &= -B_8^* \left\langle \frac{\alpha}{E_m} \right\rangle \left\langle \frac{1}{E_m} \right\rangle^{-1}, \\
B_8^* &= \left[ \left\langle \frac{1}{M} \right\rangle + \left\langle \frac{\alpha^2}{E_m} \right\rangle - \left\langle \frac{\alpha}{E_m} \right\rangle^2 \left\langle \frac{1}{E_m} \right\rangle^{-1} \right]^{-1},
\end{aligned}
\tag{A5}
$$

where $\lambda_m = K_m - (2/3)\mu$, $E_m = K_m + (4/3)\mu$, and the brackets indicate the average $\langle \zeta \rangle = d^{-1} \sum_i d_i \zeta_i$. In the unrelaxed regime, there is no interlayer flow, and the stiffnesses are

$$
\begin{aligned}
c_{66} &= c_{66}^r, \\
c_{11} - 2c_{66} &= c_{12} = 2\left\langle \frac{(E_G - 2\mu)\mu}{E_G} \right\rangle + \left\langle \frac{E_G - 2\mu}{E_G} \right\rangle^2 \left\langle \frac{1}{E_G} \right\rangle^{-1}, \\
c_{13} &= \left\langle \frac{E_G - 2\mu}{E_G} \right\rangle \left\langle \frac{1}{E_G} \right\rangle^{-1}, \\
c_{33} &= \left\langle \frac{1}{E_G} \right\rangle^{-1}, \\
c_{55} &= c_{55}^r
\end{aligned}
\tag{A6}
$$

where

$$E_G = E_m + \alpha^2 M, \quad \text{with} \quad \alpha = 1 - \frac{K_m}{K_s}, \quad M = \left( \frac{\alpha - \phi}{K_s} + \frac{\phi}{K_f} \right)^{-1} \tag{A7}$$

is the Gassmann P-wave modulus.

Because the relaxed and unrelaxed shear moduli coincide, there is no shear dissipation along and normal to layering. The qSV wave suffers attenuation because it is coupled with the qP wave, and the SH wave is not dispersive, since $c_{55} = c_{55}^r$ and $c_{66} = c_{66}^r$ imply $p_{66} = c_{66}$ and $p_{55} = c_{55}$. A set of thin layers saturated with different fluids but with the same shear modulus is isotropic. The average density of the medium is simply $\rho = \langle \rho \rangle$.

*Appendix B.2. Wave Velocities and Dissipation Factors*

For homogeneous viscoelastic waves, the complex velocities of the three waves are

$$
\begin{aligned}
v_{\mathrm{qP}} &= (2\rho)^{-1/2}\sqrt{p_{11}l_1^2 + p_{33}l_3^2 + p_{55} + A} \\
v_{\mathrm{qSV}} &= (2\rho)^{-1/2}\sqrt{p_{11}l_1^2 + p_{33}l_3^2 + p_{55} - A} \\
v_{\mathrm{SH}} &= \rho^{-1/2}\sqrt{p_{66}l_1^2 + p_{55}l_3^2} \\
A &= \sqrt{[(p_{11} - p_{55})l_1^2 + (p_{55} - p_{33})l_3^2]^2 + 4[(p_{13} + p_{55})l_1 l_3]^2},
\end{aligned}
\tag{A8}
$$

where $l_1 = \sin\theta$ and $l_3 = \cos\theta$, and $\theta$ is the angle between the symmetry axis and wavenumber vector. The phase velocity is

$$
v_p = \left[\mathrm{Re}\left(\frac{1}{v}\right)\right]^{-1},
\tag{A9}
$$

where $v$ represents the above velocities. The energy-velocity vector of the qP and qSV waves is

$$
\frac{\mathbf{v}_e}{v_p} = (l_1 + l_3 \cot\psi)^{-1}\hat{\mathbf{e}}_1 + (l_1 \tan\psi + l_3)^{-1}\hat{\mathbf{e}}_3,
\tag{A10}
$$

where

$$
\tan\psi = \frac{\mathrm{Re}(\beta^* X + \xi^* W)}{\mathrm{Re}(\beta^* W + \xi^* Z)}
\tag{A11}
$$

defines the angle between the $z$-axis and the energy-velocity,

$$
\begin{aligned}
\beta &= \sqrt{A \pm B}, \\
\xi &= \pm \mathrm{pv}\sqrt{A \mp B}, \\
B &= p_{11}l_1^2 - p_{33}l_3^2 + p_{55}\cos 2\theta,
\end{aligned}
\tag{A12}
$$

where the lower and upper signs correspond to the qSV and qP waves, respectively. Moreover,

$$
\begin{aligned}
W &= p_{55}(\xi l_1 + \beta l_3), \\
X &= \beta p_{11}l_1 + \xi p_{13}l_3, \\
Z &= \beta p_{13}l_1 + \xi p_{33}l_3,
\end{aligned}
\tag{A13}
$$

where "pv" indicates the principal value.

Moreover, the SH-wave energy velocity is

$$
\mathbf{v}_e = \frac{1}{\bar{\rho} v_p}(l_1 c_{66}\hat{\mathbf{e}}_1 + l_3 c_{55}\hat{\mathbf{e}}_3),
\tag{A14}
$$

and

$$
\tan\psi = \left(\frac{c_{66}}{c_{55}}\right)\tan\theta.
\tag{A15}
$$

The quality factor is

$$
Q = \frac{\mathrm{Re}(v^2)}{\mathrm{Im}(v^2)},
\tag{A16}
$$

and the dissipation factor is defined as the inverse of $Q$.

# References


1. Pham, N.H.; Carcione, J.M.; Helle, H.B.; Ursin, B. Wave velocities and attenuation of shaley sandstones as a function of pore pressure and partial saturation. *Geophys. Prospect.* **2002**, *50*, 615–627. [CrossRef]
2. Mavko, G.; Mukerji, T.; Dvorkin, J. *Rock Physics Handbook*; Cambridge University Press: Cambridge, UK, 2009.
3. Müller, T.M.; Gurevich, B.; Lebedev, M. Seismic wave attenuation and dispersion resulting from wave-induced flow in porous rocks—A review. *Geophysics* **2010**, *75*, 147–163. [CrossRef]

4.   Pride, S.R. Relationships between seismic and hydrological properties. In *Hydrogeophysics*; Chapter 9; Rubin, Y., Hubbard, S., Eds.; Springer: Berlin/Heidelberg, Germany, 2005; pp. 253–290.

5.   Tisato, N.; Quintal, B. Laboratory measurements of seismic attenuation in sandstone: Strain versus fluid saturation effects. *Geophysics* **2014**, *79*, WB9–WB14. [CrossRef]

6.   Tisato, N.; Quintal, B.; Madonna, C.; Grasselli, G. Seismic attenuation in partially saturated rocks: Recent advances and future directions *Lead. Edge* **2014**, 640–646. [CrossRef]

7.   Spencer, J.W.; Shine, J. Seismic wave attenuation and modulus dispersion in sandstones. *Geophysics* **2016**, *81*, D211–D231. [CrossRef]

8.   Chapman, S.; Borgomano, J.V.M.; Quintal, B.; Benson, S.M.; Fortin, J. Seismic wave attenuation and dispersion due to partial fluid saturation: direct measurements and numerical simulations based on X-Ray CT. *J. Geophys. Res. Solid Earth* **2021**, *126*, 1404–1430. [CrossRef]

9.   Biot, M.A. Theory of deformation of a porous viscoelastic anisotropic solid. *J. Appl. Phys.* **1956**, *27*, 459–467. [CrossRef]

10.  Biot, M.A. Theory of propagation of elastic waves in a fluid-saturated porous solid. I. Low frequency range. *J. Acoust. Soc. Am.* **1956**, *28*, 168–178. [CrossRef]

11.  Biot, M.A. Mechanics of deformation and acoustic propagation in porous media. *J. Appl. Phys.* **1962**, *33*, 1482–1498. [CrossRef]

12.  White, J.E.; Mikhaylova, N.G.; Lyakhovitskiy, F.M. Low-frequency seismic waves in fluid saturated layered rocks. *Phys. Solid Earth* **1975**, *11*, 654–659. [CrossRef]

13.  Brutsaert, W. The propagation of elastic waves in unconsolidated unsaturated granular mediums. *J. Geophys. Res. Solid Earth* **1964**, *69*, 243–257. [CrossRef]

14.  Garg, S.K.; Nayfeh, A.H. Compressional wave propagation in liquid and/or gas saturated elastic porous media. *J. Appl. Phys.* **1986**, *60*, 3045–3055. [CrossRef]

15.  Berryman, J.G.; Thigpen, L.; Chin, R.C.Y. Bulk elastic wave propagation in partially saturated porous solids. *J. Acoust. Soc. Am.* **1988**, *84*, 360–373. [CrossRef]

16.  Lo, W.; Sposito, G. Wave propagation through elastic porous media containing two immiscible fluids. *Water Res. Res.* **2005**, *41*, W02025. [CrossRef]

17.  Albers, B. Analysis of the propagation of sound waves in partially saturated soils by means of a macroscopic linear poroelastic model. *Transp. Porous Media* **2009**, *80*, 173–192. [CrossRef]

18.  Tuncay, K.; Corapcioglu, M.Y. Body waves in poroelastic media saturated by two immiscible fluids. *J. Geophys. Res.* **1996**, *111*, 149–159. [CrossRef]

19.  Tuncay, K.; Corapcioglu, M.Y. Wave propagation in poroelastic media saturated by two fluids. *J. Appl. Mech.* **1997**, *64*, 313–320. [CrossRef]

20.  Wei, C.; Muraleetharan, K.K. A continuum theory of porous media saturated by multiple immiscible fluids: I. Linear poroelasticity. *Int. J. Eng. Sci.* **2002**, *40*, 1807–1833. [CrossRef]

21.  Dutta N.C.; Odé, H. Attenuation and dispersion of compressional waves in fluid-filled porous rocks with partial gas saturation (White model). Part I: Biot theory. *Geophysics* **1979**, *44*, 1777–1788. [CrossRef]

22.  Pride, S.R.; Berryman, J.G.; Harris, J.M. Seismic attenuation due to wave-induced flow. *J. Geophys. Res.* **2004**, *109*, B01201.1. [CrossRef]

23.  Thovert, J.F.; Li, X.Y.; Malinouskaya, I.; Mourzenko, V.V.; Adler, P.M. Propagation of acoustic waves through saturated porous media. *Phys. Rev. E* **2020**, *102*, 023001. [CrossRef]

24.  Hamzehpour, H.; Kasani, F.H.; Sahimi, M.; Sepehrinia, R. Wave propagation in disordered fractured porous media. *Phys. Rev. E* **2014**, *89*, 023301. [CrossRef]

25.  Santos, J.E.; Corberó, J.M.; Douglas, J., Jr. Static and dynamic behaviour of a porous solid saturated by a two-phase fluid. *J. Acoust. Soc. Am.* **1990**, *87*, 1428–1438. [CrossRef]

26.  Peaceman, D.W. *Fundamentals of Numerical Reservoir Simulation*; Elsevier: Amsterdam, The Netherlands, 1977.

27.  Santos, J.E.; Savioli G.B. Long-wave equivalent viscoelastic solids for porous rocks saturated by two-phase fluids. *Geophys. J. Int.* **2018**, *214*, 302–314. [CrossRef]

28.  Qi, Q.; Müller, T.; Gurevich, M.B.; Lopes, S.; Lebedev, M.; Caspari, E. Quantifying the effect of capillarity on attenuation and dispersion in patchy-saturated rocks. *Geophysics* **2014**, *79*, WB35–WB50. [CrossRef]

29.  Liu, J.; Müller, T.M.; Qi, Q.; Lebedev, M.; Sun, W. Velocity-saturation relation in partially saturated rocks: modelling the effect of injection rate changes. *Geophys. Prospect.* **2016**, *64*, 1054–1066. [CrossRef]

30.  Cavallini, F.; Carcione, J.M.; Vidal de Ventós, D.; Engell-Sørensen, L. Low frequency dispersion and attenuation in anisotropic partially saturated rocks. *Geophys. J. Int.* **2017**, *209*, 1572–1584. [CrossRef]

31.  Scheidegger, A.E. *The Physics of Flow through Porous Media*; University of Toronto: Toronto, ON, Canada, 1974.

32.  Bear, J. *Dynamics of Fluids in Porous Media*; Dover Publications: New York, NY, USA, 1972.

33.  Douglas, J., Jr.; Furtado, F.; Pereira, F. On the numerical simulation of waterflooding of heterogeneous petroleum reservoirs. *Comput. Geosci.* **1997**, *1*, 155–190. [CrossRef]

34.  Douglas, J., Jr.; Paes-Leme, P.J.; Hensley, J.L. A limit form of the equations for immiscible displacement in a fractured reservoir. *Transp. Porous Media* **1991**, *5*, 549–565. [CrossRef]

35. Douglas, J., Jr.; Paes-Leme, P.J.; Pereira, F.; Yeh, L.M. A massively parallel iterative numerical algorithm for immiscible flow in naturally fractured reservoirs. *Int. Ser. Numer. Math.* **1991**, *114*, 75–93.
36. Chavent, G.; Jaffre, J. *Mathematical Models and Finite Element Methods for Reservoir Simulation*; Elsevier: North Holland, The Netherlands, 1986.
37. Li, K. Interrelationship between resistivity index, capillary pressure and relative permeability. *Transp. Porous Media* **2011**, *3*, 385–398. [CrossRef]
38. Krzikalla, F.; Müller, T. Anisotropic P-SV-wave dispersion and attenuation due to inter-layer flow in thinly layered porous rocks. *Geophysics* **2011**, *76*, WA135–WA145. [CrossRef]
39. Gelinsky, S.; Shapiro, S.A. Dynamic-equivalent medium approach for thinly layered saturated sediments. *Geophys. J. Int.* **1997**, *128*, F1–F4. [CrossRef]

*Article*

# Investigation of the Energy Evolution of Tectonic Coal under Triaxial Cyclic Loading with Different Loading Rates and the Underlying Mechanism

**Deyi Gao** [1,2,†], **Shuxun Sang** [2,3,4,*,†], **Shiqi Liu** [3,4,*,†], **Jishi Geng** [1], **Tao Wang** [1,2] and **Tengmin Sun** [1,2]

[1] Key Laboratory of Coalbed Methane Resources and Reservoir Formation Process, China University of Mining and Technology, Xuzhou 221008, China; TB17010005b0@cumt.edu.cn (D.G.); gengjishi@cumt.edu.cn (J.G.); TS20010128P31@cumt.edu.cn (T.W.); ts20010127p31@cumt.edu.cn (T.S.)

[2] School of Resources and Geosciences, China University of Mining and Technology, Xuzhou 221008, China

[3] Low Carbon Energy Institute, China University of Mining and Technology, Xuzhou 221008, China

[4] Jiangsu Key Laboratory of Coal-Based Greenhouse Gas Control and Utilization, China University of Mining and Technology, Xuzhou 221008, China

* Correspondence: shxsang@cumt.edu.cn (S.S.); liushiqi@cumt.edu.cn (S.L.)

† These authors contributed equally to this work.

**Abstract:** It is of great significance to ascertain the mechanical characteristics and deformation laws of tectonic coal that is under complex stress conditions for safe production, but the targeted research in this area is still insufficient at present. This paper performed triaxial tests under cyclic multi-level loading at different rates by using an MTS-815 Rock Mechanics Testing System. The strain characteristics, elastic modulus and energy evolution were obtained in order to explore the effects of the mechanism of loading rate on the evolution of deformation and energy parameters of tectonic coal. The results showed that the irreversible strain and plastic energy increased exponentially with the increase in the deviatoric stress, but the growth rate decreased with the increase in loading rate. Furthermore, the elastic strain increased linearly and the growth rate was essentially unaffected by the loading rate. During the compaction stage, the variation of each parameter was not sensitive to the loading rate; during the elastic and damage stage, the rate increase inhibited secondary defect propagation and improved rock strength. In addition, the stepwise and cumulative energy ratio was defined in order to describe the energy distribution during cyclic loading and unloading. It was found that the decrease in the loading rate was beneficial to the transformation of the total energy into plastic energy. The elastic modulus was the most sensitive to sample damage, but the energy density evolution was able to be used to describe the deformation damage process of tectonic coal in more detail. These findings provide important theoretical support for the tectonic coal deformation law and action mechanism in the damage process that occurs under complex stress conditions.

**Keywords:** tectonic coal; triaxial cycle loading; irreversible deformation; elastic modulus; energy density; energy ratio

**Citation:** Gao, D.; Sang, S.; Liu, S.; Geng, J.; Wang, T.; Sun, T. Investigation of the Energy Evolution of Tectonic Coal under Triaxial Cyclic Loading with Different Loading Rates and the Underlying Mechanism. *Energies* **2021**, *14*, 8124. https://doi.org/10.3390/en14238124

Academic Editors: Junlong Shang, Chun Zhu and Manchao He

Received: 29 October 2021
Accepted: 30 November 2021
Published: 3 December 2021

**Publisher's Note:** MDPI stays neutral with regard to jurisdictional claims in published maps and institutional affiliations.

## 1. Introduction

As it is affected by tectonism such as faulting and slipping, tectonic stress destroys the original coal structure. This results in tectonic coal, which has wide distribution globally and especially in China. Because of its gas enrichment, low permeability, and strength characteristics, tectonic coal has long been known as the major cause of coal and gas outbursts [1–3]. Previous studies on the porosity [4,5], pore structure [6], permeability, and adsorption–desorption characteristics [1,7,8] of tectonic coal have been carried out. Based on reconstituted tectonic coal samples, the preparation information for which is shown in Table 1, many scholars have performed uniaxial and triaxial tests in order to study the mechanical properties of tectonic coal [9–13]. However, in concrete engineering (especially in the tectonic coal mining processes), the change in stress is not monotonic [14–16]. The mechanical response and deformation characteristics of rocks that are under cyclic loading and unloading are completely different from those that are under monotonic loading [17–19]. Hence, important theoretical significance and engineering value exist for the study of the strength and deformation of rocks that are under complex stress conditions [20–23].

**Table 1.** Preparation information for tectonic coal.

| Particle Size (mm) | Stress | Additive | Refs. |
|---|---|---|---|
| <0.2 | 2.76–19.9 MPa | / | [9] |
| 0.18–0.425 | 100 MPa | Water | [10] |
| <0.25 | 48 MPa | Water | [11] |
| 0.38–0.83 | 100 MPa | Water | [12] |
| 0.25–0.425 | 100 MPa | Water | [13] |

Currently, cyclic loading and unloading testing in the laboratory constitutes an effective method for studying the rock deformation characteristics of complex stress paths and many scholars have conducted relevant tests and theoretical studies resulting in numerous achievements [24–29]. The results of these tests have shown that the confining pressure and the amplitude, frequency, and peak value of the axial stress are important factors affecting the mechanical behaviour of rocks [14,26,30]. Wang suggested that the damage time for coal would be delayed by an increase in frequency and that the peak stress of loading had the most significant influence on coal damage [31]. Taheri performed cyclic loading experiments on lignite and observed that, with an increase in confining pressure, the specimen damage under axial loading decreased [32]. Taking the initial properties of the coal and rock and the stress environment into account, Zhang carried out cyclic loading experiments with confining pressures and found that the Poisson's ratio and deformation of coal increased non-linearly with confining pressure [33]. Roberts, L. A. et al. carried out cyclic triaxial creep tests under various load paths and found that salt that was subjected to cyclic loading in extension was no more prone to damage than if it were subjected to compression [34]. Voznesenskii, A. S et al. established the interrelation between the acoustic emission (AE) signals and mechanical strength under cyclic loading and unloading [35]. Based on a cyclic loading experiment, Shkuratnik V. L. et al. found that stress memory in the characteristic of AE that were occurring under constant temperatures was steady [36]. Alexander Lavrov suggested that the loading rate had no significant effect on the Kaiser effect, which might be because the rate that is found in laboratory tests is higher than the in situ loading history [37]. Olovyannyy, A., and Chantsev, V. used rock salt in order to research the models of deformation and failure under various cyclic loading conditions and obtained the parameters that were needed for solving engineering problems in the context of mining [38,39].

From the perspective of thermodynamics, the essence of material damage is driven by energy, so the process of rock deformation and failure is an irreversible process of energy consumption [40]. In recent years, energy transformation and its evolution have gradually been favoured by scholars in the study of rock damage and a series of failure criteria and methods for the stability evaluation for rocks were put forward [41–44]. Zhang and Lindqvist, M. investigated the variation characteristics of rock energy dissipation in the whole process under dynamic and instantaneous dynamic effects [45,46]. Gaziev suggested that the primary element begins to fail when the strain reaches its limit value and the continuous accumulation of strain energy is a necessary condition for material failure [47]. Xie et al. and Xue et al. noted that energy dissipation leads to material deterioration and energy release results in abrupt structural collapse [48,49]. Li et al. obtained the influence behaviours of the initial confining pressure and unloading rate on the strain–energy conversion of rocks during the deformation and failure processes [50]. Peng et al. performed triaxial compression experiments on coal under different confining pressures and analysed the relationship between the failure modes and energy conversion [51]. Jiang et al. studied the AE characteristics and energy dissipation of gas-containing coal under tiered cyclic loading and unloading and established a new equation for the variable damage of coal with dissipation energy [52].

The extraction of coalbed methane from the tectonic coal reservoir is a common method that is used in order to realize the efficient utilisation and safe exploitation of resources, but borehole fractures that are caused by tectonic coal reservoir deformation are the main constraint for coal and coalbed methane production [53]. However, a majority of the research has focused on hard rocks, while research on tectonic coal is still insufficient. By utilising the MTS-815 test system, tectonic coal was subjected to a cyclic loading and unloading test under triaxial compression with different loading rates. On this basis, the damage characteristics and energy evolution of the tectonic coal during cyclic loading were explored. Moreover, the action mechanism of loading rate in the damage process was analysed. The research results provide theoretical support for methods for inducing or preventing the destruction or damage of tectonic coal.

## 2. Experimental Design and Schemes

### 2.1. Specimens

The pulverised coal that was used in this experiment came from the Huainan mining area in Anhui Province, China. The tectonic coal that was sampled underground was screened in order to obtain its particle size distribution (calculated by mass percentage), as shown in Figure 1.

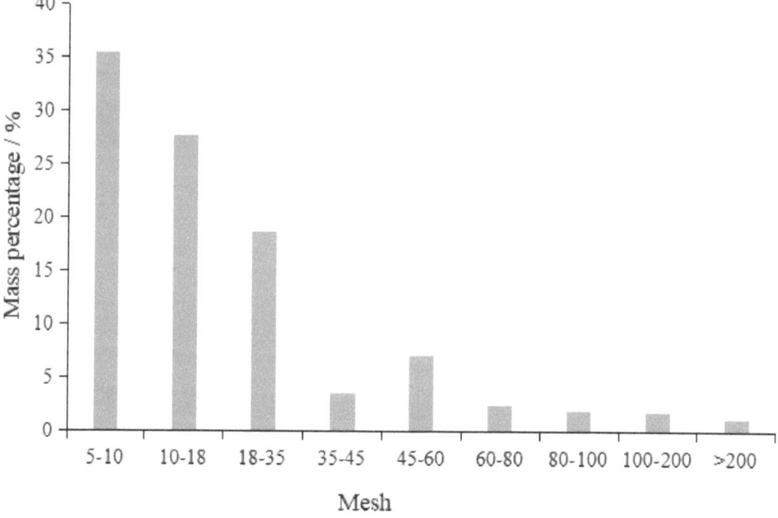


**Figure 1.** Particle size distribution of tectonic coal.

Due to its low strength and weak cohesion, tectonic coal is difficult to obtain in the form of an intact core specimen. Therefore, reconstituted coal specimens have been widely used in order to investigate the mechanical and permeability properties of tectonic coal. The main process for preparing these specimens is as follows: as shown in Figure 1, the mixed particle size pulverised coal was prepared, 10% water (mass percentage) was added, and petroleum jelly was applied to the inner wall of the mould in order to reduce the friction force. Following this, the pulverised coal was put into a steel mould and pressed at 15 MPa for 2 h in order to make a cylindrical sample with a diameter of 50 mm and a height of 100 mm, as seen in Figures 2 and 3. Formed pressure was provided by an electric–hydraulic serving compression machine. Subsequently, the samples were put in an oven and the weight of each sample was measured every 12 h until the change in weight was less than 0.2% of the initial weight. This process lasted approximately 96 h.

Uniaxial compressive strength that was measured at 0.37 MPa, tensile strength that was measured at 0.07 MPa, and an internal friction angle of 29.2° proved that the samples had the characteristics of low strength and weak cohesion.

**Figure 2.** Steel mould and electric–hydraulic serving compression machine.

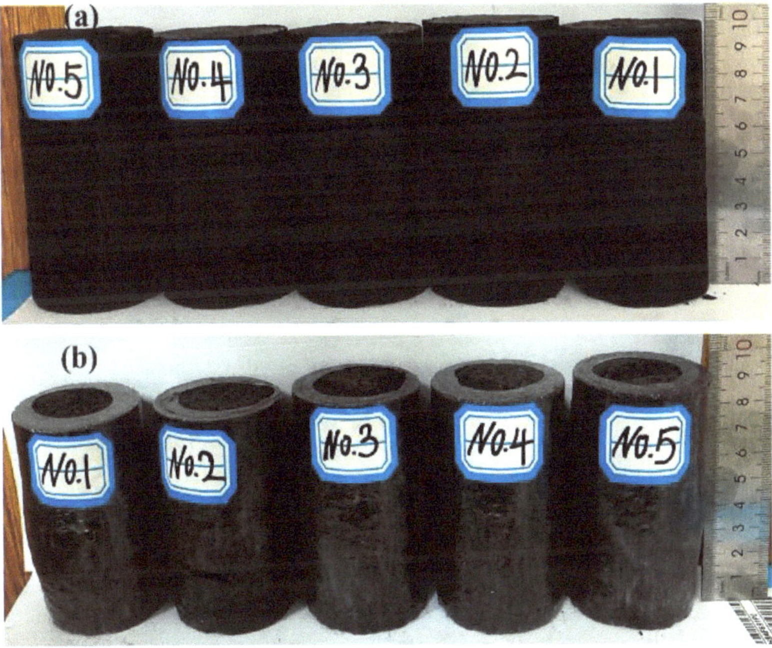

**Figure 3.** (**a**) Reconstituted tectonic coal samples, (**b**) Damaged samples.

## 2.2. Test Equipment

This study used the MTS815 test system for rock mechanics from the State Key Laboratory for Geomechanics and Deep Underground Engineering, China University of Mining and Technology, Xuzhou, China, as shown in Figure 4. The equipment consisted of three parts, including loading, testing, and control systems. The equipment could be used for loading and unloading testing under multiple control modes of force, stress, displacement, and strain. The rigidity of the test machine was $10.5 \times 109$ N/m, the maximum axial load could reach 1700 kN, the confining pressure was lower than or equal to 45 MPa, and the limiting data acquisition frequency was $2 \times 104$ times per second.

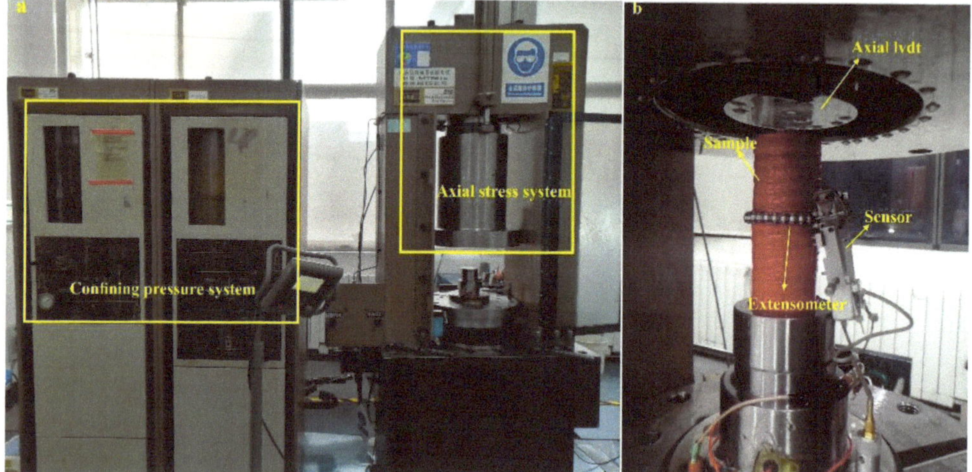


**Figure 4.** MTS-815 test system. (**a**) Loading system, (**b**) Measure system.

The CT scans were performed using the high-resolution 3D X-ray microscopy imaging system from the Advanced Analysis and Computation Center of China University of Mining and Technology, Xuzhou. The X-ray tube voltage was 30–60 KV, the X-ray power was 2–10 w, and the resolution ratio was 50 μm.

## 2.3. Test Procedure and Schemes

Triaxial cyclic loading and unloading procedures taking place under stress gradients were conducted in this test, with a confining pressure of 12 MPa. The axial upper limit of stress increased by 2 MPa per gradient and the lower limit of axial deviatoric stress remained constant at 1 MPa (as Figure 5). The cyclic loading and unloading was performed at different rates (0.05 MPa/s, 0.1 MPa/s, 0.15 MPa/s, 0.2 MPa/s, and 0.25 MPa/s) and the corresponding samples were named NO. $i$ ($i$ = 1,2,3,4,5, respectively).

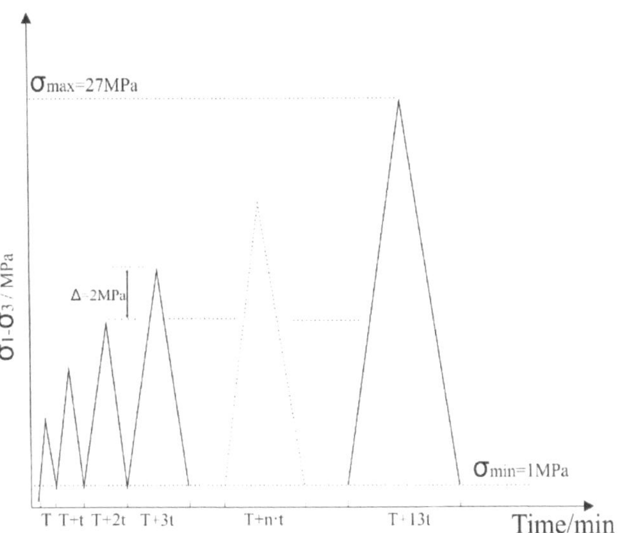


**Figure 5.** Loading path.

*2.4. Calculation of Energy Parameters*

It was assumed that there was no heat exchange between the rocks and the experimental environment during the test [54]. According to Xie et al., energy input, storage, and consumption all conform to the following formulas [48]:

$$\begin{cases} U = U_e + U_d \\ U = \int u\,dV \\ U_e = \int u_e\,dV \\ U_d = \int u_d\,dV \end{cases} \tag{1}$$

where $U$, $U_e$, and $U_d$ are the total strain energy, elastic strain energy, and plastic strain energy, respectively; $u$, $u_e$, and $u_d$ are the total strain energy density, elastic strain energy density, and plastic strain energy density of the rocks, respectively; and $V$ denotes the rock volume.

As shown in Figure 6, $u$ is equivalent to the sum of the areas that are encircled by the X-axis and loading curve; $u_e$ is equal to the sum of the areas encircled by the X-axis and unloading curve, in numerical terms; and $u_d$ is equivalent to the sum of the areas that are formed by the loading and unloading curves and abscissa axis, as per the following formulas [30]:

$$u = \int_{\varepsilon_1}^{\varepsilon_{max}} \sigma\,d\varepsilon = \sum_{i=1}^{N} \frac{(\varepsilon_i - \varepsilon_{i+1})(\sigma_i + \sigma_{i+1})}{2} \tag{2}$$

$$u_e = \int_{\varepsilon_2}^{\varepsilon_{max}} \sigma\,d\varepsilon = \sum_{n=1}^{N} \frac{(\varepsilon_n - \varepsilon_{n+1})(\sigma_n + \sigma_{n+1})}{2} \tag{3}$$

$$u_d = u - u_e = \frac{\sum_{i=1}^{N}(\varepsilon_i - \varepsilon_{i+1})(\sigma_i + \sigma_{i+1}) - \sum_{n=1}^{N}(\varepsilon_n - \varepsilon_{n+1})(\sigma_n + \sigma_{n+1})}{2} \tag{4}$$

where $\varepsilon_1$ and $\varepsilon_2$ indicate the strains that were apparent when the stress $\sigma$ equalled $\sigma_{min}$ in loading and unloading processes; and $\varepsilon_{max}$ denotes the strain when the stress $\sigma$ equalled $\sigma_{max}$ on the loading curve. Moreover, $\varepsilon_i$, $\varepsilon_{i+1}$, $\varepsilon_n$, and $\varepsilon_{n+1}$ indicate the strains at an integral step, while $\sigma_i$, $\sigma_{i+1}$, $\sigma_n$, and $\sigma_{n+1}$ denote the stresses at an integral step. The units of $\varepsilon$ and $\sigma$ are " [-]" and "MPa", respectively.

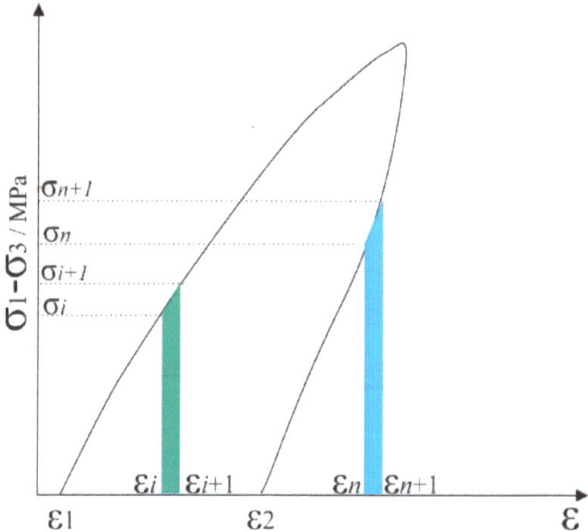


**Figure 6.** Schematic diagram for energy density calculation.

## 3. Results

### 3.1. Full Stress–Strain Curves

Figure 7 presents the curves that were measured under the cyclic loading and unloading processes. Overall, the change in behaviour was the same; as the stress gradually increased, the axial stress–strain curve experienced two stages which were described as "dense–sparse". With the increase in deviatoric stress, the strain that was generated increased in each cycle and the growth rate accelerated gradually. With the increase in the loading rate, the total axial strains at the end of the cycle were 11.67%, 9.87%, 8.47%, 7.62%, and 6.25%.

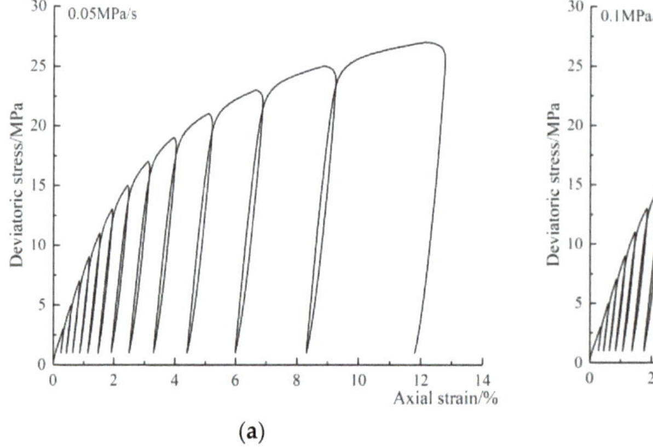


(a)

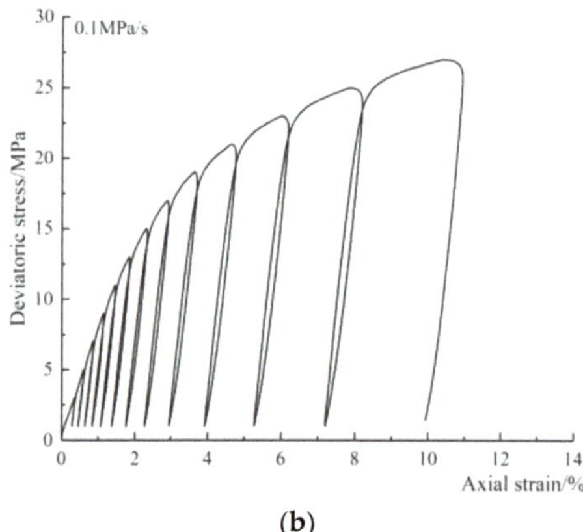


(b)

**Figure 7.** *Cont.*

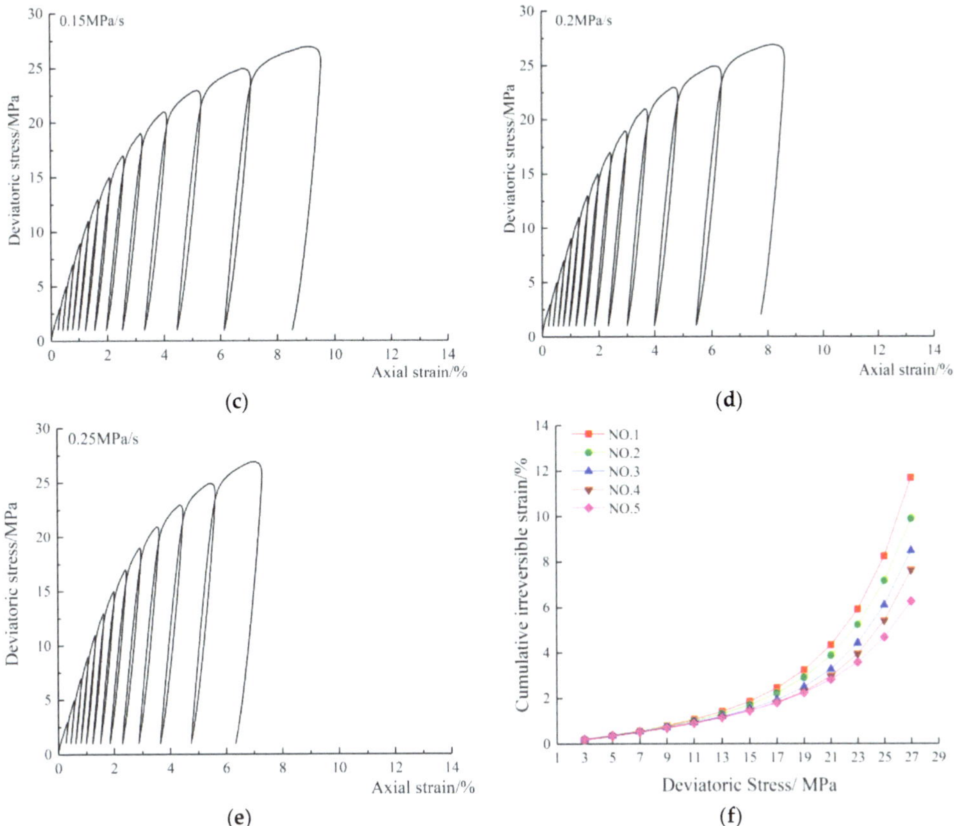

**Figure 7.** Deviatoric stress–strain curves. (**a**) Stress–strain curve of NO. 1, (**b**) Stress–strain curve of NO. 2, (**c**) Stress–strain curve of NO. 3; (**d**) Stress–strain curve of NO. 4, (**e**) Stress–strain curve of NO. 5, (**f**) Cumulative irreversible strain of all samples.

As shown in Figure 7f, the total strain in a single cycle increased with the increased deviatoric stress, but the growth rate was inversely proportional to the loading rate. From 0.05 MPa/s to 0.25 MPa/s, the strain in a single cycle increased by an average of 0.0032, 0.0027, 0.0024, 0.0023, and 0.0018 stepwise. The total strain increased exponentially with the deviatoric stress, as shown in Table 2, and the correlation coefficients were all greater than 0.99. However, with the increase in loading rate, the fitting coefficients decreased from 0.16 to 0.13.

**Table 2.** Fitting relationships of total strain with cycle numbers under different loading rates.

| Loading Rate (MPa/s) | Fitting Expression | $R_2$ |
|---|---|---|
| 0.05 | $\varepsilon = 0.0016\exp 0.1601\sigma'$ | 0.99 |
| 0.1 | $\varepsilon = 0.0017\exp 0.1505\sigma'$ | 0.99 |
| 0.15 | $\varepsilon = 0.0016\exp 0.1457\sigma'$ | 0.99 |
| 0.2 | $\varepsilon = 0.0015\exp 0.1461\sigma'$ | 0.98 |
| 0.25 | $\varepsilon = 0.0018\exp 0.1319\sigma'$ | 0.98 |

Note: $\varepsilon$ = Total strain; $\sigma'$ = Deviatoric stress.

### 3.2. Elastic Strain and Plastic Strain

Due to the effects of tectonism, the intact coal was destroyed, causing it to be severely crushed or even pulverised [55]. Therefore, even under the lower levels of stress, the tectonic coal exhibited plastic deformation. In this experiment, the irreversible strain was equal to the strain at the end of the cycle minus the strain at the beginning of the cycle and the restorable strain was the strain which rebounded during unloading.

As described in the previous section, the characteristics of the stress–strain curves of all of the samples were consistent. The NO. 4 specimen was taken as an example to illustrate the changing laws of elastic strain and plastic strain in a single cycle, as shown in Figure 8. When the deviatoric stress was less than 5 MPa, the axial plastic strain was slightly higher than the elastic strain. This is because the original sample contained several voids, which compressed rapidly under the low axial stress that was applied during the compaction. As the voids were gradually compacted by deviatoric stress in the range of 7 MPa to 21 MPa, the elastic strain was greater than the plastic strain, therefore accounting for a higher percentage of the total strain. As the axial stress increased above 21 MPa, the growth rate of the plastic strain was much faster than that of the elastic strain. When the deviatoric stress increased from 21 MPa to 27 MPa, the plastic strain increased by 230%, while the elastic strain only increased by about 25%. Finally, the irreversible strain and elastic strain were measured at 2.22% and 0.96%. The sharp increase in the plastic strain that was observed in this stage was caused by the continuous expansion of secondary defects.

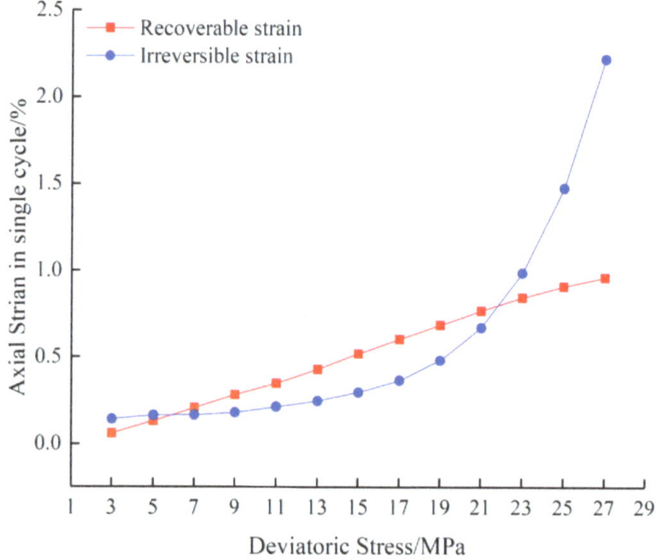


**Figure 8.** Recoverable strain and irreversible strain in single cycle of the NO. 4 sample.

As presented in Table 3 and Figure 9, the plastic strain increase that occurred with the increase in the deviatoric stress can be fitted with an exponential function, the correlation coefficient of which was greater than 0.98. The fitting coefficient decreased from 0.16 to 0.13 with the increase in the loading rate, indicating that the growth rate of the cumulative plastic strain decreased with the increased loading. An increase in the loading rate inhibited the growth rate of the plastic strain. The elastic strain increased linearly and the correlation coefficients were greater than 0.99. Compared with that of the plastic strain, the fitting coefficient of the elastic strain did not change significantly with the loading rate and the effect was weaker than that of the plastic strain. However, relatively speaking, when the loading rate was less than 0.15 MPa/s, the fitting coefficient was larger, and vice versa.

Therefore, for this experiment, 0.15 MPa/s can be regarded as the critical value to define low- or high-speed loading.

**Table 3.** Fitting relationships of stepwise elastic strain and cumulative plastic strain with cycle numbers under different loading rates.

| Loading Rate MPa/S | Fitting Expression | | | |
| --- | --- | --- | --- | --- |
| | Stepwise Elastic Strain | | Cumulative Plastic Strain | |
| | Fitting Expression | $R^2$ | Fitting Expression | $R^2$ |
| 0.05 | $\varepsilon_E = 0.039\sigma' - 0.0376$ | 0.99 | $\varepsilon_P = 0.158\exp 0.1601\sigma'$ | 0.99 |
| 0.10 | $\varepsilon_E = 0.0418\sigma' - 0.0529$ | 0.99 | $\varepsilon_P = 0.171\exp 0.1505\sigma'$ | 0.99 |
| 0.15 | $\varepsilon_E = 0.0409\sigma' - 0.0604$ | 0.99 | $\varepsilon_P = 0.163\exp 0.1457\sigma'$ | 0.99 |
| 0.20 | $\varepsilon_E = 0.0388\sigma' - 0.0627$ | 0.99 | $\varepsilon_P = 0.148\exp 0.1461\sigma'$ | 0.98 |
| 0.25 | $\varepsilon_E = 0.0389\sigma' - 0.0666$ | 0.99 | $\varepsilon_P = 0.184\exp 0.1319\sigma'$ | 0.98 |

Note: $\varepsilon_R$–Elastic strain; $\varepsilon_I$–Cumulative plastic strain.

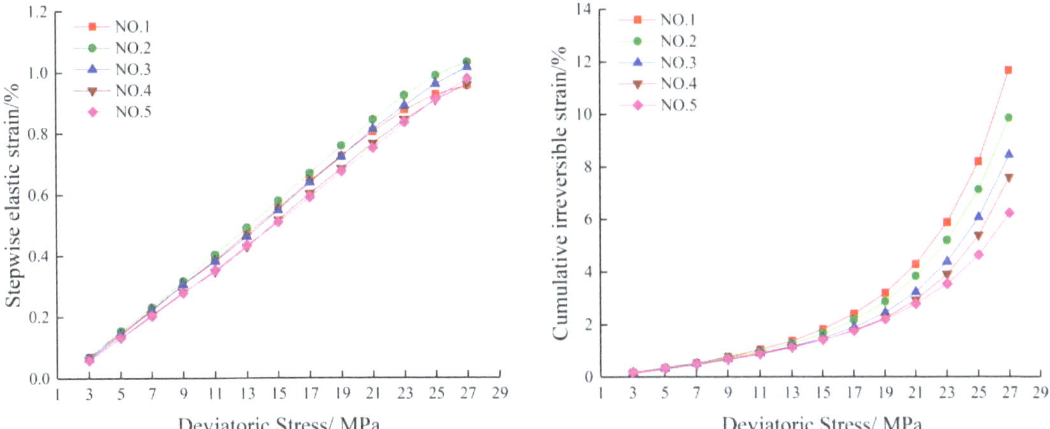


**Figure 9.** Growth curves of stepwise elastic strain and cumulative plastic strain with different loading rates.

### 3.3. Evolution Trends of the Elasticity Modulus

The elastic modulus can be used to reflect the ability of a material to resist deformation under external forces [56]. In the compression process, generally, the strain state of the rock can be divided into compaction, elastic, plastic, and damage stages [57]. Commonly, the slope of the linear variation section of the stress–strain curve is taken as the elastic modulus of the rock. In this experiment, the stress–strain curve that was gathered during the loading process can be divided into an approximate elastic stage and approximate plastic stage. Therefore, the secant slope of the approximate elastic stage in the loading process was regarded as the elastic modulus.

As presented in Figure 10, the elastic modulus evolution has the following characteristics: first, the voids were compressed during early loading and the elastic modulus increased gradually. Second, as the loading rate changed from slow to fast, the corresponding elastic modulus peaks were measured at 2.18 GPa, 2.20 GPa, 2.27 GPa, 2.44 GPa, and 2.44 GPa. Furthermore, the change curves of the elastic modulus under the different loading rates were approximately parallel. In the compaction stage, there was no obvious regularity in the elastic modulus change with the loading rate. Only after the elastic modulus approached the peak did it show a positive correlation with the loading rate. In addition, the elastic modulus experienced a rising period after which it reached the maximum value, then declined immediately when the loading rate was lower than 0.15 MPa/s. However, when the loading rate was higher than 0.15 MPa/s, the change in the elastic modulus could be divided into three stages. These were the rising stage, stable stage and decay stage. At the point at which the deviatoric stress reached 9 MPa, the elastic modulus was stable at about 2.42 GPa; that was until the deviatoric stress was 17 MPa, at which point the elastic modulus began to decrease gradually.

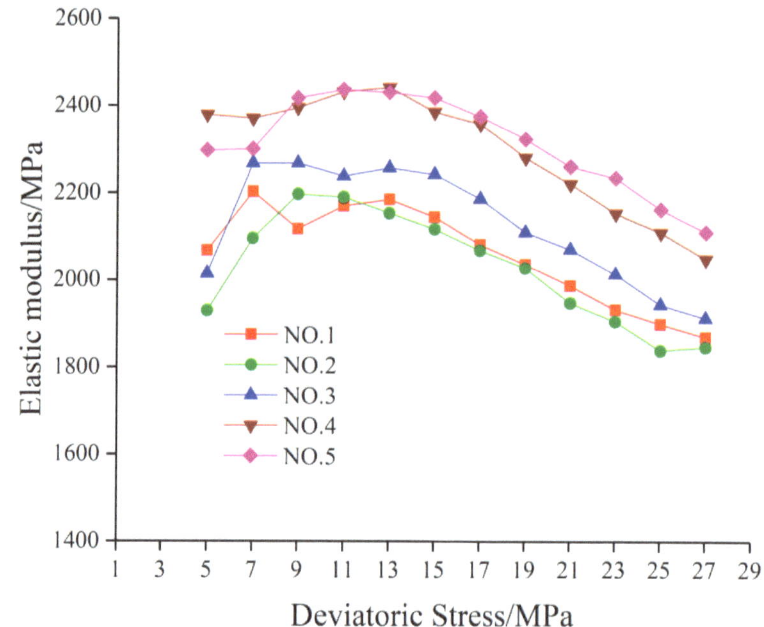


**Figure 10.** Laws of the elastic modulus of the samples with different loading rates.

If the stress corresponding to a significant decrease in elastic modulus was defined as the damage intensity, then 0.15 MPa/s can be considered to be the critical value. The ultimate damage strengths under low-speed loading were 19 MPa, 21 MPa, and 21 MPa; while under high-speed loading, they reached 25 MPa and 27 MPa. Under the low-speed loading conditions, the sample strength did not significantly improve with the increase in loading rate. However, for high-speed loading, the strength of the tectonic coal significantly increased; this is a finding which is consistent with the previous conclusion that high-frequency loading can enhance the strength of rock and inhibit damage propagation.

## 4. Discussion

### 4.1. Growth Laws of Strain Energy Density with Different Loading Rates

Through the use of Formulas (2)–(4), $u$, $u_e$, and $u_d$ were obtained. Based on the analysis of the strain and elastic moduli, 0.15 MPa/s was used as the demarcation point to define the loading rate. Therefore, the samples with loading rates of 0.05 MPa/s, 0.15 MPa/s, and 0.25 MPa/s were selected to represent low-speed, medium-speed, and high-speed loading, respectively.

The strain energy density evolution curves were plotted using the deviatoric stress as the $X$-axis and the three strain energy densities as the $Y$-axis, as shown in Figure 11. Although the loading rates of the tectonic samples were different, the $u$, $u_e$, and $u_d$ of the samples rose with the upper limit of the deviatoric stress throughout the whole cyclic loading process. By fitting the above test results through the use of previously mentioned functions, it was found that the total strain energy density and plastic strain energy density increased exponentially with the deviatoric stress, while the elastic strain energy density increased linearly, with a correlation coefficient that was greater than 0.9724, as shown in Table 4. In addition, the growth coefficients of $u$ and $u_d$ gradually decreased as the loading rate increased, while the growth coefficient of $u_e$ had no obvious response to the loading rate.

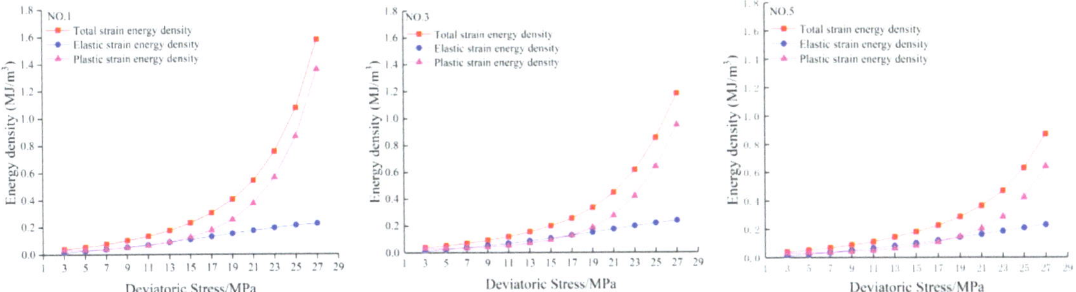



**Figure 11.** Growth curves of the energy densities. (**left**) Energy densities of NO. 1, (**middle**) Energy densities of NO. 1, (**right**) Energy densities of NO. 1.

**Table 4.** Variation in strain energy density with loading rate.

| Specimen | Total Energy Density | $R^2$ | Plasticity Energy Density | $R^2$ | Recoverable Energy Density | $R^2$ |
|---|---|---|---|---|---|---|
| NO. 1 | $u = 0.0233\exp0.1525\sigma'$ | 0.99 | $u_d = 0.0111\exp0.1691\sigma'$ | 0.99 | $u_E = 0.0093\sigma' - 0.0287$ | 0.99 |
| NO. 3 | $u = 0.0233\exp0.1425\sigma'$ | 0.99 | $u_d = 0.0106\exp0.1563\sigma'$ | 0.97 | $u_E = 0.0097\sigma' - 0.0333$ | 0.99 |
| NO. 5 | $u = 0.0249\exp0.1292\sigma'$ | 0.99 | $u_d = 0.0123\exp0.1357\sigma'$ | 0.97 | $u_E = 0.0093\sigma' - 0.0337$ | 0.99 |

Furthermore, there was evidence that both $u_e$, and $u_d$ increased slowly at the initial stage and the gap between them was not obvious. However, as the stress gradually increased, the plastic energy density began to increase rapidly and inflection points appeared in the 15 MPa, 17 MPa, and 19 MPa levels of deviatoric stress.

The above analysis shows that an increase in the loading rate not only inhibits the average growth rate of plastic strain energy density, but also elevates the deviatoric stress value that is required for the accelerated growth of the plastic strain energy density. This explains why higher loading rates in cyclic loading and unloading can improve rock strength from the perspective of energy dissipation.

*4.2. Laws of Energy Storage and Dissipation*

Li et al. studied the energy evolution characteristics of granite under different loading and unloading paths and found that the ratio of dissipated energy could be used to describe the degree of rock deformation and failure [58]. Wang et al. [59,60] established a rockmass damage indicators based on plastic strain work and energy release, by plotted contours of energy components at different damage stages of rockmass in numerical models. To further explore the internal relationships among the three energy densities, the stepwise plastic strain energy ratio (SPER)—$\lambda_d^i$, stepwise elastic strain energy ratio (SEER)—$\lambda_e^i$, cumulative elastic strain energy ratio (CEER)—$\lambda_e^n$, and cumulative plastic strain energy ratio (CPER)—$\lambda_d^n$ were used. These were defined according to the following formulas:

$$\begin{cases} u = u_e + u_d \\ \lambda_e^i = \dfrac{u_e^i}{u^i} \times 100\% \\ \lambda_d^i = \dfrac{u_d^i}{u^i} \times 100\% \end{cases} \tag{5}$$

$$\begin{cases} \lambda_e^n = \dfrac{\sum\limits_{i=1}^{n} u_e^i}{\sum\limits_{i=1}^{n} u^i} \times 100\% \\[4mm] \lambda_d^n = \dfrac{\sum\limits_{i=1}^{n} u_d^i}{\sum\limits_{i=1}^{n} u^i} \times 100\% \end{cases} \tag{6}$$

where both $i$ and $n$ are the number of cycles. The stepwise energy ratio reflects the percentage of elastic strain and plastic strain energy for the total strain energy in a single cycle; and the cumulative energy ratio describes the distribution of the total strain energy input from the first cycle to the end of the current cycle, i.e., it is a quantitative characterisation of energy distribution for the current state.

It can be seen from Figure 12 that $\lambda_d^i$ increased with the increase in deviatoric stress in the first stage and then decreased with the increase in stress in the second stage; and that the demarcation point for the deviatoric stress became 11 MPa. In the first stage, $\lambda_d^i$ had no obvious correlation with the loading rate, but it did have a significant positive correlation with the loading rate in the second stage. I.e., the higher the loading rate was, the higher the recoverable energy ratio. For example, for samples NO. 1, NO. 3, and NO. 5, the peak values of $\lambda_d^i$ were 48.40%, 56.05%, and 57.16%, respectively. At the end of the experiment these values reached low points of 13.97%, 18.30%, and 26.18% respectively. The total strain energy density was equal to the sum of the plastic and elastic strain energy densities, so the change mechanism of the plastic strain energy ratio was not found to have been repeated.

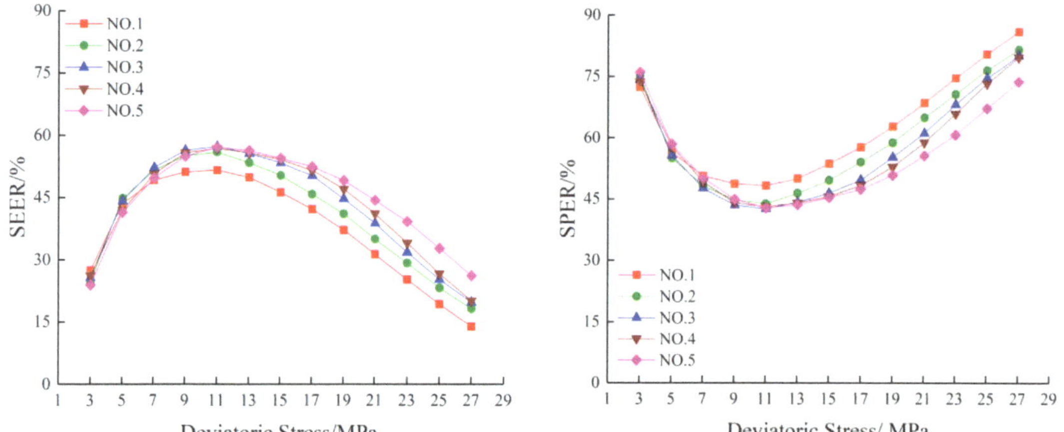


**Figure 12.** Behaviour of $\lambda_e^i$ and $\lambda_d^i$ with deviatoric stress under different loading rates.

Figure 13a–c shows the relationship between $\lambda_e^i$ and $\lambda_d^i$ of the NO. 1, NO. 3, and NO. 5 samples. It can be seen that $\lambda_e^i$ had increased from 20% to 50% as the deviational stress rose to 7 and that it reached its peak value when the deviatoric stress was 11 MPa. These evolution laws seems to be independent of the loading rate. However, with the increase in the loading rate, $\lambda_e^i$ occupied a higher proportion for a longer period. For the NO. 1 sample, $\lambda_e^i$ decreased to 50% when the deviatoric stress was 13 MPa; while for the NO. 3 and NO. 5 samples the corresponding values of deviatoric stress were 17 MPa and 19 MPa, respectively. The evolution law of $\lambda_d^i$ was opposite to that of $\lambda_e^i$. The measurements of $\lambda_d^i$ for the NO. 1, NO. 3, and NO. 5 samples were 86.03%, 80.19%, and 73.82%, respectively, when the cyclic loading ended.

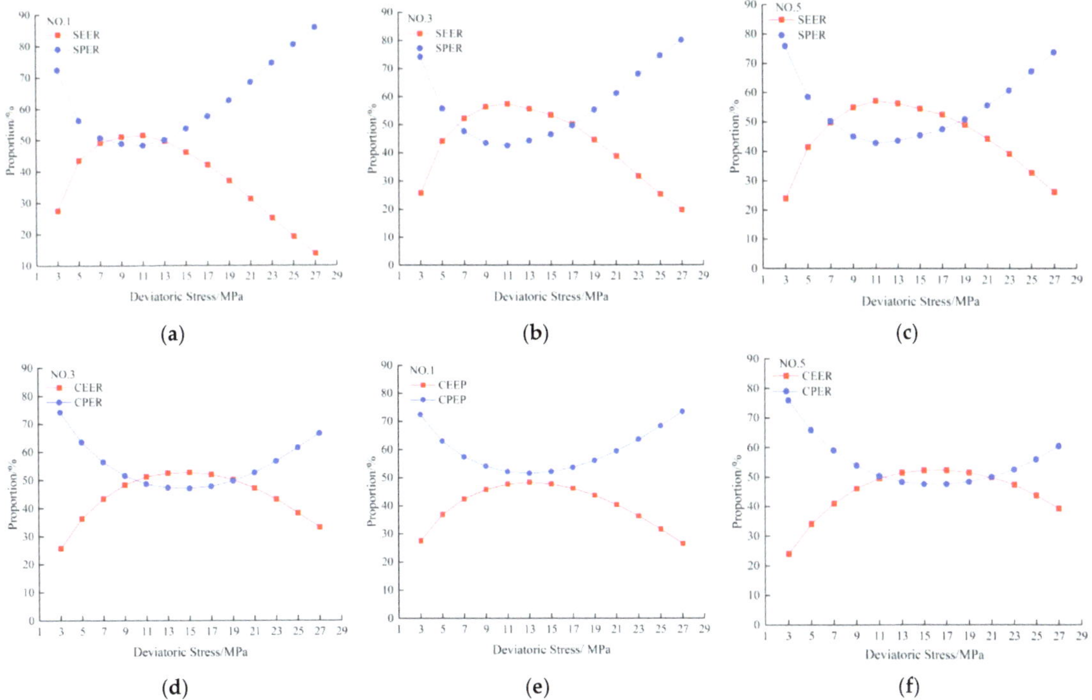

**Figure 13.** CEER and CPER evolution curves. (**a**) Stepwise strain energy ratio of NO. 1, (**b**) Stepwise strain energy ratio of NO.3, (**c**) Stepwise strain energy ratio of NO. 5; (**d**) Cumulative strain energy ratio of NO. 1, (**e**) Cumulative strain energy ratio of NO. 3, (**f**) Cumulative strain energy ratio of NO. 5.

Figure 13d–f shows the relationship between $\lambda_e^n$ and $\lambda_d^n$. The results indicate that, when the loading rate was 0.05 MPa/s, $\lambda_e^n$ was always less than 50%; however, the peak values of $\lambda_d^n$ for the NO. 3 and NO. 5 samples were 52.83% and 52.41%, respectively. At the end of the experiment, $\lambda_d^n$ was measured at 73.51%, 66.62%, and 60.550%. With the increase in loading rate, a significant downward trend was present.

### 4.3. Mechanism of the Loading Rate Action

Through parallel comparisons of the total plastic strain energy density, upper limit damage strength and total strain, some results can be obtained. As presented in Figure 14, the damage strength and total strain had a linear relationship with the loading rate, with a unit growth factor of 40, −26.2 and a fitting degree of 0.926, 0.985. However, the relationship between the total plastic strain energy density and loading rate can be fitted by power function and the fitting coefficient was 0.981. Therefore, even though the natural inhomogeneity of the samples was taken into account, the increase in loading rate exhibited a significant influence on the inhibition of the growth of the total plastic strain density and plastic strain energy density.

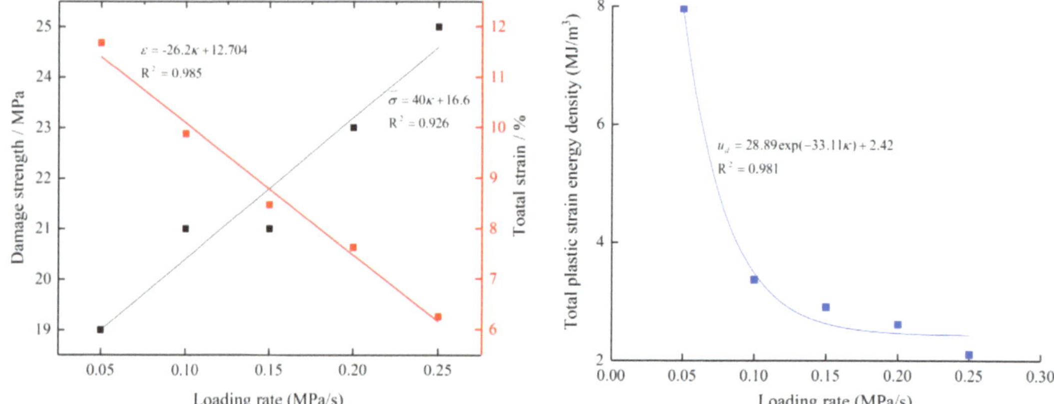

**Figure 14.** Fitting relationships of total plastic energy density, upper limit damage strength and total strain with loading rate.

Reconstituted tectonic coal samples are traditionally composed of coal particles with different particle sizes and other fine particles, such as cement. The initial defects of the samples mainly include the inter-granular spaces and the micro-cracks in the particles, as seen in Figure 15. In the initial loading stage, these voids and cracks were compressed, so the input energy was largely converted into plastic strain properties and the sample strength increased instead of incurring damage.

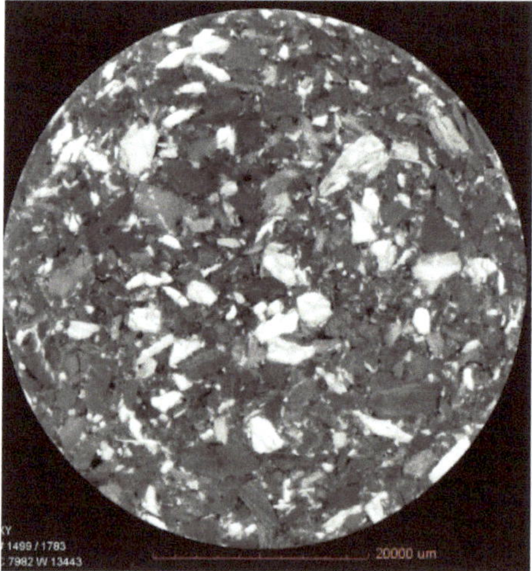

**Figure 15.** CT image of tectonic coal sample.

The strength of a reconstituted sample is affected by its cementation strength and the confining pressure, so there are two damage thresholds. The first threshold occurs when the axial stress exceeds the cementation strength, resulting in relative displacement between the particles. Peng suggested that the increase in the loading rate essentially meant a shortened disturbance time for the external force and a greatly decreased rotation time of the crystals about the weak crystal face, hence a hardening effect was generated [22]. For the reconstituted sample, as the rate increased the rotation time of the particles was shortened, thereby inhibiting the propagation of secondary defects. However, under a lower loading rate the secondary defects were fully expanded and the damage degree increased with the plastic strain at the same stress level. At this stage, although the secondary defects began to be generated, the damage degree was always at a low level and elastic strain still occupied a high proportion of the total strain. The plastic strain and plastic strain energy began to increase, but the rate of their increase was still relatively slow.

The second threshold occurs when the axial stress exceeds the peak strength and the samples lose their bearing capacity through gradual damage accumulation. At this point, the samples broke through the confining pressure and expanded rapidly in the transverse direction, leading to macroscopic damage. Additionally, when the loading rate was low, the dislocation and climbing of the internal crystal grains in the rock were more plentiful [60]. Therefore, the plastic strain and the total plastic strain energy increased rapidly and the total strain energy was converted into plastic strain energy which drove the samples' failures. In a finding that is consistent with the conclusion drawn in reference [59], the rapid change of elastic strain energy, plastic strain energy, and total strain energy can be used as indicators of rockburst. In this stage, the particles moved faster at a higher stress level, so the time effect that was caused by the decrease in the loading rate became more significant. The rock hardening effect was eventually replaced by a softening effect.

## 5. Conclusions

1. As the stress gradually increased, the axial stress–strain curve exhibited two stages, which can be described as "dense–sparse". The total strain and irreversible strain increased exponentially with the increase in the deviatoric stress; however, the elastic strain increased linearly. However, when the axial deviatoric stress was the same as the loading rate of increase, the three strains were gradually decreased.
2. The elastic modulus increased with the increase in the loading rate and rapidly increased in the compressed stage. After reaching its peak, it began to decrease gradually. Higher loading rates led to a slower decrease.
3. The total energy density and plastic energy density increased exponentially with the increase in deviatoric stress, but the elastic strain and elastic energy density increased linearly. At the preliminary stage, the difference between the $u_e$, and $u_d$ was small. As the cyclic loading continued, $u_d$ rose at an obviously faster rate than $u_e$, and the difference between them enlarged. In addition, the loading rate had a significant effect on the increase in the plastic energy density; the lower that this rate was, the higher the growth of the plastic energy density. However, there was a limited effect on the change in the recoverable energy density.
4. The relative displacement of the coal particles that was driven by axial stress was the main way in which energy was consumed. A lower loading rate was beneficial to the damage and movement of particles; i.e., the higher loading rate will inhibit the conversion of input energy to plastic energy. Increasing of loading rate results in the decreases of expansion degree of secondary defects under the same stress level, which improved the energy storage capacity and strength of structural coal.

**Author Contributions:** Conceptualization, D.G., S.S. and S.L.; Methodology, D.G., S.S. and S.L.; Validation, S.S., S.L. and J.G.; Formal analysis, D.G., S.S., S.L. and J.G.; Resources, S.S.; Data curation, D.G., J.G., T.W. and T.S.; Writing—Original draft, D.G.; Writing—Review and editing, S.S., S.L., T.W. and T.S.; Visualization, D.G.; Supervision, S.S. and S.L.; Project administration, S.S. and S.L.; Funding acquisition, S.S. and S.L. The first three authors contributed equally to this work. All authors have read and agreed to the published version of the manuscript.

**Funding:** This research was funded by National Major Scientific Research Instrument Development Project (No. 41727801), the National Natural Science Foundation of China Grant (No. 42030810), the Dominant discipline support project of Jiangsu Province (No. 2020CXNL11), the National Natural Science Foundation of China (No. 41972168), and the Foundation of Jiangsu Key Laboratory of Coal-based Greenhouse Gas Control and Utilization (No. 2020ZDZZ01D).

**Data Availability Statement:** The author of the paper declares that all data comes from the original experiment.

**Acknowledgments:** We would like to thank the technicians who helped during the experiments and the anonymous reviewer for their constructive comments. Thanks for the laboratory of the School of Resources and Geosciences, China University of Mining and Technology; State Key Laboratory for Geomechanics and Deep Underground Engineering, China University of Mining and Technology.

**Conflicts of Interest:** The authors declare no conflict of interest.

## References


1. Lu, S.; Cheng, Y.; Li, W.; Wang, L. Pore structure and its impact on CH4 adsorption capability and diffusion characteristics of normal and deformed coals from Qinshui Basin. *Int. J. Oil Gas Coal Technol.* **2015**, *10*, 94–114. [CrossRef]
2. Li, W.; Liu, H.; Song, X. Multifractal analysis of Hg pore size distributions of tectonically deformed coals. *Int. J. Coal Geol.* **2015**, *144*, 138–152. [CrossRef]
3. Pan, J.; Zhu, H.; Hou, Q.; Wang, H.; Wang, S. Macromolecular and pore structures of Chinese tectonically deformed coal studied by atomic force microscopy. *Fuel* **2015**, *139*, 94–101. [CrossRef]
4. Jin, K.; Cheng, Y.P.; Liu, Q.Q.; Zhao, W.; Wang, L.; Wang, F.; Wu, D.M. Experimental Investigation of Pore Structure Damage in Pulverized Coal: Implications for Methane Adsorption and Diffusion Characteristics. *Energy Fuels* **2016**, *30*, 10383–10395. [CrossRef]
5. Ju, Y.; Li, X. New research progress on the ultrastructure of tectonically deformed coals. *Prog. Nat. Sci.* **2009**, *19*, 1455–1466. [CrossRef]
6. Guo, D.; Li, C.; Zhang, Y. Contrast Study on Porosity and Permeability of Tectonically Deformed Coal and Indigenous Coal in Pingdingshan Mining Area, China. *Earth Sci.-J. China Univers. Geosci.* **2014**, *39*, 1600–1606.
7. Lu, S.; Cheng, Y.; Qin, L.; Wei, L.; Zhou, H.; Guo, H. Gas desorption characteristics of the high-rank intact coal and fractured coal. *Int. J. Min. Sci. Technol.* **2015**, *25*, 819–825. [CrossRef]
8. Xu, H.; Tang, D.; Zhao, J.; Li, S.; Tao, S. A new laboratory method for accurate measurement of the methane diffusion coefficient and its influencing factors in the coal matrix. *Fuel* **2015**, *158*, 239–247. [CrossRef]
9. Skoczylas, N.; Dutka, B.; Sobczyk, J. Mechanical and gaseous properties of coal briquettes in terms of outburst risk. *Fuel* **2014**, *134*, 45–52. [CrossRef]
10. Xu, J.; Ye, G.; Li, B.; Cao, J. Experimental study of mechanical and permeability characteristics of moulded coals with different binder ratios. *Rock Soil Mechan.* **2015**, *36*, 104–110, (In Chinese with English Abstract).
11. Tu, Q.; Xue, S.; Cheng, Y.; Zhang, W.; Shi, G.; Zhang, G. Experimental study on the guiding effect of tectonic coal for coal and gas outburst. *Fuel* **2022**, *309*, 1–12. [CrossRef]
12. Li, X.; Yin, G.; Zhao, H.; Wang, W.; Jing, X. Experimental study of Mechanical Properties of Outburst coal Containing gas under Triaxial Compression. *Chin. J. Rock Mechan. Eng.* **2010**, *29*, 3350–3358.
13. Zhang, D. *Experimental Study on Mechanical Characteristics and Seepage Characteristics of Coal Containing Methane under the Coupling Effect between Stress and Thermal*; Chongqing University: Chongqing, China, 2011; (In Chinese with English Abstract).
14. Meng, Q.; Zhang, M.; Han, L.; Pu, H.; Nie, T. Effects of Acoustic Emission and Energy Evolution of Rock Specimens Under the Uniaxial Cyclic Loading and Unloading Compression. *Rock Mech. Rock Eng.* **2016**, *49*, 3873–3886. [CrossRef]
15. Martino, J.B.; Chandler, N.A. Excavation-induced damage studies at the Underground Research Laboratory. *Int. J. Rock Mech. Min.* **2004**, *41*, 1413–1426. [CrossRef]
16. Shen, B.; King, A.; Guo, H. Displacement, stress and seismicity in roadway roofs during mining-induced failure. *Int. J. Rock Mech. Min.* **2008**, *45*, 672–688. [CrossRef]
17. Heap, M.J.; Vinciguerra, S.; Meredit, P.G. The evolution of elastic moduli with increasing crack damage during cyclic stressing of a basalt from Mt. Etna volcano. *Tectonophysics* **2008**, *471*, 153–160. [CrossRef]
18. Ma, L.; Liu, X.; Wang, M.; Hong, F.; Rui, P.; Peng, X.; Shen, R.; Guo, A.; Qi, K. Experimental investigation of the mechanical properties of rock salt under triaxial cyclic loading. *Int. J. Rock Mech. Min. Sci.* **2013**, *62*, 34–41. [CrossRef]
19. Yang, S.; Tian, W.; Ranjith, P.G. Experimental investigation on deformation failure characteristics of crystalline marble under triaxial cyclic loading. *Rock Mech. Rock Eng.* **2017**, *50*, 2871–2889. [CrossRef]
20. Amann, F.; Button, E.A.; Evans, K.F.; Gischig, V.S.; Bluemel, M. Experimental Study of the Brittle Behavior of Clay shale in Rapid Unconfined Compression. *Rock Mech. Rock Eng.* **2011**, *44*, 415–430. [CrossRef]
21. Baud, P.; Zhu, W.; Wong, T.F. Failure mode and weakening effect of water on sandstone. *J. Geophys. Res.* **2000**, *105*, 16371–16389. [CrossRef]
22. Peng, K.; Zhou, J.; Zou, Q.; Song, X. Effect of loading frequency on the deformation behaviours of sandstones subjected to cyclic loads and its underlying mechanism. *Int. J. Fatigue* **2020**, *131*, 1–12. [CrossRef]
23. Shapoval, V.; Shashenko, O.; Hapieiev, S.; Khalymendyk, O.; Andrieiev, V. Stability assessment of the slopes and side-hills with account of the excess pressure in the pore liquid. *Min. Miner. Depos.* **2020**, *14*, 91–99. [CrossRef]
24. Chen, J.; Du, C.; Jiang, D.; Fan, J.; He, Y. The mechanical properties of rock salt under cyclic loading-unloading experiments. *Geomech. Eng.* **2016**, *10*, 325–334. [CrossRef]
25. Duan, S.; Jian, Q.; Xu, D.; Liu, G. Experimental Study of Mechanical Behavior of Interlayer Staggered Zone under Cyclic Loading and Unloading Condition. *Int. J. Geomech.* **2020**, *20*, 04019187. [CrossRef]

26. Bagde, M.N.; Petros, V. Waveform effect on fatigue properties of intact sandstone in uniaxial cyclical loading. *Rock Mech. Rock Eng.* **2005**, *38*, 169–196. [CrossRef]

27. Taheri, A.; Yfantidis, N.; Olivares, C.L.; Connelly, B.J.; Bastian, T.J. Experimental Study on Degradation of Mechanical Properties of Sandstone Under Different Cyclic Loadings. *Geotech. Test. J.* **2016**, *39*, 673–687. [CrossRef]

28. Liu, E.; Huang, R.; He, S. Effects of Frequency on the Dynamic Properties of Intact Rock Samples Subjected to Cyclic Loading under Confining Pressure Conditions. *Rock Mech. Rock Eng.* **2012**, *45*, 89–102. [CrossRef]

29. Wen, T.; Tang, H.; Huang, L.; Wang, Y.; Ma, J. Energy evolution: A new perspective on the failure mechanism of purplish-red mudstones from the Three Gorges Reservoir area, China. *Eng. Geol.* **2020**, *264*, 1–20. [CrossRef]

30. Liu, Y.; Dai, F.; Dong, L.; Xu, N.; Feng, P. Experimental Investigation on the Fatigue Mechanical Properties of Intermittently Jointed Rock Models Under Cyclic Uniaxial Compression with Different Loading Parameters. *Rock Mech. Rock Eng.* **2018**, *51*, 47–68. [CrossRef]

31. Wang, H.; Li, J. Mechanical Behavior Evolution and Damage Characterization of Coal under Different Cyclic Engineering Loading. *Geofluids* **2020**, *2020*, 8812188. [CrossRef]

32. Taheri, A.; Squires, J.; Meng, Z.; Zhang, Z. Mechanical Properties of Brown Coal under Different Loading Conditions. *Int. J. Geomech.* **2017**, *17*, 06017020. [CrossRef]

33. Zhang, Z.; Xie, H.; Zhang, R.; Zhang, Z.; Gao, M.; Jia, Z.; Xie, J. Deformation Damage and Energy Evolution Characteristics of Coal at Different Depths. *Rock Mech. Rock Eng.* **2019**, *52*, 1491–1503. [CrossRef]

34. Roberts, L.A.; Buchholz, S.A.; Mellegard, K.D.; Düsterloh, U. Cyclic Loading Effects on the Creep and Dilation of Salt Rock. *Rock Mech. Rock Eng.* **2015**, *48*, 2581–2590. [CrossRef]

35. Voznesenskii, A.S.; Kutkin, Y.O.; Krasilov, M.N.; Komissarov, A.A. Predicting fatigue strength of rocks by its interrelation with the acoustic quality factor. *Int. J. Fatigue* **2015**, *77*, 194–198. [CrossRef]

36. Shkuratnik, V.L.; Kravchenko, O.S.; Filimonov, Y.L. Stress Memory in Acoustic Emission of Rock Salt Samples in Cyclic Loading under Variable Temperature Effects. *J. Min. Sci.* **2020**, *56*, 209–215. [CrossRef]

37. Lavrov, A. Kaiser effect observation in brittle rock cyclically loaded with different loading rates. *Mech. Mater.* **2001**, *33*, 669–677. [CrossRef]

38. Olovyannyy, A.; Chantsev, V. Simulation of sample testing under compression with the help of finite-element model of rocks being broken. *Min. Miner. Depos.* **2018**, *12*, 9–21. [CrossRef]

39. Olovyannyy, A.; Chantsev, V. Numerical experiments concerning long-term deformation of rock samples. *Min. Miner. Depos.* **2019**, *13*, 18–27. [CrossRef]

40. Stefeler, E.D.; Epstein, J.S.; Conley, E.G. Energy partitioning for a crack under remote shear and compression. *Int. J. Fract.* **2003**, *120*, 563–580. [CrossRef]

41. Wen, T.; Tang, H.; Ma, J.; Liu, Y. Energy Analysis of the Deformation and Failure Process of Sandstone and Damage Constitutive Model. *Ksce J. Civ. Eng.* **2019**, *23*, 513–524. [CrossRef]

42. Ferro, G. On dissipated energy density in compression for concrete. *Eng. Fract. Mech.* **2006**, *73*, 1510–1530. [CrossRef]

43. Liu, X.; Ning, J.; Tan, Y.; Gu, Q. Damage constitutive model based on energy dissipation for intact rock subjected to cyclic loading. *Int. J. Rock Mech. Min.* **2016**, *2016*, 27–32. [CrossRef]

44. Munoz, H.; Taheri, A.; Chanda, E.K. Fracture energy-based brittleness index development and brittleness quantification by pre-peak strength parameters in rock uniaxial compression. *Rock Mech. Rock Eng.* **2016**, *49*, 4587–4606. [CrossRef]

45. Zhang, Z.; Kou, S.; Jang, L.; Lindqvist, P.A. Effects of loading rate on rock fracture: Fracture characteristics and energy partitioning. *Int. J. Rock Mech. Min. Sci.* **2000**, *37*, 745–762. [CrossRef]

46. Lindqvist, M. Energy considerations in compressive and impact crushing of rock. *Miner. Eng.* **2008**, *21*, 631–641. [CrossRef]

47. Gaziev, E. Rupture energy evaluation for brittle materials. *Int. J. Solids Struct.* **2001**, *38*, 7681–7690. [CrossRef]

48. Xie, H.; Li, L.; Peng, R.; Ju, Y. Energy analysis and criteria for structural failure of rocks. *J. Rock Mech. Geotech. Eng.* **2009**, *1*, 11–20. [CrossRef]

49. Xue, L.; Qin, S.; Sun, Q.; Wang, Y.; Lee, M.; Li, W. A Study on Crack Damage Stress Thresholds of Different Rock Types Based on Uniaxial Compression Tests. *Rock Mech. Rock Eng.* **2014**, *47*, 1183–1195. [CrossRef]

50. Li, D.; Sun, Z.; Xie, T.; Li, X.; Ranjith, P.G. Energy evolution characteristics of hard rock during triaxial failure with different loading and unloading paths. *Eng. Geol.* **2017**, *228*, 270–281. [CrossRef]

51. Peng, R.; Ju, Y.; Wang, J.G.; Xie, H.; Gao, F.; Mao, L. Energy Dissipation and Release During Coal Failure Under Conventional Triaxial Compression. *Rock Mech. Rock Eng.* **2015**, *48*, 509–526. [CrossRef]

52. Jiang, C.; Duan, M.; Yin, G.; Wang, J.G.; Lu, T.; Xu, J.; Zhang, D.; Huang, G. Experimental study on seepage properties, AE characteristics and energy dissipation of coal under tiered cyclic loading. *Eng. Geol.* **2017**, *221*, 114–123. [CrossRef]

53. Sang, S.; Zhou, X.; Liu, S.; Wang, H.; Cao, L.; Li, Z.; Zhu, S.; Liu, C.; Huang, H.; Xu, H.; et al. Research advances in theory and technology of the stress release applied extraction of coalbed methane from tectonically deformed coals. *J. China Coal Soc.* **2020**, *45*, 2531–2543.

54. Meng, Q.; Zhang, M.; Zhang, Z.; Han, L.; Pu, H. Experimental Research on Rock Energy Evolution under Uniaxial Cyclic Loading and Unloading Compression. *Geotech. Test. J.* **2018**, *41*, 717–729. [CrossRef]

55. Wang, E.Y.; Shao, Q.; Han, S.L. Mechanics Analysis of Normal Fault Formation and Control of Structure Coal. *Coal Sci. Technol.* **2009**, *9*, 103–114.

56. Voznesenskii, A.S.; Krasilov, M.N.; Kutkin, Y.O.; Tavostin, M.N.; Osipov, Y.V. Features of interrelations between acoustic quality factor and strength of rock salt during fatigue cyclic loadings. *Int. J. Fatigue* **2017**, *97*, 70–78. [CrossRef]
57. Eberhardt, E.; Stead, D.; Stimpson, B. Quantifying progressive pre-peak brittle fracture damage in rock during uniaxial compression. *Int. J. Rock Mech. Min. Sci.* **1999**, *36*, 361–380. [CrossRef]
58. Wang, F. A Numerical Study of Rockburst Damage around Excavations Induced by Fault-Slip. 2019. Available online: https://www.proquest.com/openview/a91592b37bccda7fd007627917088f96/1?pq-origsite=gscholar&cbl=18750&diss=y (accessed on 10 October 2021).
59. Wang, F.; Kaunda, R. Assessment of rockburst hazard by quantifying the consequence with plastic strain work and released energy in numerical models. *Int. J. Min. Sci. Technol.* **2019**, *29*, 93–97. [CrossRef]
60. Wang, J.; Zhang, Q.; Song, Z.; Zhang, Y. Experimental study on creep properties of salt rock under long-period cyclic loading. *Int. J. Fatigue* **2021**, *143*, 106009. [CrossRef]

*Article*

# Study of the Stability of the Surface Perilous Rock in a Mining Area

Lu Gao [†], Xiangtao Kang *,[†], Lin Gao [†] and Zhenqian Ma [†]

Mining College, Guizhou University, Guiyang 550025, China; lulu925518@163.com (L.G.); lgao@gzu.edu.cn (L.G.); zqma@gzu.edu.cn (Z.M.)
* Correspondence: xiaokangedu@163.com
† These authors contributed equally to this work.

**Abstract:** As a result of the mining of a C3 coal seam in a mine in Guizhou, perilous rock masses on the surface collapsed. In this study, the stability of perilous rock masses on the surface of the coal mine before and after mining was calculated and examined, and the movement law of the overlying strata in the goaf, the movement and deformation law, and the failure mode of perilous rock were analyzed. This study provides a theoretical basis for the treatment of unstable rock and coal seam mining, and has important guiding significance for the safe and efficient production of the mine. The results show that: (1) The perilous rock is in a basically stable state without the influence of mining. Through theoretical analysis and the construction of the collapse model of perilous rock, it is judged that perilous rocks W1, W3, W4, and W7 were basically stable, perilous rocks W2 and W5 were in an unstable state, and perilous rock W6 was stable without heavy rainfall. (2) As a result of the mining of the C3 coal seam, the cracks in the upper strata began to develop to the surface, and the longitudinal separation cracks gradually appeared between the surface perilous rock and the rock matrix. Due to the existence of these cracks, the perilous rock had a downward shear force. In addition, due to the heavy rainfall in the Guizhou area, the transient saturated zone of perilous rock is expanding and the strength of perilous rock is reduced. The seepage increases the sliding force of the perilous rock and aggravates the opening of cracks at any time. (3) The stability of the surface perilous rock mass is largely affected by the mining of underground coal mines. The simulation analysis was repeated using the method of setting coal pillars. When 45 m permanent protection coal pillars are set at both ends, and 15 m local protection coal pillars are set at 60 m, the safety of coal mining can be ensured without affecting the surface and perilous rock.

**Keywords:** perilous rock stability; coal seam mining; UDEC; rockfall; safety production

**Citation:** Gao, L.; Kang, X.; Gao, L.; Ma, Z. Study of the Stability of the Surface Perilous Rock in a Mining Area. *Energies* **2022**, *15*, 1542. https://doi.org/10.3390/en15041542

Academic Editors: Junlong Shang, Chun Zhu, Manchao He and Manoj Khandelwal

Received: 12 January 2022
Accepted: 16 February 2022
Published: 19 February 2022

**Publisher's Note:** MDPI stays neutral with regard to jurisdictional claims in published maps and institutional affiliations.

## 1. Introduction

The surface perilous rock mass in a mining area refers to the rock structure that is distributed on a steep slope or cliff at the surface of the mine. Under the action of external forces, such as natural conditions or mining engineering activities, perilous rock is in, or near, the limit of the equilibrium state in which cracks or structural planes develop [1–5]. As a result of the rapid development of China's construction sector, human engineering activities have increased rapidly in many southwest regions of China's mountains, and perilous rock collapse disasters often occur [6–8]. Guizhou Province, China, is charactered by special karst landforms, and its geological and topographical conditions are relatively complex. Thus, coal mining can easily cause accidents involving the collapse of surface perilous rock [9–12]. Therefore, the exploration of the stability of overlying strata in the goaf is a well-studied and difficult scientific and engineering technology problem in the global field of engineering geology.

Qian et al. [13] proposed the key stratum theory of strata movement and control in 1996, which challenged the limitations of traditional mining surface movement and deformation research, and opened up a new perspective for ground surface subsidence research. Xu et al. [14] studied the dynamic process of the key strata of the overlying strata on the surface subsidence. Baryarkh et al. [15] established a dynamic surface subsidence prediction method with time factors. Singh et al. [16] studied the formation mechanism and prediction method of surface subsidence basins. By analyzing the surface subsidence at the edge of the goaf and the stability of the gully slope, Li et al. [17] determined the principle of the subsection reinforcement of the gully slope under the dynamic influence of coal seam mining. Li et al. [18] used the FLAC3D5.0 program to simulate the evolution law of the overburden strata under the coal mining conditions of the study area. A mathematical model was established to predict the settlement range and displacement of the surface after coal mining. Deng et al. [19] used numerical simulation to determine the surface subsidence boundary and surface collapse volume after mining in the study area. In order to accurately predict the surface subsidence of the mining area, Cheng et al. [20] established a surface subsidence prediction model based on the influence function method. Sun et al. [21] used theoretical analysis and numerical simulation (FLAC3D5.0) to determine the coal pillar and mining widths, and to discuss the coal pillar stress distribution and surface subsidence for different mining scenarios. Chen et al. [22], based on the geomorphology, examined the chained principle for the development of the collapse in Mt. Focusing on falling perilous rocks, Tang et al. [23,24] established a method to calculate the combined stress intensity factor of the dominant fissure under the falling excitation action of perilous rock. At present, obvious deficiencies remain in the research on the special situation of perilous rock above the coal mine goaf. In view of this, in this study, the research object was the surface perilous rock of a mining area in Guizhou Province, China, and the collapse of perilous rock above the coal mine goaf was examined. By investigating the geological conditions of coal mines and the scope, scale, and shape of the perilous rock belt, this study examined the stability of the overlying strata affected by mining, and analyzed the factors affecting the stability of perilous rock, which is of great significance to the safe and efficient production of a coal mine.

## 2. Coal Mine Risk Situation

Perilous rock refers to structural planes of rock mass on steep cliffs or slopes cutting each other at large angles. In addition, gravity, weathering, earthquakes, heavy rains, or artificial engineering activities can lead to instability and underdevelopment of the rock mass, which is otherwise in a stable or limit equilibrium state. The strata in the Panjiang mining area in China are monoclinic, and mainly composed of ridge mountains of erosion and dissolution. Affected by the lithology of strata, cliffs are formed locally, and the overall terrain is uneven, with large joint cutting. The geographical coordinates of the studied coal mines are as follows: $106°53'49''$–$106°54'25''$ east longitude; $28°30'14''$–$28°31'32''$ north latitude. By observing whether there are obvious cracks in the rock area, and whether the cracks were continuously lengthened, widened, or increased, it was determined that the perilous rock belt is distributed on the cliff on the northern side of the mining industrial square. The shortest distance between the industrial square and the perilous rock belt is about 55 m, and the total length of the perilous rock belt is about 400 m. The terrain is high in the northwest and low in the southeast, and the height difference is within 145–159 m. According to a field investigation and the analysis of collected data, the perilous rock threatening the coal mine industrial square ranges from southeast to northwest, and was assigned division numbers of W1, W2, W3, W4, W5, W6, and W7. The specific distribution of the perilous rock is shown in Figure 1.

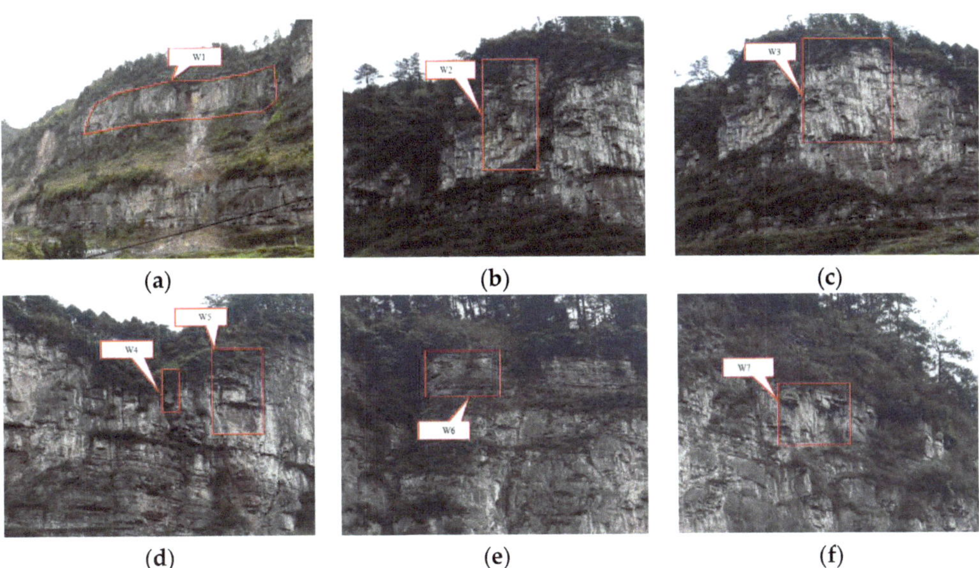

**Figure 1.** Profile of perilous rock. (**a**) W1 perilous rock; (**b**) W2 perilous rock; (**c**) W3 perilous rock; (**d**) W4, W5 perilous rock; (**e**) W6 perilous rock; (**f**) W7 perilous rock.

The details of the scope, scale, and shape of the perilous rock are shown in Table 1.

**Table 1.** Statistical table of morphological characteristics of collapse.

| Numbered | Length (m) | Thickness (m) | Height (m) | Scale and Morphological Characteristics |
|---|---|---|---|---|
| W1 | 240 | 2.5–35 | 32–38 | W1 is located at the end of the northwest side of the entire perilous rock belt, the cracks in the posterior wall have penetrated the whole perilous rock mass, and some stones are suspended on the perilous rock mass. |
| W2 | 14–17 | 1.3–16 | 32.5 | W2 is adjacent to the W1 perilous rock mass, about 15 m apart; the lower part of the perilous rock mass has undergone a block phenomenon, and the debris is a broken block. |
| W3 | 19–21 | 3–3.5 | 50 | W3 is adjacent to the W2 perilous rock mass, and the perilous rock is massive. The steeply inclined fractures on both sides of the perilous rock mass develop and penetrate the whole perilous rock mass. |
| W4 | 1.5 | 2 | 3 | W4 is developed in the wedge formed by the mutual cutting of the two fractures. The lower part of the wedge has broken away from the parent rock, and the remaining part is suspended in the two fractures. |
| W5 | 8–10 | 1.5–3 | 20–25 | W5 is about 10 m from the W3 perilous rock mass. The perilous rock mass is mainly fragmented, and the lower part has collapsed, resulting in the hanging of the upper perilous rock mass. |
| W6 | 10 | 3 | 6.5 | W6 is located at the top of the perilous rock belt. The perilous rock mass is broken into pieces, and there is a drop at the bottom of the perilous rock mass. |
| W7 | 12 | 0.5–1 | 10 | W7 is located at the end of the southeast side of the perilous rock belt, the lower part is controlled by joint fissures to form a wedge, and the rock mass at the bottom has collapsed from the parent rock. |

## 3. Stability Evaluation of Perilous Rock

### 3.1. Macroscopic Analysis of the Stability of Perilous Rock

A stereographic projection is a software-based approach used to judge the stability of a slope [25,26]. Using the geological occurrence of the plane and straight line obtained in the field, the relevant parameters of the related inclination can be obtained. Using the angular distance relationship, the perilous rock in the study area is reflected in the projection plane for analysis and processing, and the stability of perilous rock is preliminarily judged. This method can simplify the complicated calculation process, and has the advantages of being intuitive and convenient.

The macro-analysis of the W1 perilous rock mass was undertaken mainly to evaluate the overall stability of the perilous rock. According to the field investigation, there are three groups of joint cracks in the perilous rock zone of the coal mine industrial square. The relevant physical parameters are shown in Table 2:

**Table 2.** Collapsed perilous rock belt of the coal mine industry square.

| Numbered | Strike (°) | Tendency (°) | Dip Angle (°) | Word Hole (mm) | Density (g/cm$^3$) | Filling Condition | Weathering of Crack Surface |
|---|---|---|---|---|---|---|---|
| P | 340 | 330 | 27 | | | | |
| L1 | 330 | 72 | 15 | 300–500 | 2.85 | No fillings | Strong to medium weathering |
| L2 | 100 | 251 | 80 | 200–500 | 2.82 | No fillings | Strong to medium weathering |
| L3 | 80 | 188 | 69 | 100–300 | 2.74 | Filled with clay and mudstone | Strong to medium weathering |

There are three groups of joint fissures in the perilous rock zone: ① 330°∠27°; ② 72°∠15°; ③ 251°∠80°. Through field investigation, considering the scale form of the perilous rock mass, the geological analogy method was used to judge its stability from a macro perspective. The stereographic projection analysis is shown in Figure 2 below.

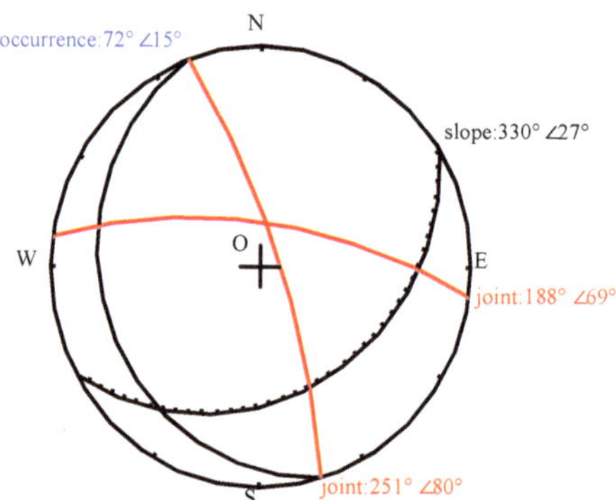


**Figure 2.** Diagram of the stereographic projection.

The analysis from the planar projection is as follows:

(1) Slope and rock occurrence analysis: From the tendency analysis, the slope inclination is 330° and the rock occurrence tendency is 72°. The inclination of the slope surface in this perilous rock zone is nearly 90° orthogonal to the attitude of the rock stratum, so the inclination of the rock stratum is conducive to the stability of the slope, and the attitude of the rock stratum belongs to the stable structural plane. From the angle analysis, the slope angle is 27° and the attitude angle of rock stratum is 15°. The dip angle of the rock is less than the slope foot, so there is no support point on the slope surface, and the rock layer at this time is not conducive to slope stability.

(2) Joint analysis: There are two groups of joints in the slope, with joint 1 having an inclination of 251°, inclination of 80°, and joint 2 having an inclination of 188°, inclination of 69°. The intersection point of joints 1 and 2 is located on the opposite side of the stereographic projection arc surface of the slope. The inclination of the combined intersection line is approximately opposite to that of the slope, so the combined surface is conducive to

slope stability. At the same time, joint 1 tends to be the main sliding direction, and joint 2 is the cutting surface. It is easier for the slope to produce more perilous rock monomers along the structural surface of joint 2.

The analysis shows that the perilous rock zone of the slope is in a relatively stable state as a whole.

### 3.2. Computational Model and Formula

The stability calculation of the perilous rock mass mainly adopted the method of field investigation and checking of the formula calculation. After analyzing the main stress conditions of the perilous rock mass, the perilous rock in the field was preliminarily classified. In addition to the main instability deformation and failure mode, the small blocks in the perilous rock zone are mostly scattered stones. The W4 perilous rock mass belongs to the dumping model, and the remainder belong to the falling model.

(1) The calculation model of dumping perilous rock is shown in Figure 3.

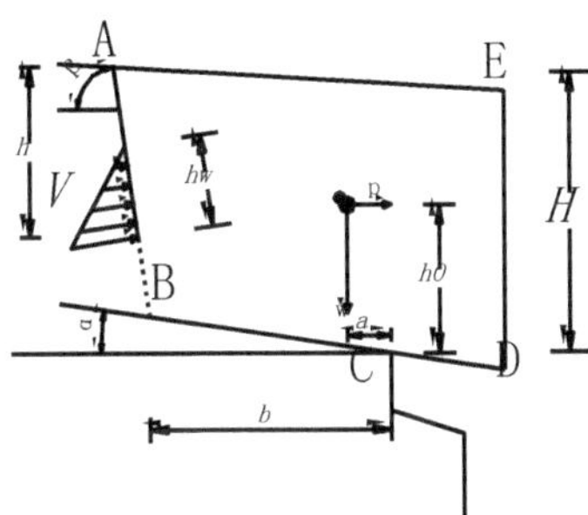

**Figure 3.** Calculation model of dumping perilous rock.

This model is considered according to the unit width without considering the tensile strength of the base. Point C is taken as the overturning point, which is the weathering outer edge point in the base rock. Then, the stability factor is:

① When the tensile strength of the trailing edge rock mass is controlled, it is calculated according to the following formula:

A: perilous rock weight heart outside the tipping point.

$$F = \frac{M_R}{M_O} = \frac{\frac{1}{2}f_{lk} \cdot \frac{H-h}{\sin\beta}\left[\left(\frac{2(H-h)}{3\sin\beta}\right) + \frac{b}{\cos\alpha}\cos(\beta - \alpha)\right]}{W_a + p_b + V\left[\frac{h_w}{3\sin\beta} + \frac{H-h}{\sin\beta} + \frac{b}{\cos\alpha}\cos(\beta - \alpha)\right]}. \tag{1}$$

B: The center of gravity of the perilous rock mass is within the overturning point, which may be C.

$$F = \frac{M_R}{M_O} = \frac{\frac{1}{2}f_{lk} \cdot \frac{H-h}{\sin\beta}\left[\frac{2(H-h)}{3\sin\beta} + \frac{b}{\cos\alpha}\cos(\beta - \alpha)\right] + W_a}{p h_0 + V\left[\frac{h_w}{3\sin\beta} + \frac{H-h}{\sin\beta} + \frac{b}{\cos\alpha}\cos(\beta - \alpha)\right]} \tag{2}$$

In the formula:

$F$: Stability coefficient;
$H_w$: Water filling height of trailing edge fracture (m);
$h$: Fracture depth of trailing edge (m);
$H$: Vertical distance from top of trailing edge fracture to bottom of unpenetrated section (m);
$F_{lk}$: Standard tensile strength of the perilous rock mass (MPa);
$a$: Horizontal distance from body weight center of perilous rock to overturning point;
$b$: Horizontal distance from bottom of unpenetrated section of trailing edge fracture to overturning point;

$h_0$: Vertical distance from perilous rock mass center to overturning point;
$\alpha$: Inclination angle of contact surface between perilous rock mass and base (°);
$\beta$: Inclination angle of trailing edge fracture (°);
$P$: seismic force (KN/m);
$V$: cleft water pressure (KN/m).

② When the tensile strength of the bottom rock mass is controlled, it is calculated according to the following formula:

$$F = \frac{\frac{1}{3}f_{lk}b^2 + W_a}{ph_0 + V\left(\frac{h_w}{3\sin\beta} + b\cos\beta\right)} \tag{3}$$

In the formula:

$F$: Stability coefficient;
$F_{lk}$: Standard tensile strength of perilous rock mass (MPa);
$H_w$: Water filling height of trailing edge fracture (m);
$a$: Horizontal distance from body weight center of perilous rock to overturning point;
$b$: Horizontal distance from bottom of unpenetrated section of trailing edge fracture to overturning point;
$h_0$: Vertical distance from perilous rock mass center to overturning point;
$\alpha$: Inclination angle of contact surface between perilous rock mass and base (°);
$\beta$: Inclination angle of trailing edge fracture (°);
$P$: seismic force (KN/m);
$V$: cleft water pressure (KN/m).

(2) The falling perilous rock calculation model is shown in Figure 4, according to the unit width considerations in the following formula:

$$F = \frac{(W\cos\beta - p\sin\beta - V)\tan\beta + c\frac{h}{\sin\beta}}{W\sin\beta + p\cos\beta} \tag{4}$$

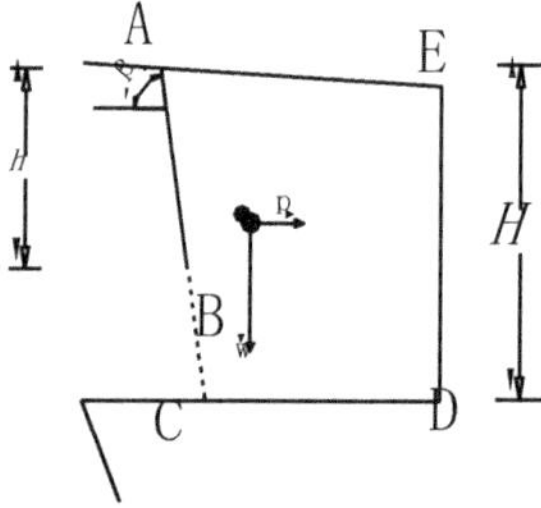


**Figure 4.** Stability calculation model of falling perilous rock.

In the formula:

$F$: Stability coefficient;
$\alpha$: Inclination angle of contact surface between perilous rock mass and base (°);
$\beta$: Inclination angle of trailing edge fracture (°);
$\varphi$: Standard value of internal friction angle of trailing edge cracks (°);
$h$: Fracture depth of trailing edge (m);
$P$: Seismic force (KN/m);
$W$: Gravity of perilous rock mass (KN/m). According to $\gamma \times S$;
$V$: cleft water pressure (KN/m);
$C$: Standard value of cohesion of trailing edge fracture (MPa).

### 3.3. Stability Calculation and Evaluation

The stability calculation and support design of the perilous rock mass do not consider the seismic action [27]. The stability calculation of the perilous rock mass in the working area during an earthquake mainly adopts the following two working conditions, and considers the reduction in the shear strength index of the rock mass structural plane under rainstorm conditions:

Condition 1: dead weight + cleft water pressure (water filling height is calculated according to 1/3 fracture height).
Condition 2: dead weight + cleft water pressure (in the case of a one-in-20-year rainstorm, water filling height is calculated according to 1/2 crack height).

According to the 'China landslide prevention engineering survey specification' (DZ/T 0218-2006), the slope stability coefficient was determined as shown in Table 3.

**Table 3.** Criteria for classification of the stability state of a perilous rock mass.

| MODE of Failure | Stable State of Perilous Rock | | | |
| --- | --- | --- | --- | --- |
| | Destabilization | Not Stable | Basically Stable | Stable |
| dumping type | $F < 1.0$ | $1.00 \leq F < 1.5$ | $1.5 \leq F < 1.8$ | $F \geq 1.8$ |
| falling type | $F < 1.0$ | $1.00 \leq F < 1.3$ | $1.3 \leq F < 1.5$ | $F \geq 1.5$ |

According to the above stability calculation formula, the stability of the perilous rock monomer and perilous rock belt was calculated, and the parameters are shown in Table 4.

**Table 4.** Stable value parameters.

| Stability Parameter | W1 | | W2 | | W3 | | W4 | | W5 | | W6 | | W7 | |
| --- | --- | --- | --- | --- | --- | --- | --- | --- | --- | --- | --- | --- | --- | --- |
| | ① | ② | ① | ② | ① | ② | ① | ② | ① | ② | ① | ② | ① | ② |
| $H$ | 16.3 | 16.3 | 32.5 | 32.5 | 50 | 50 | 1.5 | 1.5 | 22.5 | 22.5 | 6.5 | 6.5 | 10 | 10 |
| $h$ | 10.3 | 10.3 | 11.9 | 11.9 | 6.7 | 6.7 | 0.38 | 0.75 | 1.5 | 1.5 | 3.8 | 3.8 | 1.2 | 1.2 |
| $A_o$ | 0.78 | 0.78 | 1.13 | 1.13 | 2.57 | 2.57 | 0.53 | 0.53 | 2.14 | 2.14 | 0.5 | 0.5 | 2.27 | 2.27 |
| $B_o$ | 9.2 | 9.2 | 18.02 | 18.02 | 28.68 | 28.68 | 1.5 | 1.5 | 13.67 | 13.67 | 4.5 | 4.5 | 11.8 | 11.8 |
| $F_{lk}$ | 600 | 600 | 600 | 600 | 600 | 600 | 600 | 600 | 600 | 600 | 600 | 600 | 600 | 600 |
| $c$ | 120 | 80 | 120 | 80 | 120 | 80 | | | 120 | 80 | 120 | 80 | 120 | 80 |
| $\Phi$ | 32.1 | 30 | 32.1 | 30 | 32.1 | 30 | | | 32.1 | 30 | 32.1 | 30 | 32.1 | 30 |
| $\zeta$ | 0.15 | 0.15 | 0.15 | 0.15 | 0.15 | 0.15 | | | 0.15 | 0.15 | 0.15 | 0.15 | 0.15 | 0.15 |
| $p$ | 0 | 0 | 0 | 0 | 0 | 0 | 0 | 0 | 0 | 0 | 0 | 0 | 0 | 0 |
| $S$ | 19.2 | 19.2 | 72.5 | 72.5 | 155.8 | 155.8 | | | 75.98 | 75.98 | 7.98 | 7.98 | 30.26 | 30.26 |
| $V$ | | | | | | | 1.7 | 1.7 | | | | | | |
| $W$ | | | | | | | 161 | 161 | | | | | | |
| $\sin\beta$ | | | | | | | 1 | 1 | | | | | | |
| $\cos\alpha$ | | | | | | | 1 | 1 | | | | | | |
| $\cos(\beta - \alpha)$ | | | | | | | 0.2 | 0.2 | | | | | | |
| $F$ | 1.39 | 1.16 | 1.27 | 32.5 | 1.31 | 1.03 | 1.38 | 1.16 | 1.23 | 1.03 | 1.51 | 1.26 | 1.35 | 1.07 |

$H$: Vertical distance from upper part of trailing edge fracture to lower part of unpenetrated section (m); $h$: Fracture depth of trailing edge (m); $A_o$: Horizontal distance from body weight center of perilous rock to potential failure surface (m); $B_o$: The plummeting distance from the weight center of perilous rock to the centroid of potential damaged surface (m); $F_{lk}$: Standard for Tensile Strength of perilous rock mass (MPa); $c$: Standard value of cohesion of perilous rock mass (MPa); $\Phi$: Standard value of cohesion of perilous rock mass (°); $\zeta$: Calculation coefficient of bending moment of perilous rock; $p$: seismic force (KN/m); $S$: Single width area of perilous rock mass (m²); $\alpha$:Angle of contact surface between perilous rock mass and base (°); $\beta$: Inclination angle of trailing edge fracture (°); $V$: cleft water pressure (KN/m); $W$: Gravity of perilous rock mass (KN/m); $F$: stability factor.

The quantitative calculation using the formulas was carried out for the models established for the falling perilous rock and the dumping perilous rock, respectively. The calculation results are shown in Table 5.

**Table 5.** Calculation results of the stability of the perilous rock mass.

| Numbered | Mode of Failure | Calculated Work Condition | Saturated Uniaxial Compressive Strength (MPa) | Stability Factor | Steady-State |
|---|---|---|---|---|---|
| W1 | falling type | 1 | 38.50 | 1.39 | basically stable |
|  |  | 2 |  | 1.16 | not stable |
| W2 | falling type | 1 | 51.20 | 1.27 | not stable |
|  |  | 2 |  | 1.16 | not stable |
| W3 | falling type | 1 | 44.70 | 1.31 | basically stable |
|  |  | 2 |  | 1.03 | not stable |
| W4 | dumping type | 1 | 39.20 | 1.38 | basically stable |
|  |  | 2 |  | 1.16 | not stable |
| W5 | falling type | 1 | 50.10 | 1.23 | not stable |
|  |  | 2 |  | 1.03 | not stable |
| W6 | falling type | 1 | 47.30 | 1.51 | stable |
|  |  | 2 |  | 1.26 | not stable |
| W7 | falling type | 1 | 39.40 | 1.35 | basically stable |
|  |  | 2 |  | 1.07 | not stable |

## 4. Numerical Simulation Analysis of Perilous Rock Stability under Mining Influence

### 4.1. Model Establishment

The perilous rock studied in this paper has two types: falling type and dumping type. Because the W6 perilous rock is in a stable state, the remainder of the perilous rock is basically either stable and unstable. Therefore, the simulation used the W6 stable perilous rock as the research object to judge the influence of coal seam mining on the W6 perilous rock. The numerical model established a two-dimensional model of coal seam mining at the location of perilous rock W6 as a section. The boundary conditions of this study were determined according to the actual situation. The bottom and both sides of the model are zero displacement boundary conditions. The upper part of the model and the perilous rock are free boundaries, which are mainly subject to gravity stress. The gravity stress of the original rock is applied on the upper boundary [28,29], and the gradient horizontal stress is applied on the left and right boundaries of the model. Due to the small dip angle of the coal seam, the horizontal model was established as follows:

The model uses the horizontal direction as the X-axis, and the total length in the X-axis direction is 450 m. The Y-axis is along the vertical direction and the total thickness of the Y-axis is 240 m. The model has 11 inflection points, which, in the clockwise direction are (0, 0), (0, 102), (11.73, 111.82), (60.84, 133.45), (87.37, 150), (103.98, 153.36), (111.48, 228.45), (110.52, 240), (112.56, 240), (450, 240), (450, 0). The crack command was used to connect the coordinate primaries between the points with straight lines. the position of the coal seam and stratum lithology is marked in Figure 5:

The numerical simulation using the Mohr–Coulomb yield rule is:

$$f = \sigma_1 - \sigma_3 \frac{1 + \sin \varphi}{1 - \sin \varphi} = 2c \sqrt{\frac{1 - \sin \varphi}{1 + \sin \varphi}} \tag{5}$$

where:

$\sigma_1$: maximum principal stress (MPa);

$\sigma_3$: minimum principal stress (MPa);

$c$: cohesive force (MPa);

$\theta$: angle of internal friction (°).

The physical and mechanical parameters of each rock mass are shown in Table 6.

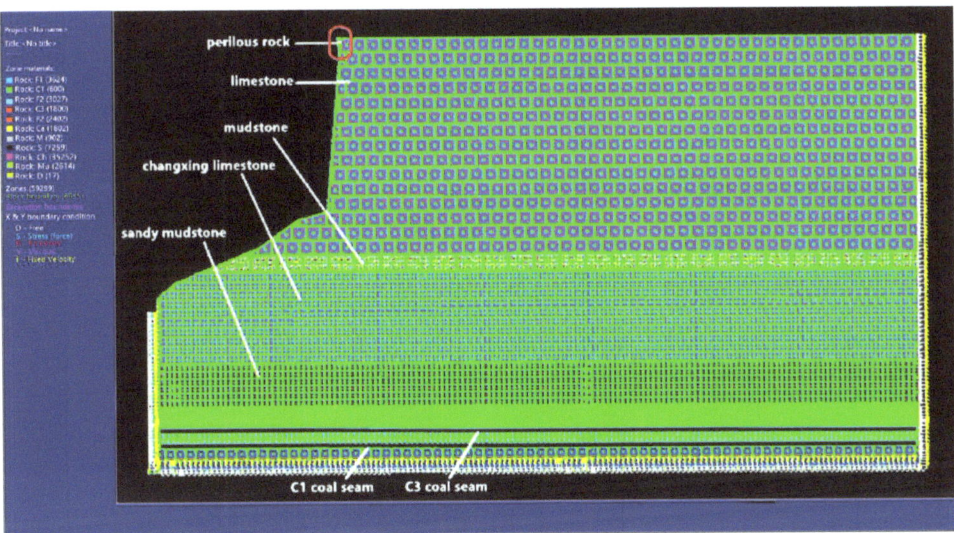


**Figure 5.** UDEC calculation model (angle 0°).

**Table 6.** Physical and mechanical parameters of rock mass.

| Number | Rock Type | Thickness /(m) | Density /(g/cm³) | Bulk Modulus/(GPa) | Shear Modulus/(GPa) | Friction Angle /(°) | Cohesion/(MPa) | Tensile Strength/(MPa) |
|---|---|---|---|---|---|---|---|---|
| 1 | limestone | 110 | 2.80 | 5.57 | 4.53 | 38 | 11.4 | 6.7 |
| 2 | mudstone | 10 | 2.70 | 2.86 | 1.4 | 39 | 2.8 | 2.48 |
| 3 | Changxing limestone | 50 | 2.43 | 11.1 | 8.3 | 35 | 2.4 | 4.4 |
| 4 | sandy mudstone | 24 | 2.25 | 10.2 | 6.1 | 30 | 1.8 | 3.2 |
| 5 | mudstone | 4 | 2.55 | 5.8 | 3.2 | 30 | 1.2 | 3.25 |
| 6 | sandy mudstone | 4.5 | 2.25 | 10.2 | 6.1 | 30 | 1.8 | 3.2 |
| 7 | carbon mudstone | 1.5 | 2.45 | 4.3 | 2.8 | 30 | 0.7 | 1.8 |
| 8 | C3 coal | 2 | 1.47 | 1.19 | 0.82 | 25 | 1.3 | 1.79 |
| 9 | Clay rock | 10 | 2.25 | 4.39 | 2.27 | 27 | 4.9 | 3.8 |
| 10 | C1 coal | 1 | 1.47 | 1.19 | 0.82 | 25 | 1.3 | 1.79 |
| 11 | mudstone | 13 | 2.55 | 5.8 | 3.2 | 30 | 1.2 | 3.25 |

*4.2. Analysis of Numerical Simulation Results*

The two-dimensional model of C3 coal seam mining was established by taking the location of the W6 perilous rock mass as the section to carry out the numerical simulation. According to the mining conditions of the C3 coal seam in the coal mine studied in this paper, 50 m coal pillars were placed at the left and right ends, and the simulation was carried out in 50 m units in front of the working face to accurately represent the mining situation of the C3 coal seam in the coal mine. According to the simulation results, two kinds of fractures occurred in the simulation: separated fractures and broken fractures [30].

As shown in Figure 6a, when the working face is mined to 50 m, coal mining only leads to local cracks above the goaf of the working face, which has little effect on the overlying strata far from the coal seam. When the working face is mined to 100 m, the fracture is still located above the goaf as a whole, but its development range is larger. The transverse separation fracture and longitudinal separation fracture are connected, and the fracture continues to develop upward, but it has little effect on the surface perilous rock, as shown in Figure 6b. When the working face advances to 150 m, due to the continuous mining, the fracture at this time has developed to the position close to the surface; however, it has not yet been connected with the surface, and is at a distance of about 10 m from the surface. At this time, the fracture is still far from the surface perilous rock, and has little influence on the perilous rock, as shown in Figure 6c. When the working face advances to

200 m, as the coal seam continues to be excavated, the fractures continue to develop in the direction toward the working face. The previous fracture zone continues to be compacted, and new fracture zones appear. At this time, above the position of mining at about 200 m, there are separate cracks connecting with the surface, and there are also upward cracks in the adjacent position. The surface at this position has shown obvious subsidence, but the influence of mining on the perilous rock mass has not been seen, as shown in Figure 6d. When the working face advances to 250 m, the fracture continues to develop in the direction of mining. At this time, the fracture is about 50 m from the horizontal distance of the goaf, and there is a local fracture at 10–40 m below the corner of the slope. This fracture is not connected with the previous fracture, and the fracture is located directly below the perilous rock. If the mining is continued, the fracture will have a great impact on the surface perilous rock. When the working face advances to 300 m, as the mining advances to the lower right side of the perilous rock, the rock strata in the goaf have fully collapsed, and collapsed to the floor with the advance in mining. The surface also has different degrees of depression. The rock strata have obvious vertical fracture cracks, and there are many fractures above the goaf and coal wall. The mining has a significant impact on the overlying strata, and the upper rock strata of the roof are greatly affected by the mining. At this point, the cracks at the foot of the slope and above the goaf have been partially penetrated, and the cracks are close to the perilous rock as a whole, as shown in Figure 6f. When the working face is advanced to 350 m, the fracture has penetrated with the surface of the perilous rock, and the perilous rock has not yet collapsed. However, the existence of the fracture has an extremely unfavorable impact on the stability of the perilous rock, as shown in Figure 6g.

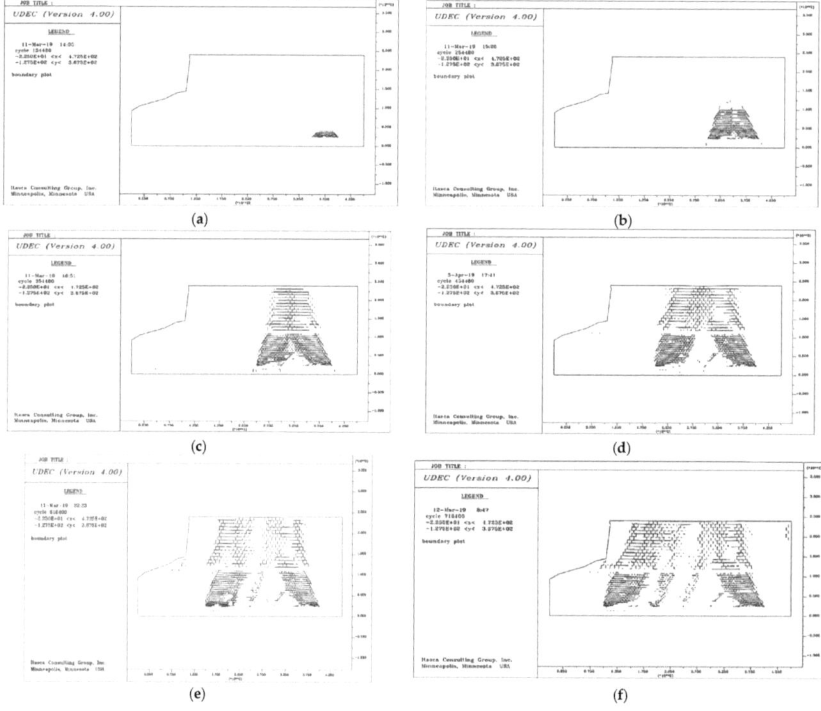

**Figure 6.** *Cont.*

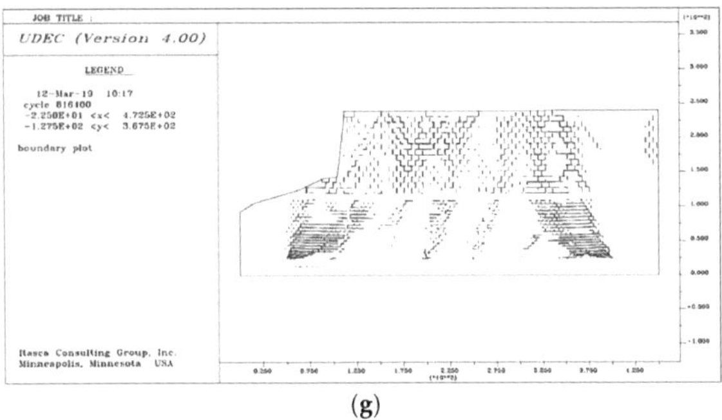

(g)

**Figure 6.** Coal seam mining effect diagram: (**a**) 50 m forward; (**b**) 100 m forward; (**c**) 150 m forward. (**d**) 200 m forward: separated fissure and surface penetration; (**e**) 250 m forward; (**f**) 300 m forward: the whole crack is close to the perilous rock; (**g**) 350 m forward: fracture and surface penetration of perilous rock.

## 5. Stability Control of Surface Perilous Rock

The surface perilous rock mass is cut by multiple sets of structural planes. The intersections of the cracks, weak layers, and faults in the rock mass determine the diversity of the failure mode of the perilous rock mass, and are the reasons for the complexity of the failure and instability. Liu et al. [31] explained the dynamic source of the failure mode of the perilous rock mass, and proposed that the main driving force of the collapse is the apparent sliding force outward from the slope. The apparent sliding force outward from the slope can overcome the shear resistance of the front edge of the perilous rock mass and the friction of the bottom surface. Xie [32] et al. analyzed the influence of mining activities on the structure and mechanical parameters of the overlying rock mass, and proposed the evaluation standard and basis for the safety degree of the cavity overlying rock mass. Aref [33] et al. proposed a flexible network locking method with an obvious effect on the treatment of the perilous rock mass, and comprehensively outlined the design principle of the flexible network locking method. Therefore, the stability control of surface perilous rock mass can be considered from two aspects of underground coal mining and surface disaster management. The mining situation of the C3 coal seam was numerically simulated by UDEC software, and the fracture of the overlying strata and the collapse of the surface perilous rock mass were theoretically analyzed. The trajectory of the perilous rock collapse was numerically simulated using Rockfall software, and the simulation was carried out from underground and at the surface. However, the actual surface collapse of the perilous rock is not only affected by coal mining, but is also related to many factors [34–38]. In addition, theoretical speculation about the range of influences on the collapse will also deviate to a certain degree from the actual situation.

Using the "support" method to set up coal pillars in the UDEC numerical simulation, the following method was finally determined to have the best mining effect: the C3 coal seam working face is pushed from right to left, with 45 m permanent protective coal pillars left at both ends, and 15 m local protective coal pillars left at 60 m. Because the UDEC software only simulates the two-dimensional plane, the width of the coal pillar can only be determined according to the field equipment. In addition, it is necessary to ensure that coal mining-related equipment can pass through the coal pillar. Finally, the simulated UDEC fracture development diagram is shown in Figure 7. For the perilous rocks W1, W2, W3, W4, W5, W6, and W7 on the surface, and the loose structure on the slope, artificial and mechanical methods can be used to remove the rock according to the current crack development of the rock mass, and the stones formed by the cutting can be transported outside and stacked at the designated location. In addition, an SNS active protection network can be used to protect the whole perilous rock belt. In this method of mining, the final fracture development is basically below the Changxing limestone section, and will not involve the surface and perilous rock.

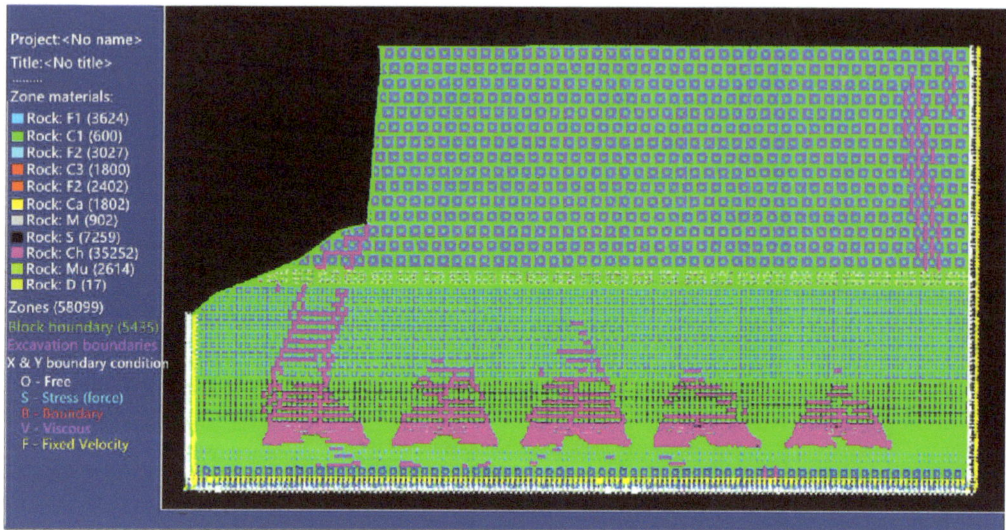



**Figure 7.** Diagram showing the effect of leaving a partial coal pillar.

## 6. Conclusions

(1) The perilous rock is basically stable without the influence of mining. The analysis of the fracture development obtained by the simulation shows that, with the mining of the C3 coal seam, the fracture of the upper strata starts to develop to the surface. The fracture gradually runs through the surface and finally causes a longitudinal separation fracture at the perilous rock. Due to the existence of the fracture, the perilous rock has a downward shear force, resulting in the collapse of the rock.

(2) The stability of the surface perilous rock mass is largely affected by the influence of underground coal mining. Through the study of the development of perilous rock fractures in the process of mining, it was concluded that when 45 m permanent protective coal pillars are left at both ends, and 15 m local protective coal pillars are left at 60 m, the safety of coal mining can be ensured and the surface and perilous rock will not be affected.

(3) The underground mining of a coal mine involves complex system engineering. In this study, UDEC software was used to simulate the influence of coal seam mining on the surface perilous rock collapse. The development of fractures was only analyzed from the two-dimensional perspective. This helps determine the significant effect of mining on the perilous rock collapse, but the weights of the influences cannot be reflected in the analysis. At present, this analysis remains in the theoretical stage, and the specific effects need to be further verified.

**Author Contributions:** Conceptualization, L.G. (Lu Gao) and X.K.; methodology, L.G. (Lin Gao); software, Z.M.; validation, Z.M., X.K. and L.G. (Lin Gao); formal analysis, X.K.; investigation, X.K.; resources, X.K.; data curation, L.G. (Lu Gao); writing—original draft preparation, L.G. (Lu Gao); writing—review and editing, X.K.; visualization, X.K.; supervision, L.G. (Lin Gao); project administration, L.G. (Lin Gao); funding acquisition, X.K. and L.G. (Lin Gao). All authors have read and agreed to the published version of the manuscript.

**Funding:** National Natural Science Foundation of China (No. 52064009; 52004073); Guizhou Science and Technology Fund Project (Qiankehe Foundation [2020] 1Y216).

**Data Availability Statement:** The data presented in this study are available on request from the corresponding author.

**Conflicts of Interest:** The authors declare no conflict of interest.

## References

1. Mu, C.; Yu, X.Y.; Zhao, B.C. The Formation Mechanism of Surface Landslide Disasters in the Mining Area under Different Slope Angles. *Adv. Civ. Eng.* **2021**, *2021*, 106–119. [CrossRef]
2. Li, H. Artificial Intelligence and Internet of Things (IoT) in Civil Engineering. *Adv. Civ. Eng.* **2021**, *2021*, 111–123.
3. Hungr, O.; Leroueil, S.; Picarelli, L. The Varnes classification of landslide types, an update. *Landslides* **2014**, *11*, 167–194. [CrossRef]
4. Liu, C.; Zhang, S.; Ding, L.; Liao, J. Identification of perilous rock mass of high slope and study of anchoring method based on laser scanning. *Chin. J. Rock Mech. Eng.* **2012**, *31*, 2139–2146.
5. Crosta, G.B.; Agliardi, F. Parametric calculation of 3D dispersion of rockfall trajectories. *Nat. Hazards Earth Syst. Sci.* **2004**, *4*, 583–598. [CrossRef]
6. Geng, J.B.; Li, Q.H.; Li, X.S.; Zhou, T.; Liu, Z.; Xie, Y. Research on the evolution characteristics of rock mass response from open-pit to underground mining. *Adv. Mater. Sci. Eng.* **2021**, *2021*, 3200906. [CrossRef]
7. Kong, D.Z.; Cheng, Z.; Zheng, S. Study on failure mechanism and stability control measures in large-cutting-height coal mining face with deep-buried seam. *Bull. Eng. Geol. Environ.* **2021**, *78*, 6143–6157. [CrossRef]
8. Bui, D.T.; Tuan, T.A.; Klempe, H.; Pradhan, B.; Revhaug, I. Spatial prediction models for shallow landslide hazards: Acomparative assessment of the efficacy of support vector machines, artificial neural networks, kernel logistic regression, and logistic model tree. *Landslides* **2016**, *13*, 361–378.
9. Cao, Z.Z.; Ren, Y.L.; Wang, Q.T. Evolution Mechanism of Water-Conducting Channel of Collapse Column in Karst Mining Area of Southwest China. *Geofluids* **2021**, *2021*, 6630462. [CrossRef]
10. Chen, H.K.; Tang, H.M.; Wang, R. Calculation method of stability for perilous rock and application to the three gorges reservoir. *Chin. J. Rock Mech. Eng.* **2004**, *23*, 614–619. [CrossRef]
11. Hou, P.; Liang, X.; Gao, F.; Dong, J.B.; He, J.; Xue, Y. Quantitative visualization and characteristics of gas flow in 3D pore-fracture system of tight rock based on Lattice Boltzmann simulation. *J. Nat. Gas Sci. Eng.* **2021**, *89*, 103867. [CrossRef]
12. Liang, X.; Hou, P.; Xue, Y.; Yang, X.J.; Gao, F.; Liu, J. A fractal perspective on fracture initiation and propagation of reservoir rocks under water and nitrogen fracturing, Fractals-Complex Geom. *Patterns Scaling Nat. Soc.* **2021**, *29*, 7.
13. Qian, M.G.; Miao, X.X.; Xu, J.L. Theoretical study of key stratumin ground control. *J. China Coal Soc.* **1996**, *21*, 225–230.
14. Xu, J.L.; Qian, M.G.; Zhu, W.B. Study on inflfluences of primary key stratum on surface dynamic subsidence, Chin. *Rock Mech. Rock. Engl.* **2005**, *24*, 787–791.
15. Baryak, A.A.; Telegin, E.A.; Samodelkina, N.A. Prediction of the Intensive Surface Subsidence in Mining Potash Series. *J. Min. Sci.* **2005**, *41*, 312–319. [CrossRef]
16. Singh, K.B. Pot-hole Subsidence in Son-Mahanadi master Coal Basin. *Eng. Geol.* **2007**, *89*, 88–97. [CrossRef]
17. Li, J.W.; Li, X.T.; Liu, C.Y. Dynamic Changes in Surface Damage Induced by High-Intensity Mining of Shallow, Thick Coal Seams in Gully Areas. *Adv. Civ. Eng.* **2020**, *2020*, 246–262. [CrossRef]
18. Li, G.; Yang, Q.H. Prediction of Mining Subsidence in Shallow Coal Seam. *Math. Probl. Eng.* **2020**, *2020*, 339–404.
19. Deng, X.; Li, L.; Tan, Y. Numerical simulation of surface subsidence induced by underground mining in mines. *Saf. Coal Mines* **2018**, *49*, 188–192.
20. Cheng, J.W.; Zhao, G.; Li, S.Y. Predicting Underground Strata Movements Model with Considering Key Strata Effects. *Geotech. Geol. Eng.* **2018**, *36*, 621–640. [CrossRef]
21. Sun, W.; Zhang, Q.; Luan, Y.Z. A study of surface subsidence and coal pillar safety for strip mining in a deep mine. *Environ. Earth Sci.* **2018**, *77*. [CrossRef]
22. Chen, H.K.; Tang, H.M.; Xian, X.F. Developing Mechanism for Collapse Disaster in Rocky Mountain Area—Taking Mt. Hongyan in the National Scenic Spots of Simianshan as an Example. *Adv. Eng. Sci.* **2010**, *42*, 1–6.
23. Tang, H.M.; Wang, Z.; Chen, H.K.; Xian, X.F. Contribution ratio of excitation action triggered by collopse of perilous rock to stability of perilous rock. *J. Vib. Shock* **2012**, *31*, 32–37.
24. Tang, H.M.; Wang, Z.; Xian, X.F.; Chen, H.K. Violent-slide rock avalanche and excitation effect of perilous rock. *J. Chongqin Univ.* **2011**, *34*, 32–37.
25. Zheng, J.; Lu, Q.; Deng, J.; Yang, X.; Fan, X.; Ding, Z. A modified stereographic projection approach and a free software tool for kinematic analysis of rock slope toppling failures. *Bull. Eng. Geol. Environ.* **2019**, *78*, 4757–4769. [CrossRef]
26. Kıncal, C. Application of two new stereographic projection techniques to slope stability problems. *Int. J. Rock Mech. Min. Sci.* **2014**, *66*, 136–150. [CrossRef]
27. Li, X.S.; Liu, Z.F.; Yang, S. Similar physical modeling of roof stress and subsidence in room and pillar mining of a gently inclined medium-thick phosphate rock. *Adv. Civ. Eng.* **2021**, *2021*, 6686981. [CrossRef]
28. Ren, W.Z.; Guo, C.M.; Peng, Z.Q.; Wang, Y.G. Model experimental research on deformation and subsidence characteristics of ground and wall rock due to mining under thick overlying terrane. *Int. J. Rock Mech. Min. Sci.* **2010**, *47*, 614–624. [CrossRef]
29. Wu, H.; Ma, D.; Spearing, A.J.S.; Zhao, G.Y. Fracture phenomena and mechanisms of brittle rock with different numbers of openings under uniaxial loading. *Geomech. Eng.* **2021**, *25*, 481–493.
30. Ma, D.; Kong, S.B.; Li, Z.H.; Zhang, Q.; Wang, Z.H.; Zhou, Z.L. Effect of wetting-drying cycle on hydraulic and mechanical properties of cemented paste backfill of the recycled solid wastes. *Chemosphere* **2021**, *282*, 131163. [CrossRef]

31. Liu, C.Z. Formation and Collapse Cause Analysis of Jiweishan perilous rock Mass in Wu long, Chongqing. *J. Eng. Geol.* **2010**, *18*, 297–304.
32. Xie, Q.D.; Ding, X.P. Dynamic Analysis of Rock Mass Stability on Cavity in East Opencast Coal Mine. *Coal Proj.* **2018**, *50*, 10–13.
33. Aref, M.; Nei, L.; Abdo, S. Causes of rock falls in Ai-lluway-shaharea, Yemen. *Glob. Geol.* **2009**, *12*, 5–12.
34. Kawamoto, T.; Ichikawa, Y.; Kyoya, T. Deformation and fracture behaviour of discontinuous rock mass and damage mechanics theory. *Int. J. Numer. Anal. Methods Geomech.* **1988**, *12*, 1–30. [CrossRef]
35. Ju, Y.; Wang, Y.; Su, C.; Zhang, D.; Ren, Z. Numerical analysis of the dynamic evolution of mining-induced stresses and fractures in multilayered rock strata using continuum-based discrete element methods. *Int. J. Rock Mech. Min. Sci.* **2019**, *113*, 191–210. [CrossRef]
36. Gao, L.; Kang, X.T.; Huang, G.; Wang, Z.Y.; Tang, M.; Shen, X.Y. Experimental Study on Crack Extension Rules of Hydraulic Fracturing Based on Simulated Coal Seam Roof and Floor. *Geofluids* **2022**, *2022*. [CrossRef]
37. Kong, D.; Xiong, Y.; Cheng, Z.; Wang, N.; Wu, G.; Liu, Y. Stability analysis of coal face based on coal face-support-roof system in steeply inclined coal seam. *Geomech. Eng.* **2021**, *25*, 233–243.
38. Donnelly, L.J.; Cruz, H.D.L.; Asmar, I.; Zapata, O.; Perez, J.D. The monitoring and prediction of mining subsidence in the Amaga, Angelopolis, Venecia and Bolombolo regions, Antioquia, Colombia. *Eng. Geol.* **2001**, *59*, 103–114. [CrossRef]

MDPI AG
Grosspeteranlage 5
4052 Basel
Switzerland
Tel.: +41 61 683 77 34

*Energies* Editorial Office
E-mail: energies@mdpi.com
www.mdpi.com/journal/energies